CARBANIONS:

MECHANISTIC AND ISOTOPIC ASPECTS

REACTION MECHANISMS IN ORGANIC CHEMISTRY

A SERIES OF MONOGRAPHS EDITED BY

C. EABORN

Professor of Chemistry,
University of Sussex, Brighton, Great Britain

and

N.B. CHAPMAN

Professor of Chemistry,
University of Hull, Great Britain

MONOGRAPH 9

Other titles in this series

1 Nucleophilic Substitution at a Saturated Carbon Atom
   by C.A. BUNTON

2 Elimination Reactions *by* D.V. BANTHORPE

3 Electrophilic Substitution in Benzenoid Compounds
   by R.O.C. NORMAN and R. TAYLOR

4 Electrophilic Additions to Unsaturated Systems
   by P.B.D. DE LA MARE and R. BOLTON

5 The Organic Chemistry of Phosphorus
   by A.J. KIRBY and S.G. WARREN

6 Aromatic Rearrangements *by* H.J. SHINE

7 Steroid Reaction Mechanisms
   by D.N. KIRK and M.P. HARTHSORN

8 Aromatic Nucleophilic Substitution
   by J. MILLER

CARBANIONS:
MECHANISTIC AND ISOTOPIC ASPECTS

E. BUNCEL

Department of Chemistry,
Queen's University,
Kingston, Ontario (Canada)

ELSEVIER SCIENTIFIC PUBLISHING COMPANY

AMSTERDAM/OXFORD/NEW YORK 1975

ELSEVIER SCIENTIFIC PUBLISHING COMPANY
335 Jan van Galenstraat
P.O. Box 211, Amsterdam, The Netherlands

AMERICAN ELSEVIER PUBLISHING COMPANY, INC.
52 Vanderbilt Avenue
New York, New York 10017

Library of Congress Card Number: 73-89148

ISBN 0-444-41190-9

With 26 illustrations and 10 tables.

Printed in The Netherlands

PREFACE

This monograph deals with some of the more important aspects of carbanion chemistry, from the viewpoint of structure and mechanism. Emphasis is given to the utilization of isotopes for elucidation of the intrinsic nature of these various processes, but the isotopic viewpoint is not intended to be confining or otherwise to provide conceptual limitations on coverage. Indeed, several important areas of carbanion chemistry are discussed extensively even though isotopic studies have contributed only in a minor way to their development and current understanding. Theoretical principles underlying the use of isotopes in organic reactions are briefly considered, and for more complete accounts, e.g. of the theory of kinetic isotope effects, the reader is referred to the several excellent published treatments.

The subject matter of carbanion chemistry is introduced at an introductory level, and only a general background in organic chemistry is assumed. Analogies are frequently drawn with areas of organic chemistry which do not involve carbanionic species, in order that a balanced account should be maintained and that the properties and reactions of carbanionic species can be placed in perspective with related chemical concepts. The presentation is intended to be useful to advanced undergraduate as well as graduate students. It is hoped that the monograph will also be of value to research workers who wish to see how mechanistic studies of carbanion chemistry may have relevance to their own fields.

In Chapter 1 the foundation is laid for discussion of the structural and electronic bases of carbanionic species, including their relationship to the carbon acid precursors, and a critical evaluation of the measurement of the pK_a values of carbon acids is presented. Chapter 2 examines stereochemical aspects of carbanions, first considering the stereochemistry of carbanions as (hypothetical) isolated species, free from any perturbing influences of the solvent medium and of the positive counterion, and then turning to the stereochemistry of carbanions as present in actual reaction systems, in which solvation and ion-pairing phenomena are of paramount importance to the stereo-chemical outcome of a given process. In Chapter 3 tautomerism, or prototropy, is discussed from the viewpoint of carbanion (enolate ion) intermediates, but for balance of presentation, emphasis is also given to acid-catalyzed reactions involving enol intermediates and to related processes such as the mechanism of halogenation. Similarly, in Chapter 4, the principles of nonclassical interactions in *carbonium ion* systems are described before consideration is given to any carbanionic counterparts. Some selected rearrangements of carbanions are discussed in Chapter 5, namely those proceeding via enolate and homoenolate ions, the relevant structural aspects having been considered in the preceding two chapters. In Chapter 6 discussion of orbital symmetry control in carbanionic systems is given within a non-mathematical framework; it was thought appropriate in this instance, for historical and pedagogical reasons, first to consider hydrocarbon rearrangements as well as carbonium ion systems. Lastly, in Chapter 7, aspects of the organometallic chemistry of some non-transition metals are examined from the viewpoint of carbanionic processes; an attempt has been made to present a broad perspective of the structural and mechanistic problems encountered in the organometallic systems, rather than to confine the account to detailed consideration of individual reactions.

It will be evident that this monograph does not attempt to present a comprehensive account of carbanion chemistry, and in particular, its role in synthesis is not discussed. While selection of the topics covered has been partly determined by the author's interests, due regard has been paid to areas which are currently of importance even though they may only be in their developmental stage. The monograph is intended to supplement rather than replace the other available texts in the general area of carbanion chemistry.

The author wishes to acknowledge the generous assistance of many colleagues, in reading and in commenting on portions of the manuscript. Selected chapters were read by M.H. Abraham, J.E. Baldwin (M.I.T.), D.J. Cram, N. Epiotis, E. Grovenstein, Jr., H.O. House, D.H. Hunter, J.R. Jones, A. Nickon, R. Stewart, A. Streitwieser, Jr., E.A. Symons, J. Warkentin and S. Wolfe, while S.I. Miller read and commented on major portions of the monograph. Professor C. Eaborn, as Editor of the series, read the entire text and not only gave his expert critical advice but also provided welcome encouragement during the preparation of the manuscript. Several of my coworkers assisted with the task of proof reading. I owe much to my colleagues at Queen's University and to the many students who, through numerous discussions, have helped to clarify my ideas. The example given by my former teachers of chemistry continues to be a source of inspiration. Lastly, my gratitude goes to my wife for her part in this partnership, as well as for her patience and understanding.

CONTENTS

Preface v

Chapter 1. Carbon acids and carbanions 1
1. Structural considerations 1
2. Equilibrium acidity of carbon acids 10
3. Kinetic acidity of carbon acids 16
4. The problem of an acidity scale 20
5. Kinetic isotope effects and the study of carbon acids 23
References 30

Chapter 2. Stereochemistry of carbanionic processes 35
1. Planar versus pyramidal configurations 35
2. Factors influencing the geometry of carbanions 42
3. Stereochemical results of hydrogen-deuterium exchange in reactions
   involving symmetrical carbanions 54
4. The chirality of carbanions adjacent to sulfur and phosphorus . . 59
References 66

Chapter 3. Carbanions and tautomerism 69
1. Keto-enol tautomerism 70
   (a) Investigation of keto-enol equilibria 70
   (b) Halogenation studies 75
   (c) Mechanism of enolization 77
   (d) Isotopic studies 80
   (e) Stereochemical consequences of keto-enol tautomerism . . . 86
2. Some further tautomeric systems 91
   (a) Nitro-*aci*-nitro tautomerism 94
   (b) Tautomerism of nitriles 97
   (c) Tautomerism of propenes 98
   (d) Tautomerism of α,β- and β,γ-unsaturated derivatives 98
   (e) The methyleneazomethine (imine) tautomeric system 100
References 102

Chapter 4. Classical and nonclassical carbanions 109
1. Introduction 109
2. Nonclassical carbonium ions: some basic principles 110
3. Homoconjugation in uncharged systems 115
4. Homoenolate ions 116
5. Homoconjugated and homoaromatic carbanions 122
6. Carbanions and antiaromaticity 129
References 138

ix

Chapter 5. Enolate and homoenolate rearrangements 143
  1. Scope of the presentation 143
  2. The Favorskii rearrangement 144
  3. Some further enolate rearrangements 155
    (a) Base-induced rearrangements of α-epoxy ketones 155
    (b) The rearrangements of halogeno amides 157
  4. Rearrangements of homoenolate ions 161
  References 167

Chapter 6. Orbital symmetry control in carbanion rearrangements 171
  1. Introduction 171
  2. Electrocyclic rearrangements 172
    (a) Rearrangement of a dienyl anion 172
    (b) Orbital symmetry and the rearrangement of 1,3,5-trienes . . . 173
    (c) Carbanion electrocyclizations 180
    (d) Carbonium ion electrocyclizations 187
  3. Sigmatropic rearrangements 188
    (a) [*1,j*] Hydrogen migrations 188
    (b) [*1,j*] Alkyl and aryl migrations 191
    (c) Sigmatropic rearrangements of the order [*i,j*] 195
  References 203

Chapter 7. Carbanions in reactions of organometallic compounds . 209
  1. Metal cations as counterions in carbanionic systems 209
  2. Scope of organometallic chemistry and of present coverage . . . 210
  3. Bond polarity in organometallic compounds and electronegativity
    considerations 211
  4. Molecular structure of organometallic compounds 213
  5. Heterolytic bond scission processes of organometallic compounds . 216
  6. Classification of S_E mechanisms 218
  7. Carbanion mechanisms in electrophilic substitution reactions of
    organomercury compounds 224
    (a) Isotopic exchange in α-carbethoxybenzylmercuric bromide . . 225
    (b) Isotopic exchange in *p*-nitrobenzylmercuric bromide 227
    (c) Protolysis of 4-pyridiomethylmercuric chloride 227
  8. Carbanions in electrophilic substitution reactions of organosilicon and
    organotin compounds 228
    (a) Alkali cleavage of (phenylethynyl)-silanes and -germanes . . . 232
    (b) Alkali cleavage of (phenylallyl)-silanes and -stannanes 233
    (c) Alkali cleavage of (benzyltrimethyl)-silanes, -germanes and
    -stannanes 234
    (d) Alkali cleavage of (phenyltrimethyl)-silanes and -stannanes . . 235
  References 238

Author Index 245

Subject Index 261

Chapter 1

CARBON ACIDS AND CARBANIONS

1. STRUCTURAL CONSIDERATIONS

A carbanion is a negatively charged ionic species in which a carbon atom formally bears unit electron charge. Typically the central carbon of a carbanion is attached to three groups through covalent bonding, and an unshared pair of electrons completes the electron octet in the valence shell. Other structural environments may be present, including a double covalent bond or a triple covalent bond. Some examples of carbanions with various hydrocarbon residues are the following:

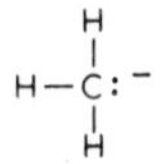

| methyl anion | 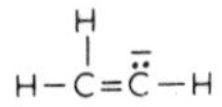 vinyl anion | 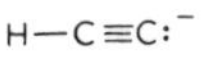 acetylide anion | 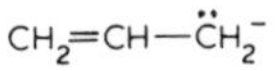 allyl anion |

| benzyl anion | 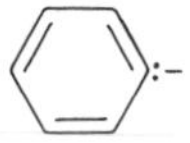 phenyl anion | 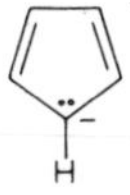 cyclopentadienyl anion | 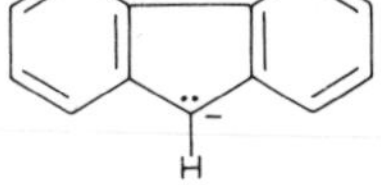 fluorenyl anion |

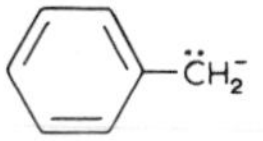

As ionic species, carbanions are, of course, formally analogous to negatively charged ions such as the chloride and hydroxide (or alkoxide) ions. However, though we can readily have a "jar-full" of hydroxide or chloride ions (in as much as jars of sodium hydroxide and sodium chloride consist essentially of the respective cations and anions packed in the crystal lattice), or we can have a "beaker-full" of aqueous solutions of these ions, it is only under exceptional circumstances that carbanions exist in large concentrations in the free state or in aqueous solution. However, small equilibrium concentrations of carbanions are known to be present in many systems and can be detected by various experimental methods. Moreover, there is definitive evidence that in

1

many other organic processes carbanions are formed as transient reaction intermediates, in concentrations too low to be observed by normal techniques.

Conceptually the simplest route to carbanions is by loss of a proton from the corresponding hydrocarbon conjugate acid. In principle, then, all compounds containing C–H bonds are potential "carbon acids" [1–8]. However, the extent of the equilibrium dissociation in aqueous solution of most carbon acids is quite small, being very much less than that of the common weak acids such as acetic acid (pK_a = 4.8) or phenol (pK_a = 9.9). Methane has an estimated pK_a of about 48; there is some uncertainty about this value, since there is no direct experimental method available for measuring acid dissociation constants of such small magnitude, and other estimates range between 40 and 58 [1, 2]. The following sections (2–4) will consider problems associated with the measurement of acidity of carbon acids and with derivation of a common pK_a scale.

The extremely small extent of dissociation of methane reflects the high degree of thermodynamic instability of the methyl anion. Since $\Delta G^\circ = -RT \ln K$, for the dissociation of methane ΔG° has a value of 65.3 kcal per mole (25°C), if the pK_a is taken as 48, whereas for the dissociation of acetic acid ΔG° = 6.5 kcal/mole; hence the former process is disfavoured to the extent of about 59 kcal/mole, and by even more if one of the higher estimated values for the pK_a of methane were chosen. The cause of the instability of the methyl anion is reasonably explained, as carbon is not an electronegative element, and hence the imposition of a unit negative charge is energetically an extremely unfavourable process.

There are various ways in which structural modification can result in increased dissociation of the hydrocarbon acid. For example, there is increased dissociation if one or more strongly electronegative elements is covalently bonded to the reaction centre, for in the resulting carbanion some of the negative charge density is then withdrawn from the central carbon by an inductive and/or field mechanism (i.e. transmitted by electrostatic action through the σ-bond network and/or through space). Trichloromethane, CCl_3H, for instance, is estimated [9] to have a pK_a of 24.

An alternative to inductive electron withdrawal for carbanion stabilization is the conjugative delocalization mechanism, whereby the

2

negative charge is spread over two or more atoms by resonance. In the benzyl anion, for instance, the charge is distributed in part over the benzene ring, as a result of contributions from resonance structures in which negative charge is borne by the *ortho* and *para* carbons [10]:

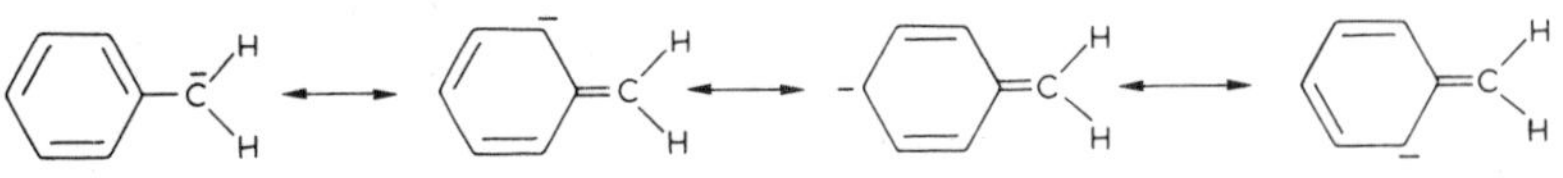

The considerably greater acidity of toluene (pK_a 41) than of methane reflects the increased carbanion stability due to charge delocalization. The situation with diphenylmethane (pK_a 34) is analogous. However, this acidity sequence does not proceed smoothly to triphenylmethane, which has a pK_a of 31.5; that is, the introduction of the third phenyl has a much smaller acidifying effect than that of the first and second. Since 10 canonical structures could potentially contribute to the overall structure of the $(C_6H_5)_3C^-$ anion, this shows that one must exercise caution in equating canonical structures with delocalization energy. Other factors, such as electronegativity and molecular geometry, must also be considered, and can, in fact, be of overriding importance.

In writing the canonical structures for the benzylic anion, as above, we imply coplanarity of the ring with the *alpha* methylene group (i.e. sp^2 hybridization for the *alpha* carbon), for it is only then that there can be effective overlap between the p-orbital of the *alpha* carbon containing the unshared pair of electrons and the p_π-orbitals of the benzene ring. Such planarity is implied, of course, in writing a double covalent bond between the *alpha* carbon and the carbon of the benzene ring. The planarity requirement for maximum p-orbital overlap is illustrated in the diagram below for the benzyl as well as for the analogous but structurally simpler case of the allyl anion:

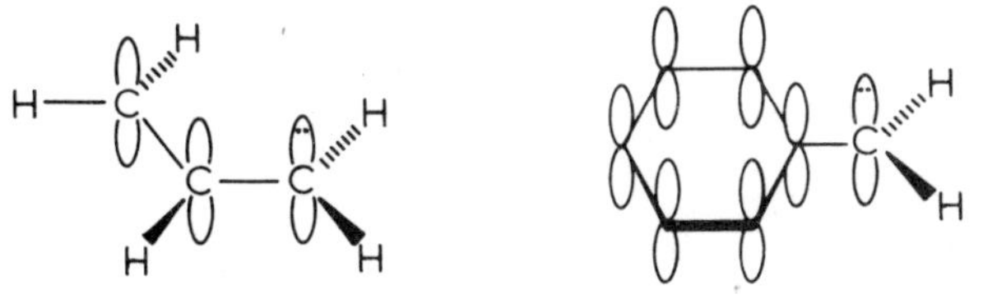

planar allyl anion

planar benzyl anion

In the triphenylmethyl (trityl) anion it is sterically not possible for the three benzene rings to be coplanar. Hence all three phenyl groups

3

cannot simultaneously act to delocalize the negative charge of the central carbon. The term "steric inhibition of resonance" is used to describe the situation, such as is found in the trityl anion, in which coplanarity is prevented by steric requirements of the system, so that the full effect of resonance cannot operate on a measured property, in this case the pK_a. An illustrative representation of the trityl anion is given below and corresponds to the generally accepted structures of the corresponding radical and carbonium ion. The "propeller" like configuration is supported by extended Hückel calculations on the anion [11], as well as by an X-ray analysis of the triphenylmethyllithium tetramethylenediamine complex [12].

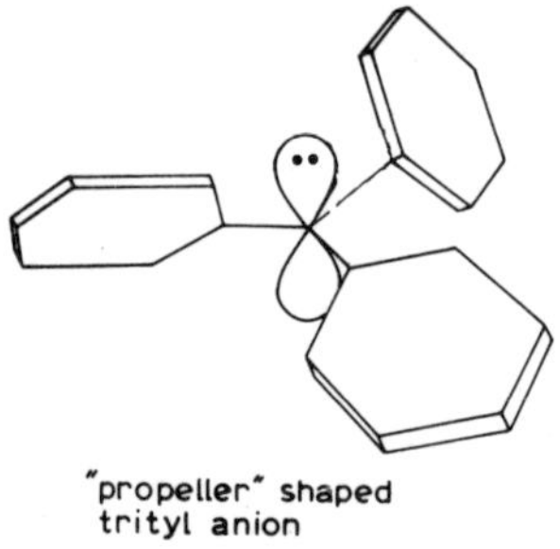

"propeller" shaped
trityl anion

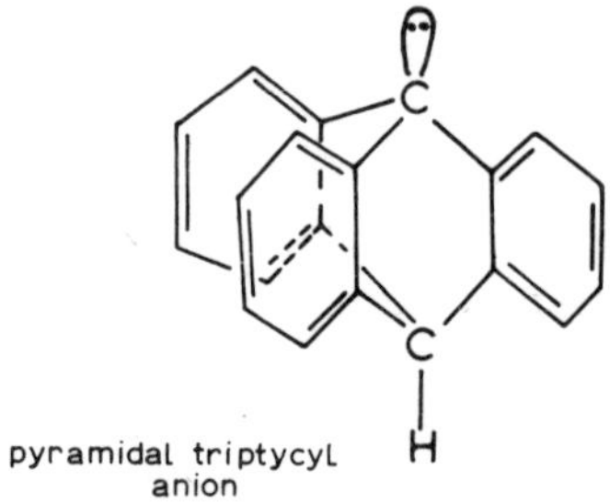

pyramidal triptycyl
anion

A more acute case of steric inhibition of resonance on hydrocarbon acidity is provided by the triptycene system [13]. Here proton abstraction from the bridgehead carbon yields the triptycyl anion which is forced by the geometry of the system to be pyramidal in configuration. Moreover, in the absence of coplanarity there is now no opportunity for charge delocalization involving the benzene rings. Despite this, the phenyls still act in stabilization of the triptycyl anion, as evidenced by the considerable acidity of triptycene (pK_a 42), which is comparable to that of benzene (pK_a 43) or ethylene (pK_a 44) and much larger than of cyclohexane (pK_a 51). It is apparent that the acidifying effect of the phenyl groups in the triptycene system must be via an inductive-field effect mechanism. Thus in the case of the trityl, and even benzyl, anions the phenyl group also stabilizes in part by inductive electron withdrawal.

An even more effective means of resonance stabilization than that in benzylic-type anions is possible for those oxygen- or nitrogen-

4

containing molecules in which resonance formulation allows the negative charge to be placed on the electronegative heteroatoms. Examples of this are given by the anions of acetonitrile, acetone, as well as nitro-methane:

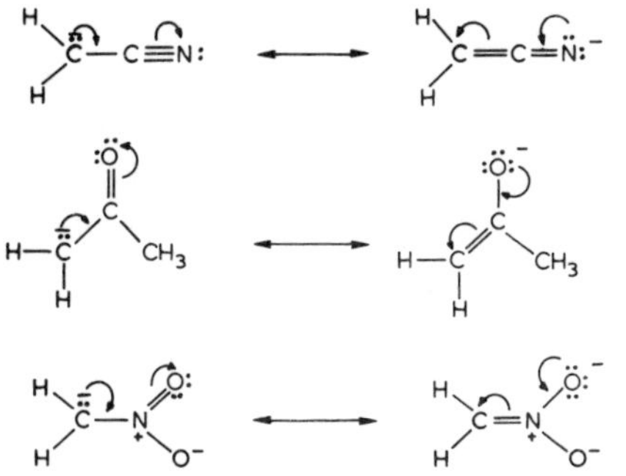

The major contributing structure in each case is the one with the formal negative charge on the heteroatom. The resulting resonance stabilization of these carbanions is reflected in the pK_a's of the parent carbon acids [1]: pK_a (CH_3CN) = 25, pK_a (CH_3COCH_3) = 20, pK_a (CH_3NO_2) = 11.

One would expect that the introduction of more than one of these substituents (CN, $COCH_3$, NO_2) would result in further large increases in acidity. The data of Table 1 show that while this expectation is qualitatively correct, the observed effects on pK_a are generally not additive.

It is seen that only for cyano substitution is there an even approximately additive effect on the pK_a. The linearity of the cyano group results in a minimal steric requirement, so that the anion derived from tricyanomethane has nearly a planar structure [14]. The much

Table 1

SUBSTITUENT EFFECTS ON pK_a [1]

	pK_a		pK_a		pK_a
CH_3CN	25	CH_3COCH_3	20	CH_3NO_2	11
$CH_2(CN)_2$	12	$CH_2(COCH_3)_2$	9	$CH_2(NO_2)_2$	4
$CH(CN)_3$	0	$CH(COCH_3)_3$	6	$CH(NO_2)_3$	0

5

bulkier acetyl and nitro groups hinder coplanarity to a much greater degree, as shown by the successively smaller decreases in pK_a values. The acidity of tricyanomethane or of trinitromethane is comparable with the "medium strong" inorganic acids, being stronger than, say, phosphoric or dilute hydrofluoric acids though weaker than the "very strong" perchloric, sulfuric, hydrochloric, and nitric acids.

The inductive electron withdrawal effect, and the delocalization of charge through resonance, are fully borne out as factors which stabilize carbanion formation and hence increase carbon acidity. To these we must now add another factor, which becomes evident in the order of carbon acidity, $RC \equiv CH > R_2C = CH_2 > R_3C\text{-}CH_3$ (for R = H, the pK_a values are 25, 44 and 49, respectively). The reader will recall that alkynes can readily be separated from alkenes and alkanes through the ease with which alkynes form metal salts. The structural or environmental factor which varies in the series alkynes, alkenes, alkanes, is hybridization of the C–H bond, which changes from sp to sp^2 to sp^3 along the series. We see that increased s character in the orbital is associated with increased C–H acidity. This is a reasonable result, since the space probability distribution functions of s- and p-orbitals require that s-electrons are, on the average, closer to the carbon nucleus than p-electrons. This leads to an s-orbital being more electronegative than a p-orbital, and hence more able to support negative charge. The relatively greater stabilization of carbanions with increased proportion of s-orbital character is referred to as the *hybridization* factor or the s-orbital character factor.

The cyclopropane system provides a clear, though perhaps unexpected, case of s-orbital carbanion stabilization. Thus theoretical calculations and a variety of experimental criteria indicate that hybridization of the C–H bonds in cyclopropane is not sp^3, as would ordinarily have been expected [15, 16]. It appears that the actual C–H bond hybridization in cyclopropane is $sp^{2.28}$, so that the s-orbital character in this system is intermediate between that in ethane and in ethylene. As a result of this exalted s-character the cyclopropane structure stabilizes carbanion formation. The comparatively large acidity of cyclopropane (pK_a 46) attests to this argument.

A different kind of stabilizing effect is responsible for the relatively high acidity of cyclopentadiene (pK_a 15) compared with other unsubstituted hydrocarbons. That the anion derived from cyclopentadiene is not simply allylic in type is seen by the pK_a values of propene (α-position) and of cycloheptatriene, both of which are considerably weaker acids:

pK_a 15 pK_a 40 pK_a 36

It is apparent that with cyclopentadiene some special effect must operate. In fact a greater number of formally analogous resonance structures may be written for the cycloheptatrienyl carbanion than for the cyclopentadienyl carbanion[*]. The special nature of the cyclopentadienyl anion is that it obeys the Hückel aromaticity rule for a cyclic system in that it contains a total of $4n + 2$ p-type electrons ($n = 1$ in this system, but in general n can be zero or any integer) [17]. The Hückel rule applies also to closed shell systems of other charge types, including neutral (e.g. benzene) and positively charged (e.g. cyclopropenyl cation) molecular systems. (The contrasting instability of the dianion of cyclobutadiene, a species for which only recently has experimental evidence become available [18], is due to other factors such as ring strain and electron repulsion forces.) The well authenticated cycloocta-

cyclopropenyl   cyclopentadienyl
cation (n = 0)   anion (n = 1)

tetraenyl dianion (1) [19], which has 10 π electrons and is formed by *addition* of two electrons to cyclooctatetraene (e.g. by means of potassium metal in ether) also falls in this category. The cyclooctatetraenyl dianion is known to be planar whereas cyclooctatetraene itself is non-planar. The cyclononatetraenide anion 2 [20, 21] and its nitrogen analogue 3 [22] have likewise been characterized as aromatic 10 π-electron systems. The methano-bridged anion 4 [23], a structure which would hardly have been expected to show any stability but for the fact

---

[*]See, however, Chapter 4, Section 6 where the possibility is discussed that the destabilization of the cycloheptatrienyl anion is a result of antiaromaticity.

that it obeys the Hückel rule, is an apt example of the general validity of this unifying principle. Similarly, in the annulene series, we have the dianions **5** and **6** which are derived by 2-electron reduction of [12]annulene and 1,5,9-tridehydro[12]annulene, respectively [24, 25];
anions with 10 π-electron systems (n = 2):

anions with 14 π-electron systems (n = 3):

A series of hydrocarbons related to cyclopentadiene is shown below, together with their pK_a values. It is seen that indene, with one benzo group fused onto cyclopentadiene is a weaker acid, and fluorene is weaker still. However, the remarkable hydrocarbon fluoradene [26] is a considerably stronger acid and we note that the anion resulting from proton loss actually possesses a "double fluorenyl" fused aromatic system.

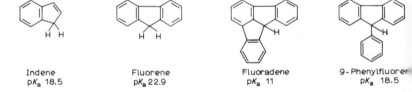

| Indene | Fluorene | Fluoradene | 9-Phenylfluorene |
| pK_a 18.5 | pK_a 22.9 | pK_a 11 | pK_a 18.5 |

Proceeding from fluorene to fluoradene, the additional benzene ring has a much greater acidifying effect on the tertiary C-9 hydrogen than with the non-fused but otherwise structurally analogous 9-phenyl-fluorene system. In the latter case the phenyl substituent acts through a combination of the conjugative and inductive effects. As would be

8

anticipated, introduction of a cyano or a carbomethoxy substituent in the 9-position of fluorene causes even larger carbanion stabilization than does the phenyl; the pK_a values of 11.4 and 12.9 respectively for 9-cyanofluorene and 9-carbomethoxyfluorene attest to this expectation [27]. Cyano-substituted cyclopentadienes are stronger acids still; pentacyanocyclopentadiene (pK_a -11) [28] has an acidity comparable to that of sulfuric or perchloric acid. Thus we have covered the full range of carbon acids, from the extremely weak, which in solution give rise to extremely small concentrations of carbanion, to the very strong which are virtually completely dissociated in solution.

Before concluding this section it is appropriate to mention the *ylid* type of structures such as 7 and 8, which are formed by reactions 1 and 2 [29, 30]:

(1) $[Ph_3\overset{+}{P}-CH_2-CO-CH_3]\,Cl^-$ $\xrightarrow{\text{base}}$ $Ph_3\overset{+}{P}-\overset{-}{C}H-CO-CH_3$

7

(2) $[MeEt\overset{+}{S}-CH_2-CO-Ph]\,ClO_4^-$ $\xrightarrow{\text{base}}$ $MeEt\overset{+}{S}-\overset{-}{C}H-CO-Ph$

8

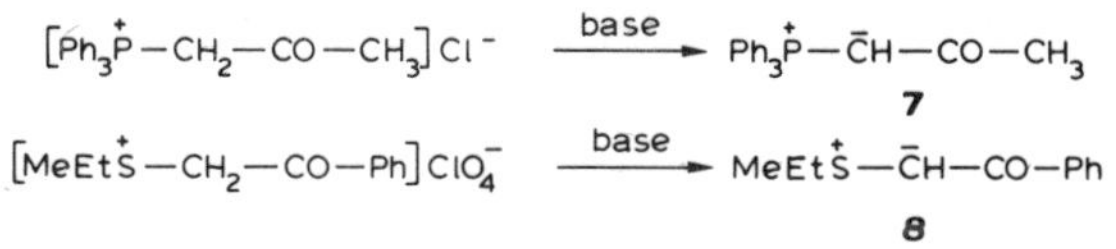

It is seen that ylids are molecules of overall electrical neutrality, though structurally they may be regarded as carbanions which are stabilized by an adjacent atom bearing a positive charge. The ylids may also be represented by other contributing resonance structures; for example in the case of 7 we have the following possibilities:

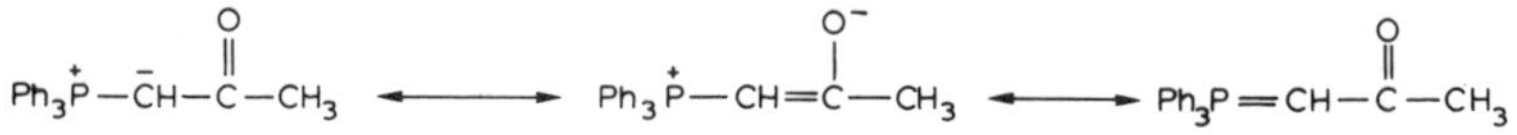

$Ph_3\overset{+}{P}-\overset{-}{C}H-\overset{O}{\overset{\|}{C}}-CH_3$ $\longleftrightarrow$ $Ph_3\overset{+}{P}-CH=\overset{O^-}{\overset{|}{C}}-CH_3$ $\longleftrightarrow$ $Ph_3P=CH-\overset{O}{\overset{\|}{C}}-CH_3$

The ylids 7 and 8 are fully stable molecules which can be isolated, recrystallized, and stored in the normal manner. Other ylids such as $(CH_3)_3\overset{+}{P}-\overset{-}{C}H_2$ are formed only as metastable species which can be detected by several techniques, including deuterium exchange (eq. 3). Ylid dianions have also been prepared [31].

(3) $(CH_3)_3\overset{+}{P}-CH_3$ $\xrightarrow{OD^-}$ $(CH_3)_3\overset{+}{P}-CH_2^-$ $\xrightarrow{D_2O}$ $(CH_3)_3\overset{+}{P}-CH_2D$

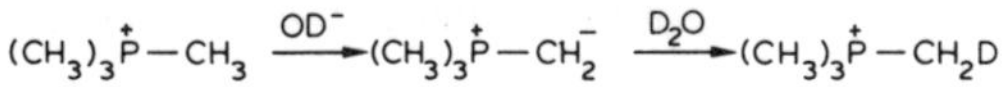

Though detailed discussion of ylids is not given in this monograph such species will be considered incidentally from time to time. Thus deuterium exchange processes of the type shown in eq. 3 are considered

9

further in Chapter 2, while ylids also appear in Chapter 6 in connection with rearrangement processes of carbanions. For further discussion of the chemistry of these interesting molecules the reader is referred to ref. 32. Also, for the chemistry of dianions, see ref. 3.

2. EQUILIBRIUM ACIDITY OF CARBON ACIDS

Of the hydrocarbon acids only few have large enough dissociation constants to allow pK_a measurement in water. The nitro- and poly-cyano- and acetyl-substituted alkanes fit into this category but, as was seen in Section 1, proton loss from these molecules yields anions with the negative charge residing largely on the electronegative heteroatoms. The strongly acidic cyano-cyclopentadienes form another obvious example. Of the unsubstituted hydrocarbon acids, only fluoradene and related structures such as 9 and 10 [33] are associated with an acidity

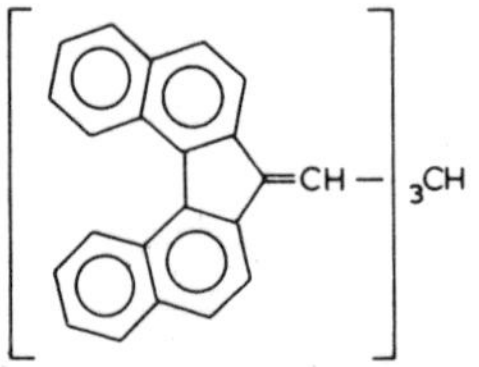

9(pK_a 5.9) 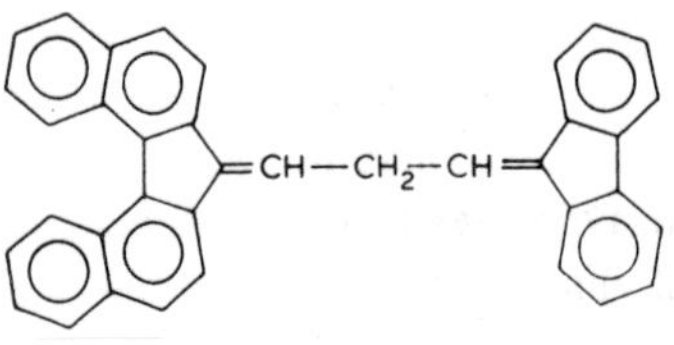   10(pK_a 9.8)

sufficient for measurement in aqueous solution. This limitation on pK_a measurement results from the fact that the strongest basic species which can exist in water is the hydroxide ion. Thus, if AH is a weaker acid than H_2O, then the conjugate base A⁻ will become protonated in aqueous solution so that the equilibrium $A^- + H_2O \rightleftharpoons AH + OH^-$ will be far to the right hand side. For *all* acids which are weaker than H_2O this equilibrium is essentially to the right, so that we have here in operation a "levelling effect" on acidity. Alternatively, we may view a very weak carbon acid in water as producing a concentration of H_3O^+ which is *less than that produced due to the autoprotolysis of water* itself, and hence not measurable.

How then are acid strengths determined for the very weakly acidic hydrocarbons such as triphenylmethane? There are two general methods for measurement of acid strengths of very weak acids: the equilibrium

10

(thermodynamic) method and the kinetic method. The present section is devoted to the first method, and the next section to the second method.

One practical way of dealing with the limitation of the aqueous system is to study acid-base equilibria in systems in which the solvent is more weakly acidic than water (pK_a 15.7). Methanol (pK_a 17.7) comes to mind first, perhaps, as another amphiprotic solvent, but the gain over the aqueous system is slight. Much greater gains are possible, however, when we go on to solvents such as dimethyl sulfoxide ($pK_a \sim 33$), ammonia ($pK_a \sim 35$), or benzene ($pK_a \sim 43$).

However, of these weakly acidic media, only the relatively polar dimethyl sulfoxide (dielectric constant, ϵ, 47) and liquid ammonia (ϵ, 17) are capable of sustaining electrochemical (potentiometric or conductivity) measurement of ionization constants. A continuous acidity scale in dimethyl sulfoxide, based on glass electrode measurements, has been recorded [34] for carbon acids covering the pK_a range 10–30, but there are as yet few potentiometric data available for carbon acids in solvent ammonia [35]. Though benzene (or cyclohexane, ether, etc.) might allow one in principle to study the ionization of even extremely weak carbon acids, the very low polarity of these solvents (ϵ, 2.3, 2.0, and 4.3, respectively) lead to other complications and limitations, including the inapplicability of electrochemical measurements.

A method which has been used with considerable success for evaluation of the pK_a of weak carbon acids in benzene or ether, as well as ammonia and related solvents such as cyclohexylamine, is based on determination of the position of equilibrium which is established between two acids R_1H and R_2H and their alkali metal salts R_1M and R_2M:

(4) $$R_1H + R_2^-M^+ \rightleftharpoons R_1^-M^+ + R_2H$$

Organometallic compounds of the simple alkanes (e.g. ethyllithium, n-butyllithium, etc.) are known to be highly associated in polymeric clusters and their solutions in ether or benzene are non-conducting [36]. These compounds are believed to be essentially covalent, with partial ionic character in the carbon-lithium bond (see Chapter 7 for discussion of bonding and molecular structure in these alkyllithium compounds). This contrasts with the phenylalkane or fluorenyl type salts, which are

11

believed to be essentially ionic but undissociated in ether or benzene solution. Now whereas in ether or benzene organometallic compounds of the arylmethanes will be undissociated and exist as tight (contact) ion pairs (R^-, M^+) and higher aggregates, in the more polar cyclohexylamine there is slight dissociation into solvent-separated ions and it is only in liquid ammonia or in dimethyl sulfoxide that dissociation into free ions will be appreciable. The possibility that the state of dissociation of the RM species varies with the structure of the substrate and the polarity of the medium suggests that caution must be exercised in the interpretation of equilibrium measurements. (For application of cyclic chelating polyethers in ion pairing, see ref. 45.)

Historically, the first pK_a scale for carbon acids, based on equilibrium measurements according to eq. (4), was developed [37, 38] for the benzene and ether solvent systems, extending to hydrocarbons of pK_a as low as ~40. More recently the cyclohexylamine-alkali metal cyclo-hexylamide $(C_6H_{11}NH_2/C_6H_{11}NHM)$ basic system has yielded more accurate data on acidity of very weak carbon acids [5, 39, 40]. Also, an acidity scale based on similar principles has been derived for carbon acids in liquid ammonia containing potassium amide [41]. Owing to the importance of the cyclohexylamine system, this will be described in more detail, including the manner in which the position of the equilib-rium of eq. (4) is determined.

Let us for convenience consider separately the ionization of two acids, R_1H and R_2H, by lithium or cesium cyclohexylamide in cyclo-hexylamine (LiCHA/CHA, CsCHA/CHA):

$$(5) \qquad R_1H + C_6H_{11}NHM \rightleftharpoons R_1^-M^+ + C_6H_{11}NH_2$$

$$(6) \qquad R_2H + C_6H_{11}NHM \rightleftharpoons R_2^-M^+ + C_6H_{11}NH_2$$

If we restrict the carbon acids under consideration to those which are stronger acids than cyclohexylamine, then the equilibria of equations (5) and (6) will lie far to the right in each case. Hence if to a mixture of R_1H and R_2H in cyclohexylamine one adds *a deficiency* of lithium cyclohexylamide, then the equilibrium of eq. (4) will be established and the position of equilibrium will depend on the relative acidities of R_1H and R_2H. Provided that these acidities do not differ by more than 2 pK units, both R_1M and R_2M will be present at equilibrium in appreciable concentration. The concentrations of the various species in the equilibrium mixture is conveniently measured spectrophotometrically when the carbanion is highly conjugated and absorbs in the ultraviolet-visible region of the spectrum.

Intensive study of the spectral characteristics of lithium and cesium carbanides in cyclohexylamine has allowed Streitwieser and coworkers [42] to distinguish two types of ion pairs: *contact* and *solvent-separated*. This dual concept had originally been recognized by Winstein [43] in order to explain the behaviour of carbonium ions produced in solvolytic processes, and has also become firmly established in carbanion chemistry, largely as a result of studies by Cram [44], and Szwarc, Smid and their coworkers [45]. In the cyclohexylamine–diethylamine medium [42] (the co-solvent is added to allow study at lower temperature), it is found that whereas fluorenylcesium exhibits visible absorption spectra which are temperature independent, fluorenyllithium gives rise to spectra in which the absorbances show appreciable temperature dependence (A_{max}^{512nm} has the value 0.69 at $-42°$ but decreases to 0.29 at $88°$, while A_{max}^{480nm} correspondingly changes from 0.95 to 0.60, and A_{max}^{447nm} from 0.73 to 0.62). These findings are interpreted on the basis of the equilibrium given in eq. (7), for which K values can be derived by analysis of the spectral data (e.g. $K \triangleq 20$ at $25°$, in 73% cyclohexylamine -27% diethylamine by weight).

$$(7) \quad R^-, Li^+ \xrightleftharpoons{K} R^- \| Li^+$$
$$\text{contact} \qquad\qquad \text{solvent separated}$$

The position of the equilibrium shifts to the right as the temperature is lowered ($\Delta H° = -8.5$ kcal/mole, $\Delta S° = -22.7$ e.u.). Fluorenyllithium in cyclohexylamine at normal temperatures exists largely as the solvent-separated ion pair species. On the other hand, fluorenylcesium in cyclohexylamine (λ_{max} 447, 472 and 504 nm) is present almost exclusively as contact ion pairs. It appears that, in general, lithium salts of delocalized carbanions are present largely as solvent-separated ion pairs, whereas cesium salts are present as contact ion pairs. The small lithium cation is preferentially solvated by one or more Lewis base solvent molecules than by the large carbanion in which charge is diffused as a result of resonance delocalization; the converse applies for the cesium cation.

The observations mentioned above have a direct bearing on the

13

determination of carbon acidity in the cyclohexylamine system. Thus equations (8) and (9) will be applicable for measurements made with lithium cyclohexylamide and cesium cyclohexylamide, respectively:

(8) $RH + Li^{+ -}NHC_6H_{11} \rightleftharpoons R^- \| Li^+ + C_6H_{11}NH_2$

(9) $RH + Cs^{+ -} NHC_6H_{11} \rightleftharpoons R^-, Cs^+ + C_6H_{11}NH_2$

Since there is greater charge separation in the $R^- \| Li^+$ species of eq. (8) than in the R^-, Cs^+ species of eq. (9), the extent of ionization of the carbon acid will, in general, be smaller for the LiCHA case. Thus triphenylmethane is known to be more acidic towards CsCHA than it is towards LiCHA. It may be concluded that the CsCHA system provides a better measure of carbon acidity than the LiCHA system (though both systems provide internal self-consistency). Cesium carbanide ion pairs should be valid models for carbanions and the nature of the ion pair should remain unchanged within a series of structurally related compounds – this is the basic assumption for the validity of the pK_a scale in cesium cyclohexylamide–cyclohexylamine. The evidence to date supports the validity of this assumption.

It is relevant to point out, before concluding this section, that CsCHA can cause a double ionization in certain cases in which both negative charges can be delocalized. An example is given by 9-benzylfluorene [40]:

(10)

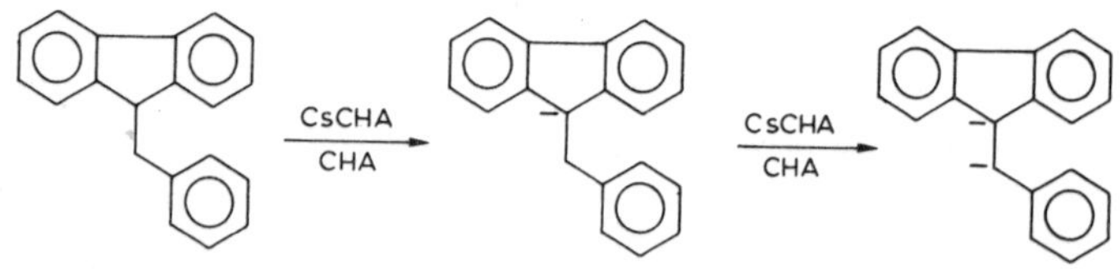

Another equilibrium method for pK_a determination of weak carbon acids is based on a discovery, made in about 1960 [46–48], that when a dipolar aprotic solvent such as dimethyl sulfoxide (DMSO) is added to an aqueous (or alcoholic) solution containing hydroxide (or alkoxide) ion, the reactivity of the base in proton abstraction processes is greatly increased. Dimethyl sulfoxide (unlike water) is capable of forming only very weak hydrogen bonds with hydroxide ion and as a result the thermodynamic activity of the latter increases tremendously as the proportion of DMSO is continuously increased. A 50% DMSO solution

14

containing 0.01 M hydroxide ion has a basicity which is 10^5 times greater than that of a purely aqueous medium, and further successive increases in basicity of 10^5 result when the DMSO content is increased to 90% and then to 99.9%!

We can now see the usefulness of the DMSO–H_2O system [49] in the study of the ionization of weak acids (eq. (11)). Thus a solution of 0.01 M OH$^-$ in 50% DMSO solution will cause half ionization of an acid of pK_a 17, in 90% DMSO an acid of pK_a 22 is half ionized, while in 99.9% DMSO an acid of pK_a 27 will be half ionized. Hydroxide or methoxide ion in pure DMSO causes the ionization of acids of pK_a close to 30 [50].

(11) $\quad$ AH $+$ OH$^-$ $\rightleftharpoons$ A$^-$ $+$ H$_2$O

In practice one measures spectrophotometrically the extent of ionization of the acid under examination in media of constant [OH$^-$] but with an increasing proportion of the dipolar aprotic component. This approach to pK_a determination may be called the *acidity function* approach [51, 52]. The method has been applied to the study of carbon acids as well as nitrogen and oxygen acids [53]. Other dipolar aprotic solvents used in this method are tetramethylenesulfone (commonly called sulfolane), and dimethylformamide (DMF) [47, 54].

Table 2 presents some selected values of the H_- function for binary mixtures of dipolar aprotic solvents and water, containing 0.01 M hydroxide ion. The H_- function was originally introduced by Hammett and gives a measure of the proton abstracting ability of a medium toward a neutral weakly acidic substrate according to eq. (11). The mathematical definition of H_- is as follows:

$$(12) \quad H_- = -\log a_{H^+} \frac{f_{A^-}}{f_{HA}} = pK_a - \log \frac{[A^-]}{[HA]}$$

In eq. (12), a_{H^+} is the hydrogen ion activity in solution, f_{A^-}, f_{HA} are activity coefficients, and $[A^-]/[HA]$ is the ionization ratio. In dilute aqueous solution $H_- = pH$, so that the purely aqueous solution corresponding to Table 2 would have a pH of 12.

Table 2

SELECTED H_- DATA FOR AQUEOUS BINARY MIXTURES WITH SEVERAL
DIPOLAR APROTIC SOLVENTS, EACH WITH 0.011 M TETRAMETHYL-
AMMONIUM HYDROXIDE [49, 54a]

Mole % dipolar aprotic component	Aqueous pyridine	Aqueous tetramethylene sulfone	Aqueous dimethyl-formamide	Aqueous dimethyl sulfoxide
20	13.75	13.22	14.20	14.48
40	14.76	14.25	15.75	16.50
60	15.31	15.56	17.34	18.50
80.78				20.68
90.07				21.98
99.59				26.59

3. KINETIC ACIDITY OF CARBON ACIDS

The equilibrium methods for determination of hydrocarbon acidity
are based, in the majority of cases, on color differences between the
hydrocarbon acid and the carbanionic species derived on proton
abstraction, through the application of quantitative spectral techniques.
Obviously this limits such measurements to hydrocarbons whose anions
form extended conjugated systems, and as a result absorb strongly in
the visible or ultraviolet region of the spectrum. However, many of the
weaker hydrocarbon acids form anions which do not absorb in these
regions: saturated hydrocarbons and aromatic hydrocarbons such as
benzene fall into this category, as do many others. The kinetic acidity
method is independent of color change and does not have the limitations
of medium polarity or dielectric constant which are necessary in
potentiometric and conductivity type measurements.

The kinetic acidity method is dependent on isotopic exchange between
the isotopically normal acidic substrate and a deuterium- or tritium-
labelled medium. (It is of course equally valid in principle to use the
isotopically labelled substrate and to follow the rate of exchange with
isotopically normal solvent.) The exchange is catalyzed by a basic
reagent (B^- in eq. (13)), which is generally the lyate ion of the solvent
medium:

(13a) $\qquad AH + B^- \xrightarrow{\text{slow}} A^- + BH$

(13b) $\qquad A^- + BD \underset{\text{slow}}{\overset{\text{fast}}{\rightleftharpoons}} AD + B^-$

16

Proton abstraction by base B^- is normally the slow, rate-determining, step and the anion A^- is then protonated by BD, which forms the deuterium containing solvent pool, in a rapidly attained equilibrium process. Hydrogen-tritium or deuterium-tritium exchange are directly analogous. Some commonly used B^-/BD combinations are OD^-/D_2O, CH_3O^-/CH_3OD, $t\text{-BuO}^-/t\text{-BuOD}$, ND_2^-/ND_3 and $C_6H_{11}ND^-/C_6H_{11}ND_2$ where C_6H_{11} is the cyclohexyl moiety.

The deuteroxide ion-deuterium oxide basic system is applicable to proton abstraction from relatively strong carbon acids such as nitro-alkanes and ketones, while methoxide ion-deuterated methanol suffices with more weakly acidic carbon acids such as fluorene, without resorting to stringent reaction conditions. Now as was seen in the previous section, the reactivity of the hydroxide or alkoxide ion can be further increased by partial replacement of the hydroxylic component by a dipolar aprotic component such as dimethyl sulfoxide (cf. Table 2). The use of pure dimethyl sulfoxide with methoxide ion or t-butoxide ion leads to even more strongly basic systems which contain in part the very strongly basic dimethylsulfinyl ("dimsyl") anion, formed as a result of the equilibrium [55—57]:

$$(14) \qquad RO^- + CH_3SOCH_3 \rightleftharpoons ROH + CH_3SOCH_2^-$$

The basicity of the $t\text{-BuO}^-$/DMSO system is comparable to that of the amide ion-liquid ammonia or the cyclohexylamide anion-cyclohexyl-amine system. Some of the weakest hydrocarbon acids can be examined by use of the kinetic acidity method in these basic systems.

What is the relationship between the rate of proton exchange for an acid AH and its equilibrium dissociation constant? Intuitively one would predict that the stronger an acid in its equilibrium ionization, the easier will the proton be transferred to a base in a kinetic process, and hence the greater the rate constant for proton abstraction. Quantitatively the relationship was first proposed by Brönsted and Pedersen [58] and is known as the Brönsted catalysis law:

$$(15) \qquad \log k = \alpha \log K_a + \text{const.}$$

In eq. (15), k is the rate constant for proton transfer from AH to B, K_a the thermodynamic ionization constant of AH, and α is known as the Brönsted exponent. The interpretation given to α traditionally (cf.

17

ref. 59) is that it represents the degree to which the proton is transferred from acid AH to base B in the transition state of the process. However this interpretation is brought into question by recent findings [60–62] that in certain systems the values of α may range outside the limits of 0 and 1.

From eq. (13) we note that the rate determining step for hydrogen exchange is proton abstraction, for which the transition state is $A \cdots H \cdots B$. Hence, in the context of the Brönsted relationship (eq. (15)), the rate constant for proton transfer is directly given by the rate constant for hydrogen exchange. Of course, proton transfer can also be measured by criteria other than isotopic exchange (see refs. 60–64 and Chapter 3), but at present we are concentrating on this particular aspect.

According to the Brönsted catalysis law, for a series of structurally similar carbon acids the logarithm of the rate constant for proton exchange, as determined with a standard base B under given experimental conditions, should be directly proportional to the pK_a of the carbon acids. Since eq. (15) makes no allowance for variation in solvation, conjugation and steric requirements, it is presumed that such factors remain constant throughout a given series of acids. A linear relationship between log k (exchange) and pK_a has in fact been observed for various series of carbon acids [1, 59, 62]. It follows that for an acid of unknown ionization constant, measurement of the rate of hydrogen exchange should provide a direct estimate of the pK_a, by substitution in eq. (15) for that series of carbon acids. [A number of studies of Brönsted catalysis have examined proton transfer *from a given carbon acid to a series of Brönsted bases*. A parallel relationship to eq. (15) will hold, i.e. log k (exchange) is proportional to log K_b for the series of structurally related bases (see refs. 62–65 for relevant examples.)]

The application of the hydrogen isotope exchange method to pK_a determination of some of the stronger carbon acids such as nitroalkanes is simple in principle and, indeed, hydrogen exchange in nitroalkanes was studied over 30 years ago [66]. However, standard electrochemical methods for pK_a determination will suffice here (ref. 61 and others cited therein). Of more direct interest are the data obtained by application of the base-induced isotopic exchange method to the series of carbon acids in the following listing:
ketones (with OD^-/D_2O [67, 68] or $Et_3N/D_2O/DMF$ [69]), sulfones (with CH_3O^-/CH_3OD [70]), alkynes (with $Et_3N/D_2O/DMF$ [71]),

monohydrofluoroalkanes such as CF_3H or $(CF_3)_3CH$ (with $CH_3O^-/$ CH_3OD [72, 73]), haloforms and mixed halomethanes (with OD^-/D_2O [74]), halogen-substituted methyl esters (with OD^-/D_2O [75]), polynitrobenzenes (with $OD^-/D_2O/DMF$ [76]), polyfluorobenzenes (with CH_3O^-/CH_3OT [77a]), unsubstituted polycyclic aromatic hydrocarbons such as naphthalene or phenanthrene (with $C_6H_{11}ND^-/C_6H_{11}ND_2$ [78]), fluorenes (with CH_3O^-/CH_3OT [79]), triphenylmethane and diphenylmethane (with CH_3O^-/CH_3OT [79] or ND_2^-/ND_3 [80]), toluene and the xylenes (with t-BuO$^-$/tritiated DMSO [81], $C_6H_{11}ND^-/C_6H_{11}ND_2$ [82], or ND_2^-/ND_3 [80]), benzene and ethylene ($C_6H_{11}ND^-/C_6H_{11}ND_2$ [83] or ND_2^-/ND_3 [80]), saturated hydrocarbons such as isopentane and cyclohexane (with ND_2^-/ND_3 [80]), cyclopropane and higher cycloalkanes ($C_6H_{11}ND^-/C_6H_{11}ND_2$ [84]) and, lastly, the parent carbon acid, methane ($C_6H_{11}ND^-/C_6H_{11}ND_2$ [85]).

The picture that emerges from the data summarised above is that even the simple saturated hydrocarbons can be regarded as acids according to the criterion of kinetic acidity. It should be born in mind, though, that with these extremely weak acids quite drastic experimental conditions are required to bring about exchange. For example cyclohexane with 1 M potassium amide in deuterated liquid ammonia exchanges on average only one of its hydrogens with the medium *after 180 hours at 120°*!

In concluding this section, we may indicate more explicitly how the current estimates of the acidities of methane, ethylene, and cyclopropane have been arrived at. Deuterium exchange with cesium cyclohexylamide in cyclohexylamine at 50° shows that methane is 3.1×10^{-2} as reactive as cyclopropane (per hydrogen) and 2.2×10^3 as reactive as cyclohexane [85]. The most recent estimate of the kinetic acidity of cyclohexane is given [86]* as 51, so that extrapolation yields, approximately, the pK_a of methane as 48, and the pK_a of cyclopropane as 46. In making these estimates one assumes that a Brönsted relationship (slope $\sim$1.0) holds between rate of proton exchange and pK_a. This is a hazardous assumption in this case since the hybridization at the reacting C–H does not remain constant. Nevertheless, it is likely that the pK_a values are not

---

*The kinetic acidity of cyclohexane is eight powers of 10 less than of benzene (pK_a 43); also the pK_a of toluene is 41 on a scale in which triphenylmethane is 31.5 [86].

further out than about $1-2$ pK units, and that the formerly accepted value for the pK_a of methane of 40 was on the low side. Similarly, the kinetic acidity of ethylene, relative to that of benzene, has been determined toward cesium cyclohexylamide in cyclohexylamine [83b], and has yielded the rate constant ratio, $k(C_2H_3D)/k(C_6H_5D) = 0.08$. Thus ethylene has a kinetic acidity which is about 1 pK unit less than that of benzene. The pK_a of benzene has been estimated as 43 (extra-polation of kinetic and equilibrium measurements for fluorinated benzenes in CsCHA/CHA) [77b], which yields for ethylene a pK_a of 44.

A summary of the pK_a values of the hydrocarbons which have been considered in this and the previous sections is given in Table 3.

Table 3

ACIDITIES OF SOME HYDROCARBONS[a]

Compound	pK_a	Compound	pK_a
Cyclohexane	51	Diphenylmethane	34
Ethane	49	Triphenylmethane	31.5
Methane	48	Acetylene	25
Cyclopropane	46	Fluorene	22.9
Ethylene	44	9-Phenylfluorene	18.5
Benzene	43	Indene	18.5
Triptycene	42	Cyclopentadiene	15
Toluene	41	Fluoradene	11
Propene	40	Compound 10	9.8
Cycloheptatriene	36	Compound 9	5.9

[a] See text for discussion of dependence of pK_a value on method of measurement and the consequent uncertainties in the derived pK_a values, particularly in the case of the very weak carbon acids given in the left column. No correspondence with an absolute pK_a scale is meant to be implied by this Table.

4. THE PROBLEM OF AN ACIDITY SCALE

It is appropriate at this stage to ask whether there is mutual consistency among the various methods for pK_a determination which were considered in the previous sections. Thus, if a given hydrocarbon acid were examined by more than one method, would a single, "absolute", value be obtained? The answer to this question must be a qualified negative. Though the agreement may usually be satisfactory (i.e. within 1 pK unit) for moderately weak acids (p$K_a \leqslant 20$), it is generally

20

unsatisfactory with the very weak acids ($pK_a \geqslant 25$). In other words a single absolute pK_a scale that would cover hydrocarbon acids which are moderately strong ($pK_a \sim 15$) and extremely weak ($pK_a \sim 40$) is not yet available. Actually, we should not expect that measurements made in dilute aqueous medium would give results which are interchangeable with those made in, say, pure dimethyl sulfoxide, ammonia, cyclohexyl-amine, or in benzene. The standard state in each case is the pure solvent. The solvation of the ions and neutral molecules will not remain constant as the solvent is varied, and thus the activities of the various species will not be invariant. Let us consider the problem from a quantitative view-point.

The *intrinsic acidity* of a carbon acid is given by its tendency to ionize according to eq. (16); the equilibrium constant for this process is given the symbol K_{abs}, denoting the absolute acidity constant:

$$(16) \quad RH \rightleftharpoons R^- + H^+$$

$$(17) \quad K_{abs} = \frac{a'_{R^-} \times a'_{H^+}}{a'_{RH}}$$

The a'_i terms in eq. (17) refer to the absolute activities which can be defined as equivalent to mole fractions in the gas phase. Measurement of *gas-phase acidities* is actually feasible, by means of the recently developed method of ion cyclotron resonance [87–91]. To date the measurements reported have been primarily for binary inorganic acids, as well as for the common aliphatic alcohols, and normally yield relative orders of gas phase acidities (e.g., $H_2S > AsH_3 > PH_3 > SiH_4 > H_2O > NH_3 > CH_4$). Where relative orders of acidity of carbon and oxygen acids have been estimated, some rather startling observations have resulted, such as that toluene and propene are *stronger* acids than water in the gas phase! It is clear that information derived by this method may potentially reveal intrinsic electronic effects free of solvation effects. Correlation of theoretical acidity parameters with experimental results should also become more meaningful then.

The aqueous medium provides traditionally the reference state for the definition of thermodynamic acidity constants. The protolytic equilibrium in water involves the transfer of a proton to a solvent molecule which acts as a Brönsted base (eq. (18)); the thermodynamic acidity constant is then given by eq. (19), since a_{H_2O} is

21

(18) $\quad$ RH + H$_2$O $\rightleftharpoons$ R$^-$ + H$_3$O$^+$

(19) $\quad K_a = \dfrac{a_{R^-} \times a_{H_3O^+}}{a_{RH}} = \dfrac{[R^-][H_3O^+]}{[RH]} \cdot \dfrac{f_{R^-} \times f_{H_3O^+}}{f_{RH}}$

taken as constant. The f_i terms are activity coefficients, which tend to unity at infinite dilution of the solute for the aqueous standard state.

As we have noted, for weak carbon acids the dissociation in aqueous medium is not measurable, and a more basic medium must be used. In such a solvent, S, the acid dissociation constant, K_S, will be given by

(20) $\quad K_S = \dfrac{a_{R^-} \times a_{H^+}}{a_{RH}} = \dfrac{[R^-][H^+]}{[RH]} \cdot \dfrac{f_{R^-}^S \times f_{H^+}^S}{f_{RH}^S}$

We may therefore relate acidity constants in the aqueous medium and in solvent S by the expression

(21) $\quad K_a = K_S \cdot \dfrac{f_{R^-} \times f_{H_3O^+} \times f_{RH}^S}{f_{R^-}^S \times f_{H^+}^S \times f_{RH}}$

The quotients $f_{R^-}/f_{R^-}^S$, etc. (sometimes denoted as $^\circ\gamma_i^S$) give a measure of the free energy of transfer of one mole of the species i from the standard state in water to the standard state in solvent S and may be termed the "medium effect". Such quotients have been evaluated for a number of systems of interest [92–96], but they are not available for ionization of the weak carbon acids, with which we are primarily concerned.

Thus, in practice, evaluation of pK's of weak carbon acids is based on extrapolative methods which use a standard acid as the point of reference. However, the choice of the standard acid becomes of crucial importance. Carbon acids whose ionization constants can be measured in water include 9-cyanofluorene (pK_a 11.4) and malononitrile (pK_a 11.2); these are actually used as reference compounds in the acidity function approach for study of weaker acids [27]. This procedure leads to a secondary reference acid, given by 9-phenylfluorene, for which the acidity function method in aqueous or alcoholic DMSO systems yields a pK_a of 18.5. The latter is then used as the "anchor" for pK measurement in the CsCHA/CHA systems [39]. However, the fact that acidities may not be constant as the medium is changed, is illustrated by measurement of the pK of 9-t-butyl-fluorene in Me$_4$NOH—DMSO—H$_2$O which yields the value [97] 23.4, while in CsCHA—CHA the value 24.3 is obtained [40]. This difference in measured acidity reflects the different sensitivity of the species involved in the ionization equilibria in the two media to solute—solute and solute—solvent interactions. Since differential medium effects are to be expected generally, it becomes apparent that a uniform pK scale is unlikely to result when the measurements must, by necessity, be made in basic systems of widely diverging characteristics.

As stated previously, the first comprehensive attempt at the construction of an acidity scale was made by McEwen, in 1936 [38]. This scale was based on equilibrium measurements, in benzene-ether, of the type RH + R′M $\rightleftharpoons$ RM + R′H. Nearly 30 years [2] later another acidity scale was constructed by Cram which represented a combination of the data obtained by McEwen, Streitwieser, Applequist, and Dessy. The result, known as the MSAD scale, was based on thermodynamic as well as kinetic acidity measurements. It was realized [2] that further work would lead to modifications of this scale and the studies carried out in the last decade have borne this out. Moreover, as our discussion has shown, the idea of an all-encompassing acidity scale has now receded into the background, at least until such time as new experimental methods are perfected and our understanding of medium effects becomes greatly advanced.

5. KINETIC ISOTOPE EFFECTS AND THE STUDY OF CARBON ACIDS

An important aspect of the kinetic acidity of carbon acids remains to be discussed, namely the kinetic isotope effect in proton transfer processes. A brief and much simplified account is given here; for full and rigorous treatments the reader is referred to Melander's text [98] and to several review articles [99—103].

We consider in the present context the reaction (22) in which a carbon-hydrogen, or carbon-deuterium or -tritium, covalent bond is ruptured in a slow, rate-controlling, step:

$$(22) \quad -\underset{|}{\overset{|}{C}}-H \,(D,T) \,+\, B^- \xrightarrow{\text{slow}} -\underset{|}{\overset{|}{C}}{}^- \,+\, B-H(D,T)$$

One may represent this process by the potential energy profile shown in Fig. 1, which leads us to the *zero-point-energy* explanation of the

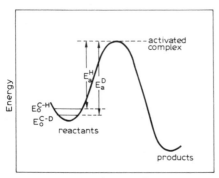

Figure 1. Potential energy-reaction coordinates profile illustrating the zero point energy origin of the kinetic isotope effect for carbon-hydrogen versus carbon-deuterium bond rupture.

kinetic isotope effect. In this figure, the C–H bonds are shown in the zeroth vibrational level, i.e. the vibrational quantum number, v, is zero. This lowest energy level, known as the zero point energy, is the vibrational energy retained even at $0°K$. Most (~99%) molecules will be in this vibrational level at room temperature owing to the magnitude of the transition to the next higher level ($v = 1$). The zero point energy is dependent on nuclear mass and is therefore different for the C–H and C–D bonds. On the other hand the overall nature of the potential energy curve is unchanged on isotopic substitution since the electron distribution in the molecule, which governs the potential energy surface, is to a good approximation independent of the mass numbers of the constituent isotopes.

The zero point energy is given by $E_0 = 1/2(hv)$, where h is Planck's constant and v is the vibrational frequency of the bond concerned. Now $v \propto 1/\sqrt{\mu}$, with μ, the reduced mass, given by $\mu = m_1 m_2/(m_1 + m_2)$ where m_1, m_2 are the masses of the particles linked by the covalent bond. It will be seen that the ratio μ_H/μ_D approaches 2 for bonds to hydrogen and deuterium as the mass of the fragment to which the proton is attached becomes large; correspondingly, the frequency ratio v_H/v_D will approach $\sqrt{2}$. This relationship can readily be verified by measurement of infrared stretching frequencies of isotopically substituted molecules. From the values of $v_{C–H}$ and $v_{C–D}$ one may calculate $E_0^{C–H}$ and $E_0^{C–D}$ as 4.15 and 3.00 kcal/mole, respectively.

24

For bond rupture in the rate-determining step, it is seen from Fig. 1 that the activation energy required for rupture of the C–D bond is larger than for C–H rupture, and the difference between the two quantities is equal to the zero point energy difference. Thus the kinetic isotope effect is

$$(23) \quad \frac{k_H}{k_D} = e^{-(E_a^H - E_a^D)/RT}$$

$$= e^{(E_o^H - E_o^D)/RT} = e^{\Delta E_o/RT}$$

Values of k_H/k_D obtained from this equation are 8.3 at 0°C, 6.9 at 25°C, and 4.3 at 100°C. Similarly, one may calculate kinetic isotope effect values for rupture of bonds involving other isotopes. For example, for carbon-tritium bond rupture one obtains $k_H/k_T = 20$ at 0°C. Of interest, also, is the relationship which one can derive between deuterium and tritium isotope effects [104]:

$$(24) \quad \frac{k_H}{k_T} = \left(\frac{k_H}{k_D}\right)^{1.44}$$

$$(25) \quad \frac{k_H}{k_D} = \left(\frac{k_D}{k_T}\right)^{2.26}$$

A more complete treatment of kinetic isotope effects would be based on the theory of absolute reaction rates and would use as the starting point relevant equations for rate constants of reaction of the two isotopically substituted molecules. However, the expressions that result from such treatment can only be solved by making a number of simplifying assumptions which, in the most extreme case, reduce to the equation derived in the zero point energy method, eq. (23). Actually, one may work quite usefully with this equation for many purposes, provided its limitations are recognized and allowed for. Thus it is possible to arrive at an improved state of understanding of the application of the isotope effect criterion in proton transfer processes by taking into account some additional factors relevant to such systems.

In the first place, it was assumed in the above discussion that only a single C–H stretching vibration need be considered, or in other words that non-reacting bonds are not affected in any way, whereas in actual

25

reactions this C—H vibration may often be coupled to other vibrational modes. (Further exploration of this aspect could lead us to the topic of *secondary isotope effects* [105], i.e. the effect of isotopic substitution at non-reacting atomic centres, whereas our discussion is restricted to primary isotope effects.) Secondly, for polyatomic molecules one should also consider C—H bending modes; their inclusion via the zero point energy term could augment the kinetic isotope effect by about half as much again from the values indicated above. Thirdly, the possibility of quantum mechanical tunneling [106] has been neglected. Thus there is a finite probability that an isotope of hydrogen will "tunnel through" the energy barrier instead of passing over the hump of the potential energy-reaction coordinate curve as indicated in Fig. 1. Since the possibility of tunneling is more favourable for the lighter isotope, this will lead to abnormally high values of the hydrogen isotope effect. Hydrogen-deuterium isotope effects as large as 25—30, and hydrogen-tritium isotope effects as large as 70—80, have been observed and ascribed to the tunneling phenomenon [107]. A relevant case is that of proton transfer from 2-nitropropane to a series of pyridine bases, where k_H/k_D increases markedly with steric hindrance at the reaction site, reaching a maximum value of 24.3 for 2,4,6-trimethylpyridine; the tritium isotope effect is 79.1 [107]. Other cases of abnormally large isotope effects have also been recorded (e.g. [108]).

Now we come to examine an underlying assumption of the zero point energy treatment as given above, namely that the transition state or activated complex has no residual vibrational frequency which could affect the magnitude of the isotope effect via a contribution from a zero point energy term. However, in a proton abstraction process involving an acid AH, the proton actually becomes transferred to a base B, so that one must consider a three-centred transition state, A· · ·H· · ·B, rather than simply A· · ·H. Now if the three atomic centres A.H.B are assumed to be collinear then there will be two possible vibrational modes, 11 and 12, characteristic of the activated complex:

$$\overset{\leftarrow}{A} \cdots \vec{H} \cdots \overset{\leftarrow}{B} \qquad\qquad \overset{\leftarrow}{A} \cdots H \cdots \vec{B}$$

| 11 | 12 |

The asymmetric stretching vibration 11 corresponds to motion along the reaction coordinate; it is an imaginary vibration and has no zero point energy term associated with it. The symmetric stretching vibration **12**, however, is retained in the transition state, so that we need actually write:

$$(26) \qquad \overset{\leftarrow}{A} - \vec{H} + B \longrightarrow [\overset{\leftarrow}{A} \cdots H \cdots \vec{B}]^{\ddagger}$$

Now if the transition state were truly symmetrical then there would be no motion of the hydrogen in **12** and hence deuterium substitution would result in no zero point energy contribution. However, if the hydrogen atom is unequally bonded to A and B an asymmetric transition state would arise, $A \cdot \cdot H \cdot \cdot \cdot \cdot B$ or $A \cdot \cdot \cdot \cdot H \cdot \cdot B$, in which case a residual zero point energy term due to the vibration corresponding to **12** would be maintained in the transition state. This would partially cancel the zero point energy contribution of the ground state and hence result in a decreased kinetic isotope effect compared to the value expected from eq. (23), as seen from the following relationship:

$$(27) \qquad \frac{k_H}{k_D} = e^{[(E_0^H - E_0^D) - (E_0^{\ddagger H} - E_0^{\ddagger D})]/RT}$$

$$= e^{(\Delta E_0 - \Delta E_0^{\ddagger})/RT}$$

It should be emphasized that the possibility of unsymmetrical transition states in proton transfer processes is a very real one. It is probable that only in the case that A and B have equal base strengths, and are structurally similar, that one would expect equal bonding of the hydrogen to A and B. In fact the available evidence indicates that for $\Delta pK = 0$ a maximum isotope effect is obtained [109–111], while smaller values result when this condition does not hold. Model calculations of hydrogen isotope effects on such systems have been performed and have been extended also to non-linear transition states [112–114].

Small primary hydrogen isotope effects may also have their origin in another kind of explanation, based on a wholly different phenomenon, that of "internal return", which is the result of ion pairing and the diffusion among hydrogen-bonded species in solution. Thus proton abstraction by base gives first a hydrogen-bonded carbanion which may return to the initial carbon acid at a faster rate than it dissociates to free carbanion, so that proton abstraction is no longer rate-controlling. The

27

kinetic scheme for internal return is indicated in the following equation, with the condition that $k_{-1} \gg k_2$:

(28) $\quad C - D + {}^-OR \underset{k_{-1}}{\overset{k_1}{\rightleftarrows}} [C^- \cdots DOR] \xrightarrow{k_2} C^- + DOR$

Some of the reactions in which internal return is believed to predominate are the following: proton abstraction from 2-phenylbutane and 2-phenylbutane-2-d with potassium t-butoxide in dimethyl sulfoxide containing t-butyl alcohol [115], $k_H/k_D = 3$; also, for 1-phenylmethoxyethane [115], $k_H/k_D = 1.7$ and for toluene (α-position) [81] $k_H/k_D = 0.6$. Internal return can also be important for proton transfer processes in the more dissociating hydroxylic media. For instance [77a], for the polyfluorobenzenes undergoing proton exchange with sodium methoxide in methanol, $k_H/k_D = 1.0$. Proton exchange by the haloforms is similarly associated with rather small isotope effects ($k_H/k_D = 2-3$) [9]. However, it is essential in such cases to have good evidence that the origin of the small isotope effect values is in fact internal return, rather than a reactant-like or product-like transition state. This is an inherent ambiguity in the interpretation of small isotope effects which it may not always be possible to eliminate.

Now we may return to the discussion of kinetic acidity (Section 3) from the viewpoint of the isotope effect criterion. Consider, therefore, the experimental design in a typical isotopic exchange reaction, as implied in equations (13a, b). The experimental limitation of this exchange system is that it does *not*, in fact, allow a valid k_H/k_D isotope rate constant ratio to be measured *directly*. For instance, in the case of deuteroxide ion catalyzed isotopic exchange where $B^- = OD^-$ and $BD = D_2O$, we may measure the rate of exchange for the $AH/OD^-/D_2O$ combination and compare it with the rate for the $AD/OH^-/H_2O$ combination. However, we now have a superposition of a *substrate isotope effect* on top of a *solvent isotope effect* since we are changing two variables simultaneously. The problem can be solved in the following way. Suppose we measure the rate of isotopic exchange for the two systems $AD/OH^-/H_2O$ and $AT/OH^-/H_2O$. In the former case we would be measuring hydrogen-deuterium exchange and in the latter case hydrogen-tritium exchange. In either case one could monitor the isotopic composition of the substrate or of the reaction medium, at various times as reaction proceeds. The isotope rate constant ratio k_D/k_T is then

28

obtained directly from the two sets of measurements since the medium effect variable has been kept constant. Having measured k_D/k_T we can calculate k_H/k_D by means of eq. (25). We may illustrate with several examples in which isotope effect measurements have been related to kinetic acidity, selecting mainly those studies which are mentioned elsewhere in this chapter.

The kinetic acidity of hydrocarbons has been extensively investigated by means of proton exchange in the lithium (or cesium) cyclohexyl-amide—cyclohexylamine system and we consider the application of the isotope effect criterion in this system in the first instance. For the isotopic exchange reaction of toluene-α-d and toluene-α-t with lithium cyclohexylamide in cyclohexylamine [116], the measured isotope rate constant ratio is $k_D/k_T = 2.8$, from which one calculates $k_H/k_D = 10$. Similarly [117] for proton exchange of benzene-d and benzene-t with cesium cyclohexylamide $k_D/k_T = 2.5$, leading to $k_H/k_D = 8$. The fluorenyl system has been recurrent in our discussion of carbanions and it is noteworthy that fair-sized isotope effects have been obtained in the exchange of fluorene-9-d and fluorene-9-t with two basic media: with liquid ammonia as the base [80] $k_D/k_T = 1.9$ ($k_H/k_D = 5$), while with sodium methoxide in methanol [5] $k_D/k_T = 2.2$ ($k_H/k_D = 6$). The isotope effect results for these several systems constitute strong evidence for a large degree of carbon-hydrogen bond weakening in the transition states of the proton exchange processes, in accord with equations (13a, b).

$$\frac{k(C_6H_5CD_3)}{k(C_6H_5CT_3)} = 2.8 \text{ (with } C_6H_{11}NHLi/C_6H_{11}NH_2)$$

$$\frac{k(\text{fluorene-9-}d)}{k(\text{fluorene-9-}t)} = 2.2 \text{ (with } CH_3ONa/CH_3OH)$$

In conclusion of this section, it can be noted that the isotope effect criterion has clearly been of great value in the study of proton transfer from carbon acids. However, while the finding of normal isotope effects provides clear evidence for rate-determining proton transfer, the interpretation of small primary isotope effects may not always be unambiguous. The observation of abnormally large isotope effects is indicative of the tunneling effect.

REFERENCES

1   R.P. Bell, "The Proton in Chemistry", 2nd edition, Cornell University Press, Ithaca, New York, 1973.
2   D.J. Cram, "Fundamentals of Carbanion Chemistry", Academic Press, New York, 1965, Chapter 1.
3   E.M. Kaiser and D.W. Sloacum, in "Organic Reactive Intermediates", ed. S.P. McManus, Academic Press, New York, 1973.
4   H.F. Ebel, "The Acidity of CH Acids", Georg Thieme Verlag, Stuttgart, 1969.
5   A. Streitwieser, Jr. and J.H. Hammons, Prog. Phys. Org. Chem., 3 (1965) 41.
6   H. Fischer and D. Rewicki, Prog. Org. Chem., 7 (1968) 116.
7   A.I. Shatenshtein and I.O. Shapiro, Russ. Chem. Revs., 37 (1968) 845.
8   J.R. Jones, Quart. Rev. Chem. Soc. (London), 25 (1971) 365; "The Ionization of Carbon Acids", Academic Press, London, 1973.
9   J. Hine, R. Wiesbock and R.G. Ghirardelli, J. Amer. Chem. Soc., 83 (1961) 1219; Z. Margolin and F.A. Long, J. Amer. Chem. Soc., 95 (1973) 2757.
10 G. Fraenkel, J.G. Russell and Y.H. Chen, J. Amer. Chem. Soc., 95 (1973) 3208.
11 R. Hoffmann, R. Bissell and D. Farnum, J. Phys. Chem., 73 (1969) 1789.
12 J.J. Brooks and G.D. Stucky, J. Amer. Chem. Soc., 94 (1972) 7333.
13 A. Streitwieser, Jr., R.A. Caldwell and M.R. Granger, J. Amer. Chem. Soc., 86 (1964) 3578; A. Streitwieser, Jr., M.J. Maskornick and G.R. Ziegler, Tetrahedron Lett., No. 42 (1971) 3927.
14 C. Bugg, R. Desiderato and R.L. Sass, J. Amer. Chem. Soc., 86 (1964) 3157.
15 C.A. Coulson and W. Moffitt, Phil. Mag., 40 (1949) 1.
16 H.A. Bent, Chem. Rev., 61 (1961) 275.
17 E. Hückel, Z. Physik., 70 (1931) 204; 76 (1932) 628.
18 J.S. McKennis, L. Brener, J.R. Schweiger and R. Pettit, Chem. Commun., 365 (1972).
19 T.J. Katz, J. Amer. Chem. Soc., 82 (1960) 3784, 3785.
20 T.J. Katz and P.J. Garratt, J. Amer. Chem. Soc., 86 (1964) 5194.
21 E.A. LaLancette and R.E. Benson, J. Amer. Chem. Soc., 87 (1965) 1941.
22 R.T. Seidner and S. Masamune, Chem. Commun., 149 (1972).
23 P. Radlick and W. Rosen, J. Amer. Chem. Soc., 88 (1966) 3461; 89 (1967) 5308.
24 J.F.M. Oth and G. Schröder, J. Chem. Soc. (B), 904 (1971).
25 P.J. Garratt, N.E. Rowland and F. Sondheimer, Tetrahedron, 27 (1971) 3157.
26 H. Rapoport and G. Smolinsky, J. Amer. Chem. Soc., 82 (1960) 934.
27 K. Bowden and R. Stewart, Tetrahedron, 21 (1965) 261; K. Bowden and A.F. Cockerill, J. Chem. Soc. (B), 173 (1970).
28 O.W. Webster, J. Amer. Chem. Soc., 88 (1966) 3046.
29 J.D. Taylor and J.F. Wolf, Chem. Commun., 876 (1972).
30 D. Darwish and R.L. Tomilson, J. Amer. Chem. Soc., 90 (1968) 5938.
31 P.A. Grieco and C.S. Pogonowski, J. Amer. Chem. Soc., 95 (1973) 3071.
32 A.W. Johnson, "Ylid Chemistry", Academic Press, New York, 1966.
33 R. Kuhn and D. Rewicki, Ann., 706 (1967) 250; Angew. Chem., 79 (1967) 648.

34   C.D. Ritchie and R.E. Uschold, J. Amer. Chem. Soc., 89 (1967) 1721, 2752; 90 (1968) 2821.

35   J. Badoz-Lambling, M. Herlem and A. Thiebault, Analyt. Letters, 2(1) (1969) 35; M. Herlem and A. Thiebault, Bull. Soc. Chim. France, 383 (1970).

36   T.L. Brown, Advan. Organometal. Chem., 3 (1965) 365.

37   J.B. Conant and G.W. Wheland, J. Amer. Chem. Soc., 54 (1932) 1212.

38   W.K. McEwen, J. Amer. Chem. Soc., 58 (1936) 1124.

39   A. Streitwieser, Jr., E. Ciuffarin and J.H. Hammons, J. Amer. Chem. Soc., 89 (1967) 63.

40   A. Streitwieser, Jr., C.J. Chang and D.M.E. Reuben, J. Amer. Chem. Soc., 94 (1972) 5730.

41   J.H. Takemoto and J.J. Lagowski, Inorg. Nucl. Chem. Letters, 6 (1970) 315.

42a A. Streitwieser, Jr., C.J. Chang, W.B. Hollyhead and J.R. Murdoch, J. Amer. Chem. Soc., 94 (1972) 5288.

42b P.C. Mowery and A. Streitwieser, Jr., in "Ions and Ion Pairs in Organic Reactions", edited by M. Szwarc, Interscience, New York, in press.

43   S. Winstein and G.C. Robinson, J. Amer. Chem. Soc., 80 (1958) 169.

44   D.J. Cram, F. Hauck, K.R. Kopecky and W.D. Nielsen, J. Amer. Chem. Soc., 81 (1959) 5767.

45a M. Szwarc, in "Ions and Ion-Pairs in Organic Reactions", Vol. 1, edited by M. Szwarc, Interscience, New York, 1972.

45b J. Smid, in "Ions and Ion-Pairs in Organic Reactions", Vol. 1, edited by M. Szwarc, Interscience, New York, 1972.

46   D.J. Cram, J.L. Mateos, F. Hauck, A. Langemann, K.R. Kopecky, W.D. Nielsen and J. Allinger, J. Amer. Chem. Soc., 81 (1959) 5774; D.J. Cram, B. Rickborn and G.R. Knox, ibid., 82 (1960) 6412.

47   C.H. Langford and R.L. Burwell, J. Amer. Chem. Soc., 82 (1960) 1503.

48   A.J. Parker, Quart. Rev. Chem. Soc. (London), 16 (1962) 163.

49   D. Dolman and R. Stewart, Can. J. Chem., 45 (1967) 911.

50   E.C. Steiner and J.M. Gilbert, J. Amer. Chem. Soc., 85 (1963) 3054; 87 (1965) 382; E.C. Steiner and J.D. Starkey, ibid., 89 (1967) 2751.

51   L.P. Hammett, "Physical Organic Chemistry", 2nd Edition, McGraw Hill, New York, 1970, Chapter 9.

52   C.H. Rochester, "Acidity Functions", Academic Press, London, 1970.

53   J.R. Jones, Prog. Phys. Org. Chem., 9 (1972) 241.

54a E. Buncel, E.A. Symons, R. Stewart and D. Dolman, Can. J. Chem., 48 (1970) 3354.

54b D. Bethell and A.F. Cockerill, J. Chem. Soc. (B), 913 (1966).

55   E. Buncel, E.A. Symons and A.W. Zabel, Chem. Commun., 173 (1965).

56   A. Albagli, R. Stewart and J.R. Jones, J. Chem. Soc. (B), 1509 (1970).

57   J.I. Brauman, J.A. Bryson, D.C. Kahl and N.J. Nelson, J. Amer. Chem. Soc., 92 (1970) 6679.

58   J.N. Brönsted and K.J. Pedersen, Z. Physikal. Chem., 108 (1924) 185.

59   M. Eigen, Angew. Chem. Int. Ed., 3 (1964) 1.

60 F.G. Bordwell, W.J. Boyle, Jr., J.A. Hautala and K.C. Lee, J. Amer. Chem. Soc., 91 (1969) 4002.
61 M. Fukuyama, P.W.K. Flanagan, F.J. Williams, Jr., L. Frainier, S.A. Miller and H. Shechter, J. Amer. Chem. Soc., 92 (1970) 4689.
62 A.J. Kresge, Chem. Soc. Revs. (London), 2 (1973) 475.
63 J.E. Dixon and T.C. Bruice, J. Amer. Chem. Soc., 92 (1970) 905; M.J. Gregory and T.C. Bruice, ibid., 89 (1967) 2327.
64 R.G. Pearson and R.L. Dillon, J. Amer. Chem. Soc., 75 (1953) 2439.
65 F. Hibbert and F.A. Long, J. Amer. Chem. Soc., 94 (1972) 2647.
66 K.F. Bonhoeffer, K.H. Geib and O. Reitz, J. Chem. Phys., 7 (1939) 644.
67 J.R. Jones and R. Stewart, J. Chem. Soc. (B), 1173 (1967).
68 J. Warkentin and C. Barnett, J. Amer. Chem. Soc., 90 (1968) 4629.
69 H.W. Amburn, K.C. Kauffman and H. Shechter, J. Amer. Chem. Soc., 91 (1969) 530.
70 Ref. 60, footnote 5.
71 R.E. Dessy, Y. Okuzumi and A. Chen, J. Amer. Chem. Soc., 82 (1962) 2899.
72 S. Andreades, J. Amer. Chem. Soc., 86 (1964) 2003.
73 K.J. Klabunde and D.J. Burton, J. Amer. Chem. Soc., 94 (1972) 5985.
74 J. Hine, "Physical Organic Chemistry", McGraw Hill, New York, 1962, p. 487.
75 J. Hine, L.G. Mahone and C.L. Liotta, J. Amer. Chem. Soc., 89 (1967) 5911.
76 E. Buncel and E.A. Symons, J. Org. Chem., 38 (1973) 1201; E. Buncel and A.W. Zabel, J. Amer. Chem. Soc., 89 (1967) 3082.
77a A. Streitwieser, Jr., J.A. Hudson and F. Mares, J. Amer. Chem. Soc., 90 (1968) 648.
77b A. Streitwieser, Jr., P.J. Scannon and H.M. Niemeyer, J. Amer. Chem. Soc., 94 (1972) 7936.
78 A. Streitwieser, Jr. and R.G. Lawler, J. Amer. Chem. Soc., 87 (1965) 5388.
79a A. Streitwieser, Jr., J.I. Brauman, J.H. Hammons and A.H. Pudjaatmaka, J. Amer. Chem. Soc., 87 (1965) 384.
79b D.J. Cram and W.D. Kollmeyer, J. Amer. Chem. Soc., 90 (1968) 1791.
80 A.I. Shatenshtein, Adv. Phys. Org. Chem., 1 (1963) 155; Tetrahedron, 18 (1962) 95.
81 J.E. Hofmann, R.J. Muller and A. Schriesheim, J. Amer. Chem. Soc., 85 (1963) 3002; J.E. Hofmann, A. Schriesheim and R.E. Nickols, Tetrahedron Lett., 22 (1965) 1725.
82 A. Streitwieser, Jr. and H.F. Koch, J. Amer. Chem. Soc., 86 (1964) 404.
83a A. Streitwieser, Jr., R.G. Lawler and C. Perrin, J. Amer. Chem. Soc., 87 (1965) 5383.
83b M.J. Maskornick and A. Streitwieser, Jr., Tetrahedron Lett., No. 17 (1972) 1625.
84 A. Streitwieser, Jr., R.A. Caldwell and W.R. Young, J. Amer. Chem. Soc., 91 (1969) 529.
85 A. Streitwieser, Jr. and D.R. Taylor, Chem. Commun., 1248 (1970).
86 A. Streitwieser, Jr., private communication (1973).
87 J.D. Baldeschwieler and S.S. Woodgate, Accounts Chem. Res., 4 (1971) 114.

88 J.L. Beauchamp, Ann. Rev. Phys. Chem., 22 (1971) 527.

89 J.I. Brauman and L.K. Blair in "Determination of Organic Structures by Physical Methods", Vol. 5, edited by F.C. Nachod and J.J. Zuckerman, Academic Press, New York, 1973; J.I. Brauman, C.A. Lieder and M.J. White, J. Amer. Chem. Soc., 95 (1973) 927.

90 L.B. Young, E. Lee-Ruff and D.K. Bohme, Can. J. Chem., 49 (1971) 979; D.K. Bohme, E. Lee-Ruff and L.B. Young, J. Amer. Chem. Soc., 94 (1972) 5153.

91 R.T. McIver, Jr., J.A. Scott and J.M. Riveros, J. Amer. Chem. Soc., 95 (1973) 2706.

92 R.G. Bates, in "Solute-Solvent Interactions", edited by J.F. Coetzee and C.D. Ritchie, M. Dekker, New York, 1969.

93 R.H. Boyd in ref. 92.

94 C.D. Ritchie in ref. 92.

95 B.W. Clare, D. Cook, E.C.F. Ko, Y.C. Mac and A.J. Parker, J. Amer. Chem. Soc., 88 (1966) 1911.

96 M.K. Chantooni, Jr. and I.M. Kolthoff, J. Phys. Chem., 77 (1973) 527.

97 K. Bowden, A.F. Cockerill and J.R. Gilbert, J. Chem. Soc. (B), 179 (1970).

98 L. Melander, "Isotope Effects on Reaction Rates", The Ronald Press, New York, 1960.

99 J. Bigeleisen and M. Wolfsberg, Adv. Chem. Phys., 1 (1958) 16.

100 F.H. Westheimer, Chem. Rev., 61 (1961) 265.

101 W.H. Saunders, Jr., in "Investigation of Rates and Mechanisms of Reactions", Technique of Organic Chemistry, Vol. 8, Part 1, edited by S.L. Friess, E.S. Lewis and A. Weissberger, Interscience, New York, 1961.

102a W.A. Van Hook, in "Isotope Effects in Chemical Reactions", edited by C.J. Collins and N.S. Bowman, A.C.S. Monograph 167, Van Nostrand Reinhold, New York, 1970.

102b E.K. Thornton and E.R. Thornton in ref. 102a.

103 M. Wolfsberg, Accounts Chem. Res., 5 (1972) 225.

104 C.G. Swain, E.C. Stivers, J.F. Reuwer, Jr. and L.J. Schaad, J. Amer. Chem. Soc., 80 (1958) 5885.

105 E.R. Thornton, Ann. Rev. Phys. Chem., 17 (1966) 349.

106 E.F. Caldin, Chem. Rev., 69 (1969) 135; M.D. Harmony, Chem. Soc. Revs., 1 (1972) 211.

107 E.S. Lewis and L.H. Funderburk, J. Amer. Chem. Soc., 89 (1967) 2322; E.S. Lewis and J.K. Robinson, J. Amer. Chem. Soc., 90 (1968) 4337.

108 H. Wilson, J.D. Caldwell and E.S. Lewis, J. Org. Chem., 38 (1973) 564.

109 R.P. Bell, W.H. Sachs and R.L. Tranter, Trans. Faraday Soc., 67 (1971) 1995; R.P. Bell and B.G. Cox, J. Chem. Soc. (B), 194 (1970); 783 (1971).

110a A.F. Cockerill, S. Rottschaefer and W.H. Saunders, Jr., J. Amer. Chem. Soc., 89 (1967) 901; A.F. Cockerill, J. Chem. Soc. (B), 964 (1967).

110b L. Melander and N.-A. Bergman, Acta Chem. Scand., 25 (1971) 2264.

111a J.L. Longridge and F.A. Long, J. Amer. Chem. Soc., 89 (1967) 1292.

111b A.J. Kresge and Y. Chiang, J. Amer. Chem. Soc., 91 (1969) 1025.

112 N.-A. Bergman, W.H. Saunders, Jr. and L. Melander, Acta Chem. Scand., 26 (1972) 1130.
113 R.A. More O'Ferrall and J. Kouba, J. Chem. Soc. (B), 985 (1967).
114 R.A. More O'Ferrall, J. Chem. Soc. (B), 785 (1970).
115 D.J. Cram, C.A. Kingsbury and B. Rickborn, J. Amer. Chem. Soc., 83 (1961) 3688.
116 A. Streitwieser, Jr., W.C. Langworthy and D.E. Van Sickle, J. Amer. Chem. Soc., 84 (1962) 251.
117 A. Streitwieser, Jr. and R.A. Caldwell, J. Amer. Chem. Soc., 87 (1965) 5394.

Chapter 2

STEREOCHEMISTRY OF CARBANIONIC PROCESSES

1. PLANAR VERSUS PYRAMIDAL CONFIGURATIONS

The stereochemical course of reactions proceeding by way of carbanion intermediates is of considerable interest and importance in our understanding of known processes and in the prediction of others. Let us examine in this light the molecular geometry of carbanionic species.

The methyl carbanion, as structurally the simplest carbanionic species, is considered first. The question is, what is the spatial orientation of the three coordinated hydrogens, and of the unshared pair of electrons, with respect to the carbon nucleus?

The two most probable configurations involve the following hybridization states for carbon: sp^2 and sp^3. Since sp^2 hybrid orbitals are coplanar and trigonally oriented, CH_3^- in this hybridization state would have a planar structure, with the unshared electron pair occupying a p orbital in a plane perpendicular to that containing the C—H bonds (Fig. 1a). On the other hand, sp^3 hybrid orbitals are tetrahedrally directed, so in this case CH_3^- would have a pyramidal structure with the electron pair in a sp^3 orbital. We note (Fig. 1b) that *two equivalent* pyramidal structures are possible; these are interconvertible in principle through the act of *pyramidal inversion*. The topic of pyramidal inversion has been reviewed by a number of workers [1–3] *.

A third possible alternative, shown in Fig. 1c, would involve the $2s2p_x2p_y2p_z$ valence shell electron configuration for carbon, with the unshared electron pair occupying the s orbital (spherical symmetry) and the C—H bonds orthogonally directed. This configuration, involving non-directional distribution for the unshared electron pair, is not supported by any of the available evidence. The concept of the

---

*For a historical account of the stereochemical capabilities of carbanions see ref. 4a. The first evidence that carbanions could exist in asymmetric environments was reported in 1955 [4b]; the first evidence for intrinsically dissymmetric carbanions was given in 1960 [4c].

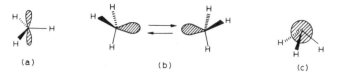

Figure 1. Illustration of various possible configurations of the methyl anion. The shaded orbital contains the unshared electron pair in each case.

directionality of lone electron pairs forms an important part of most current theories which set out to explain the stereochemical nature of molecular species, as for example the valence shell electron pair repulsion theory [5a] (see, however, [5b]). The possibility shown in Fig. 1c will not be considered further in this discussion. Differentiation between the configurations shown in Fig. 1a and Fig. 1b is much more difficult, and in fact there is evidence for both structural types. The evidence to be considered is both of a theoretical nature as well as the experimental approach.

The results of *ab initio* quantum mechanical calculations performed for the CH_3^- system [6a] are illustrated in Fig. 2, in the form of a cross-section of the potential energy surface as a function of the out of plane angle θ. It is seen that the potential energy surface exhibits two minima, corresponding to $\theta = 23°$, and these minima are joined by an energy barrier of almost 6 kcal/mole occurring at $\theta = 0°$. (In other calculations, using different assumptions and basis sets, the energy barrier varies somewhat though the general character of the curves, including the energy minima, remain unchanged [6b].) The configuration with $\theta = 0°$ is of course equivalent to the planar carbanion structure of Fig. 1a, while the two configurations with $\theta = 23°$ are equivalent to the pyramidal structures (the calculated value of the out of plane angle for tetrahedral configurations is $109.5° - 90° = 19.5°$). Thus the calculations support the pyramidal interconverting species of Fig. 1b as the more stable carbanion configurations. In this approach then the planar structure of Fig. 1a actually corresponds to a transition state (energy maximum) configuration in the interconversion of the two pyramidal structures. Calculations [7] on the ammonia molecule (which is isoelectronic with CH_3^-) also point to pyramidal interconverting species and the calculated inversion barrier (5.1–5.6 kcal/mole) is in fair agreement with the observed value (5.8 kcal/mole) obtained from microwave spectroscopy.

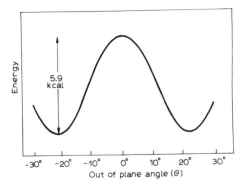

Figure 2. Cross section of potential energy surface for methyl anion showing energy as function of out of plane angle [6a].

The *ab initio* or non-empirical (SCF–LCAO–MO) calculations allow one to reach certain conclusions concerning the physical origin of inversion barriers. Thus the calculated total energy of the system (E_T) may be separated into its component terms, which consist of the nuclear–electron attraction (V_{ne}), the nuclear–nuclear repulsion (V_{nn}), the electron–electron repulsion (V_{ee}), and the kinetic energy of the electrons (T). An inversion barrier may be *attractive dominant* or *repulsive dominant*, depending on whether the attractive term (V_{ne}) or the repulsive term ($V_{nn} + V_{ee} + T$) shows the larger variation on going from the ground state (GS) to the transition state (TS). In the case of ammonia, for example, the pyramidal form is the more stable with respect to the repulsive terms but the planar form is more stable with respect to the attractive term. However, the net change (GS → TS) in the repulsive terms exceeds that in the attractive term, hence the pyramidal structure is stabilized relatively and the barrier is repulsive dominant [7]. Thus, by evaluating the various components of the total energy, one gains greater insight into the inversion process than would be attained in ascribing the inversion barrier simply to the energy associated with the hybridization change, from sp^3 to p of the lone-pair orbital, and from sp^3 to sp^2 of the bonding orbitals. Unfortunately, due to computational limitations, the *ab initio* method is limited to molecules of small size. Several semi-empirical methods of calculation have considerably extended the scope in this regard [8, 9]. Generally, the barrier to inversion is obtained as the difference between the calculated

37

energies of the ground state and of the planar transition state, using appropriate parameters which have been tested with molecules of known energy barriers. Data obtained by the MINDO and CNDO/2 methods are given in Table 1, which includes also results for inversion at atomic centres other than carbon, for comparative purposes.

It will be apparent from the above account that the process of pyramidal inversion must occur without rupture and formation of covalent bonds. The usual mechanism for inversion of a XY_3 species may be represented by

$$
\mathrm{Y} \cdots \overset{\ddot{X}}{\diagup}\hspace{-0.3em}\diagdown\, \mathrm{Y} \quad \rightleftharpoons \quad \mathrm{Y}\diagdown \overset{Y}{\diagup} \underset{X}{\diagup} \mathrm{Y}
$$

Thus inversion occurs along a normal bending vibrational mode (parallel to the axis of the pyramid) in a thermally activated process. The energy barrier for inversion may be calculated, using an appropriate potential energy function, which is based on an input of vibrational (symmetric stretching and bending) frequencies, and on changes in the bond lengths and bond angles, from their equilibrium values, on passing to the planar configuration. Such calculations have been performed for a fair number of molecules, with considerable success [10]. Some of the results of this treatment have been included in Table 1.

The classical traversal over the energy barrier, as outlined above, may give way to non-classical penetration of the energy barrier, in which case the *electron-pair tunneling* mode would be operative. In practice the tunneling mode will be important only for relatively low values of the inversion barrier ($<$ca. 5 kcal/mole) and for light substituent atoms, protons mainly. The ammonia molecule is a case where inversion occurs by quantum mechanical tunneling, as shown by the observed characteristic splitting of the rotation–vibration energy levels in the microwave spectrum. However, in the majority of cases, and especially for the molecules of interest to us, it may be assumed that the classical inversion mode is predominant. Actually it should be stated that for discussion of relative magnitudes of inversion barriers it is unimportant whether the classical or the non-classical inversion mode is predominant. On the other hand, when rates of inversion are used to calculate inversion barriers, then the nature of the inversion process does become important

38

Table 1

INVERSION BARRIERS OF PYRAMIDAL SPECIES

Molecule	Measured barrier [a] kcal/mole	Calcd. barrier [b] kcal/mole	Method	Ref.
CH_3^-		5.6	*ab initio*	5, 6
		20.2	MINDO	8
		10.9	vibr. model	10
$CH_2=CH^-$		31.1	MINDO	8
		38	vibr. model	10
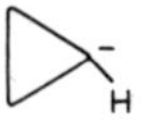 ▷⁻ H		20.8	*ab initio*	3
		14.2	CNDO/2	9
		36.6	MINDO	8
 ▷⁻ H		52.3	*ab initio*	3
		39.3	CNDO/2	9
NH_3	5.8	5.1	*ab initio*	7
		3.2	CNDO/2	9
		3.7	MINDO	8
		5.6	vibr. model	10
$(CH_3)_3N$	6.0	6.9	CNDO/2	9
		7.5	vibr. model	10
$PhCH_2(Et)(Me)N$	10.8			
NF_3	50	62.6	(INDO)	3
		42	vibr. model	10
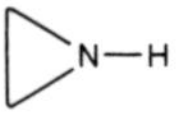 ▷N—H	12.0	18.3	*ab initio*	3
		21.4	CNDO/2	9
		13.8	MINDO	8
SiH_3^-		39.6	*ab initio*	3
PH_3		37.2	*ab initio*	3
		27	vibr. model	10
PD_3		28	vibr. model	10
$Me_2(Ph)P$	32.1	28.2	CNDO/2	9
H_3O^+		0	CNDO/2	9
		1.7	vibr. model	10
D_3O^+		0.9	vibr. model	10
H_3S^+		30	*ab initio*	3
		29.3	CNDO/2	9
$CH_3-S(O)-CH_3$	39.7	37.5	CNDO/2	9
$CH_3-S(O)-OCH_3$		53.3	CNDO/2	9
$CH_3-S(O)-SCH_3$	23	36.4	CNDO/2	9
t-BuEtMeS⁺ClO₄⁻	25			
MeEtŚ-C̄HCOPh	23.3			

[a] Data taken from refs. 1–3.

[b] In a few cases the value given in the table has been selected from a range of values given in the original reference.

39

since the two modes are characterized by different energy-rate relationships.

It should be pointed out that though the *configurational integrity* of stereoisomers is directly dependent on the barrier heights to inversion, the *detectability* of such isomers is dependent on the time scale relevant to a given operating technique [10]. Thus, using normal laboratory techniques, two pyramidal isomers will be separately observable, and their interconversion subject to ordinary kinetic study, provided that the barrier to inversion is in the range 20—40 kcal/mole. On the other hand the use of dynamic nuclear magnetic resonance spectroscopy allows energy barriers of 10—20 kcal/mole to be measured, while microwave spectroscopy takes us into the region of ca. 5 kcal/mole barriers. Finally, theoretical methods (of which there are a number) have no limitation as to life-time (or existence) of molecular species, though it is necessary that a particular theoretical approach should be tested on model molecules before reliance can be placed on the results of such calculations. At present theoretical methods are used when possible in conjunction with experimental observations.

How does one verify experimentally whether theory correctly predicts the configuration of the methyl carbanion? Unlike ammonia, the methyl carbanion is not observable experimentally. It is known (see Chapter 7, Section 4) that methyllithium exists in solution as the tetramer, $(MeLi)_4$, and the hexamer, $(MeLi)_6$, and that the structures of these units are characterized by delocalized bonding. Ethyllithium and butyllithium are also polymeric. Thus information on carbanion configuration which can be obtained from the study of such metal alkyls is limited[*].

Our task in the first instance is to find carbanions which can be investigated experimentally unambiguously. Secondly, we have to have valid criteria for the study of the stereochemistry of the carbanions. Thirdly, we must realize that a pyramidal carbanion may not be experimentally discernable unless its configurational stability is assured by an appreciable energy barrier to inversion (see, however, the previous comments relating configurational stability and energy barriers to the timescale of the technique).

---

[*]For discussion of stereomutation in organometallic compounds, and of pertaining experimental measurements, see refs. 3, 11.

To date the criterion of carbanion configuration which has been used most extensively is the stereochemical one, with chiral substrates, particularly when combined with hydrogen-deuterium exchange. We note in this connection that for a hypothetical optically active carbanion $R_1R_2R_3C^-$ the act of pyramidal inversion results in racemization, since the two pyramidal configurations are chiral and thus related to each other as mirror-image, or enantiomeric forms (see Fig. 1b).

Consider then the stereochemical consequences of allowing proton abstraction from a chiral carbon acid substrate to occur in a deuterium containing medium:

$$(1) \qquad R_1-\underset{\underset{R_3}{|}}{\overset{\overset{R_2}{|}}{C}}-H \quad \underset{BH}{\overset{B^-}{\rightleftarrows}} \quad R_1-\underset{\underset{R_3}{|}}{\overset{\overset{R_2}{|}}{C^-}} \quad \underset{B^-}{\overset{BD}{\rightleftarrows}} \quad R_1-\underset{\underset{R_3}{|}}{\overset{\overset{R_2}{|}}{C}}-D$$

If the carbanion were *planar* then deuterium exchange would be accompanied by *complete racemization*. Similarly, if the carbanion had a pyramidal configuration but were undergoing pyramidal inversion at a rate which was greater than the rate at which it was being re-protonated, then again deuterium exchange would be accompanied by racemization. However, if the carbanion were *pyramidal and inversion were slow* relative to protonation, then *retention* of configuration (predominant or complete depending on the relative rates) would be the result of isotopic exchange.

This theme will be explored in depth in the following sections, together with further discussion of factors which influence geometry of carbanions (planar vs. pyramidal) and increase the energy barrier to pyramidal inversion. In the first part of the discussion, solvation of carbanions in solution will be assumed, though not shown explicitly, but the subsequent section considers this aspect in some detail. It will also become apparent then that a quantitative evaluation of inversion barriers from the types of experiments under consideration is exceedingly complicated as a result of the solvation requirements of carbanions in the media used to generate these species. Thus, theoretical examination of the energetics of carbanionic processes, as indicated in this Section (see also Section 4), provides valuable insight in our understanding of intrinsic properties of such species, in the absence of complicating effects which relate to actual reacting systems. Some of the difficulties

41

relating to the derivation of inversion barriers from experimental measurements are discussed in refs. 1–3 and 12.

2. FACTORS INFLUENCING THE GEOMETRY OF CARBANIONS

In order to study experimentally the geometry of carbanions we have to go to more complex systems than afforded by the hypothetical methyl carbanion, which hitherto has been investigated only by the theoretical approach. In considering structural modification of CH_3^- we clearly must aim in the direction of introducing substituents (or other structural characteristics) which have a stabilizing effect on carbanion formation. Now although substituents often act through a combination of electronic effects, we can generally select one factor as being dominant in a given system. As carbanion stabilizing factors, we have already recognized the inductive effect exerted by electron-withdrawing groups, the conjugative effect of groups capable of interacting with the carbanionic center through resonance, and the s-orbital effect which relates carbanion stability to the s-orbital character of the orbital containing the unshared electron pair.

A priori one might expect that a pyramidal anion would be more favoured by the s-character effect than a planar anion, since in the former case the electron pair is in a sp^3 orbital with 25% s-character whereas in the latter case the electron pair is in a pure p orbital. However, many other factors, both electronic and steric, could counterbalance the s-character effect and it would therefore be incorrect to come to a premature decision before considering the problem in some depth. In fact a more useful approach is one which recognizes that there are factors or structural characteristics which favour *planar stability*, just as there are those which favour *pyramidal stability* [13].

How do the carbanion stabilizing (or destabilizing) structural factors influence carbanion geometry — planar or pyramidal and, in the latter case, what is the effect on the barrier to pyramidal inversion?

The s-orbital stabilizing effect and its bearing on the geometry of carbanions is discussed first, and it is instructive in this regard to consider the situation in the vinyl (sp^2) anionic system, since here the stereochemical criterion can be applied relatively unambiguously. In this case, depending on its configurational stability, the carbanion once formed can in principle reprotonate to give *either* the *cis or* the *trans* isomer.

42

Now it has been found that *cis*-1,2-dichloroethylene undergoes deuteroxide ion catalyzed hydrogen-deuterium exchange *without isomerization* (eq. (2)) and the same is true in the case of the *trans* isomer [14].

$$(2) \quad \underset{Cl}{\overset{H}{}}C=C\underset{Cl}{\overset{H}{}} \quad \underset{HOD}{\overset{OD^-}{\rightleftharpoons}} \quad \underset{Cl}{\overset{H}{}}C=C\underset{Cl}{\overset{\ominus}{}} \quad \underset{OD^-}{\overset{D_2O}{\rightarrow}} \quad \underset{Cl}{\overset{H}{}}C=C\underset{Cl}{\overset{D}{}}$$

These observations clearly show that vinyl carbanions retain their geometry. The authors estimated the activation energy for isomerisation of 1,2-dihalovinyl anions to be of the order 25–35 kcal/mole, representing a high degree of configurational stability to rotation or inversion. Since rotation about the double bond would require an energy of the order of 40–60 kcal/mole, in rupture of the π bond, this type of process can be ruled out in this case. Another possibility to be considered is that of "addition–elimination", i.e. RCH=CHR $\xrightarrow{R'O^-}$ R$\bar{C}$H–CHR(OR') $\xrightarrow{R'OD}$ RCHD–CHR(OR') $\xrightarrow{-R'OH}$ RCD=CHR. In this scheme a vinyl anion is not formed at all; on the other hand such a process would not be expected to occur with retention of configuration. Hence these possible alternative explanations for reaction (2) may be ruled out. The inversion process for a vinylic anion occurs by an in-plane wagging vibrational mode, via a linear transition state.

More recently, the configurational stability of the vinyl anion has been elegantly examined by use of the chiral molecule 2,2,4,6,6-penta-methylcyclohexylideneacetonitrile (1) [15a]. Comparison of the rate of isotopic exchange of the vinyl hydrogen (k_e) with the rate of racemization (k_α), in methanolic sodium methoxide at 90°, shows a high degree (99.3%) of retention of configuration in the vinyl anion. Thus the vinyl anion is formed 140 times as frequently as it is racemized.

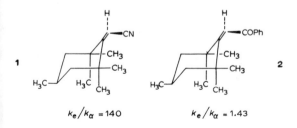

1        $k_e/k_\alpha = 140$ $k_e/k_\alpha = 1.43$ 2

43

In contrast, the keto analog **2** exhibits only a moderate degree (30%) retention of optical activity [15b]. The contrasting observations can be accounted for by the greater charge delocalizability of the carbonyl function, relative to the cyano function, which results in a greater tendency to form a planar (symmetric) anion in the former system compared to the latter:

(3) 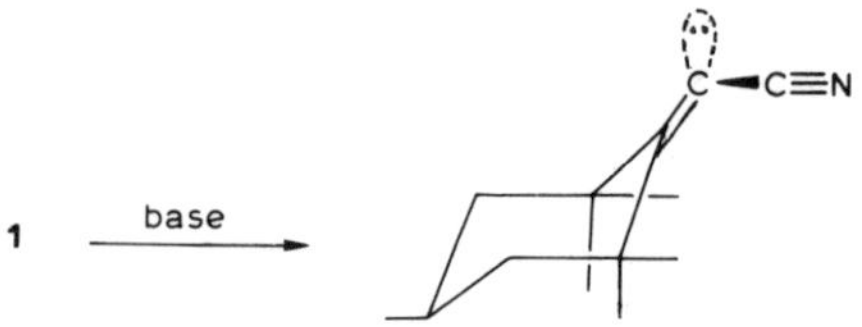 **1** $\xrightarrow{\text{base}}$

asymmetric anion

(4) 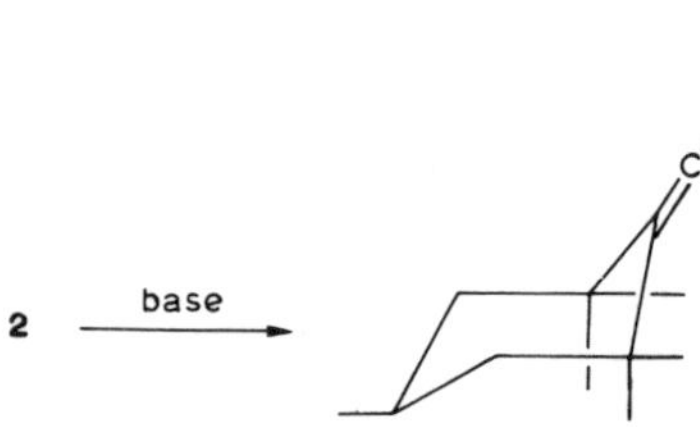 **2** $\xrightarrow{\text{base}}$

symmetric anion

In Chapter 1 we referred to the relatively large acidity of cyclo-propane, which results from the exalted s-character of the exocyclic C–H bonds. Now we turn to consider the geometry of the cyclopropyl anion: is the configuration of the cyclopropyl carbanion planar or pyramidal and, if the latter, is pyramidal inversion facile or not? It should be noted that though cyclopropane constitutes a strained ring system, the transition state for inversion *increases* the strain energy further still. This follows from a consideration of angles of strain. The deviation from the normal bond angle of an acyclic system, which is $109° - 60° = 49°$ in the ground state, becomes $120° - 60° = 60°$ in the planar transition state for inversion. Hence the cyclopropane system provides an in-built energy barrier to pyramidal inversion compared to open-chain systems. That the cyclopropyl carbanion does indeed possess configurational stability is shown by the fact that optically active 2,2-diphenylcyclopropyl cyanide (**3**) undergoes hydrogen exchange with

44

methoxide ion in methanol-O-*d* at the *alpha* hydrogen with net *retention* of configuration [k(exchange)/k (racemization) = 910 at 90°)] [15c].

(5)   **3**

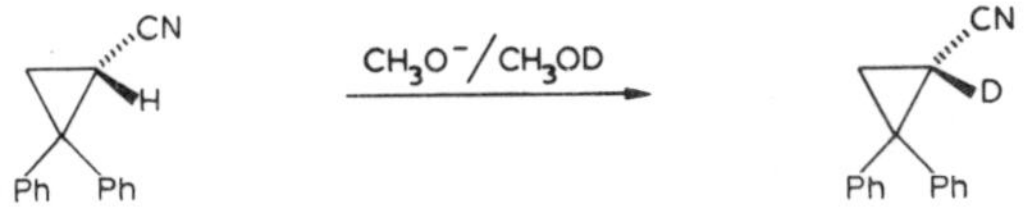

We conclude from this result that the cyclopropyl carbanion (in this system at least) is pyramidal with a considerable barrier to inversion (cf. Table 1).

Data are available which allow comparison, in corresponding systems, between the configurational stability of the cyclopropyl anion and the vinyl anion, on the one hand, and an acyclic tetrahedrally hybridized molecule on the other hand. Results for the three relevant cyano carbon acids are given in Table 2 for the exchange and racemization processes in methanolic sodium methoxide [15a]. Now *if* the hybridization state (s-character) in the anions were the only factor to be considered, then one would have expected the vinyl anion to be more configurationally stable than the cyclopropyl anion. However, it is seen that in practice

Table 2

COMPARISON OF CARBANIONS*

Compound		Hybrid	k_e/k_α	% retention of configuration
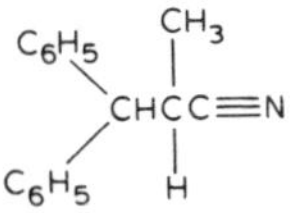	**4**	sp^3	1.87	47
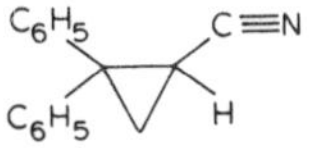	**3**	sp$^{2.28}$	913	99.9
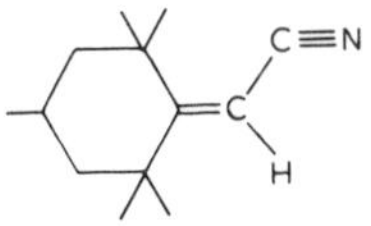	**1**	sp^2	140	99.3

*Reprinted with permission from H.M. Walborsky and L.M. Turner, J. Amer. Chem. Soc., 94 (1972) 2273. Copyright by the American Chemical Society.

45

the reverse applies, as judged by the larger k_e/k_α value in the cyclopropane derivative. One can explain this result as arising from the strain factor in the cyclopropane ring which opposes the planarity of the transition state for racemization, and it appears that the ring strain factor is the more important, in this case at least. The partial retention of configuration of the acyclic molecule is surprising at first sight, since the delocalized carbanion would have been expected to be symmetrical. A possible explanation for this result is that steric hindrance to rotation in the carbanion favours the asymmetric configuration **5**, relative to the symmetric configuration **6** [15c]:

(6)

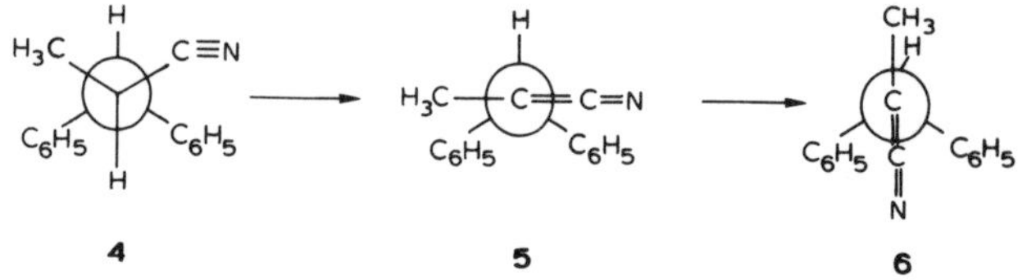

4   **5**   **6**

The problem of the stereochemical result of hydrogen-deuterium exchange in symmetrical carbanions is considered more fully in the next section, at which time asymmetric solvation will be advanced as one of the main operative mechanisms.

We now consider the conjugative effect as a carbanion stabilizing factor. This may well have been expected to be a rather clear-cut case, in that we can readily see that the conjugative effect will be most effective when the system is coplanar. For instance, in the case of the benzyl carbanion the *alpha* carbon must be sp^2 hybridized, for it is then that maximum overlap can occur with the π-electron system of the benzene ring (see also Chapter 1). Independent evidence for charge delocalization in the benzylic carbanion derives from the observation that when 2-phenyl-2-propylpotassium is treated with DCl in ether, ring deuteration occurs in addition to deuteration at the *alpha* benzylic carbon [16]; for *para* deuteration,

(7)

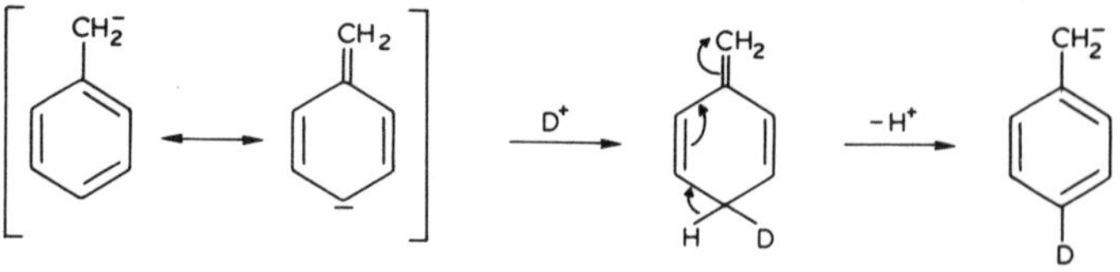

46

Similarly, the carbanions with NO_2, CN or $COCH_3$ as the stabilizing groups are also coplanar, as has been noted previously. Now the corollary statement is also largely true; that if these carbanions were not essentially planar then the substituent stabilizing effect should be considerably smaller. However, our discussion is incomplete, since the inductive effect has been ignored. The NO_2, CN, and $COCH_3$ groups will also withdraw electronic charge inductively, as a result of the dipoles $N^{\delta+}-O^{\delta-}$, $C^{\delta+}-N^{\delta-}$, and $C^{\delta+}-O^{\delta-}$ induced by the electronegative oxygen and nitrogen. Moreover, the phenyl group is also known to act in part as an inductively electron withdrawing substituent (see the discussion on the triptycene system in Chapter 1)[*].

The inductive electron withdrawal of substituents as a factor in stabilizing carbanion formation is readily demonstrated, as for example in the facile hydrogen-deuterium exchange of chloroform as catalyzed by deuteroxide ion in D_2O [21]. However, that this stabilizing effect is much smaller quantitatively than the conjugative effect is seen by the fact that chloroform is a weaker acid than $CH(NO_2)_3$ or $CH(CN)_3$ by at least 15 pK units.

Now chlorine and other heteroatoms also affect the geometry of carbanions by influencing the energy barrier to pyramidal inversion. A particularly interesting example is given by the case of bromochloro-fluoromethane, CHFClBr, which is one of the least complex compounds structurally capable of exhibiting optical activity. It has been shown that the ⁻CFClBr carbanion retains its chirality and is the simplest carbanion to do so [22]. The chirality of ⁻CFClBr follows from the fact that HCFClBr is obtained in optically active form in a reaction known to proceed by a carbanionic mechanism:

(8) $CH_3-\overset{\overset{O}{\|}}{C}-CFClBr \xrightarrow{\ ^-OH\ } CH_3-\overset{\overset{O}{\|}}{C}-OH + {}^-CFClBr \xrightarrow{\ H_2O\ }$ HCFClBr

$[\alpha]_D = +0.39°$ $\qquad\qquad\qquad\qquad\qquad\qquad [\alpha]_D = +0.25°$

Clearly the ⁻CFClBr carbanion must be pyramidal in configuration. Now if the halogen substituents were acting merely as electron with-

---

[*]For further discussion of the problems encountered in the separation of electronic effects the reader is referred to the classic chapter by Taft [17] and to recent texts [18–20].

drawing groups it would seem unlikely that the barrier to pyramidal inversion would be raised appreciably from the relatively small (calculated) value in CH_3^- to allow the ^-CFClBr carbanion to retain its chirality. The available evidence indicates that heteroatomic substitution in general has a retarding effect on inversion both at carbon and at nitrogen. For instance NF_3, which is isoelectronic with CF_3^-, has a barrier to inversion of about 50 kcal/mole. The reason why heteroatoms retard pyramidal inversion is not fully understood although several explanations have been proposed [1–3].

There is diverse evidence, however, which indicates that the halogens may not act in carbanion-forming processes simply as inductively electron withdrawing groups. Various workers have proposed other concurrently operating influences, including a polarizability factor, d-orbital participation, and opposing conjugative mechanisms which can have *stabilizing or destabilizing* influence on carbanion formation. Only limited discussion is presented here of these interesting topics, with emphasis on inductive and the opposing conjugative mechanisms. Also, our discussion concentrates on fluorine since the various influences on carbanion stability and geometry are manifested most prominently with this element, and also more data are available for evaluation with fluorine than with the other halogens. Finally, the fluorine case is also the most controversial for the halogens, and its discussion presents a more challenging prospect.

The fluoroalkanes provide an interesting series for discussion in this context. Thus the acidities of monohydrofluoroalkanes, as measured by the rates of proton exchange with CH_3O^-/CH_3OD, or CH_3O^-/CH_3OT, follow the increasing order CF_3H (pK_a 31), $CF_3(CF_2)_5CF_2H$ (pK_a 30), $(CF_3)_2CFH$ (pK_a 20), $(CF_3)_3CH$ (pK_a 11) [23]. It is seen that three CF_3 substituents are vastly more effective in carbanion stabilization than are three fluorines. Now the 9 fluorines attached to the *beta* carbons could conjugatively interact with the anionic center in the following manner:

7a   7b

This type of carbon-fluorine conjugative interaction is described as double bond–no bond resonance [24], or negative hyperconjugation by analogy with the hyperconjugative interaction of β-hydrogens with a carbonium ion centre ($^+$C–C–H $\leftrightarrow$ C=C H$^+$)*.

However, this interpretation of negative hyperconjugation is brought into question by the observation that the fluoro-substituted bicyclo-heptane derivative 8 undergoes proton exchange at the bridgehead carbon with sodium methoxide in tritiated methanol at a *faster* rate (by a factor of 2) than does $(CF_3)_3CH$ [26].

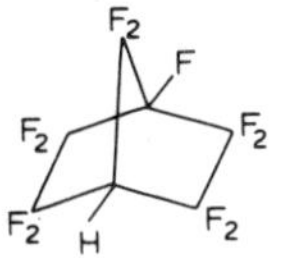

8 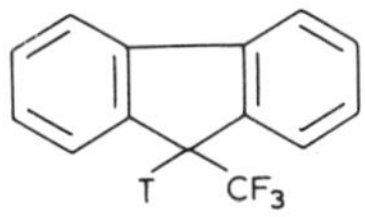   9

The carbanion intermediate derived from 8 is *constrained to be pyramidal* and it is difficult to see how overlap of the type implied in the hyperconjugative mechanism could be effective to any extent given the geometry of this system. If carbon-fluorine resonance is *not* important in this system, there would remain the normal inductive effect to account for the high acidity of 8 and presumably also of $(CF_3)_3CH$. By analogy, the carbanion derived from $(CF_3)_3CH$ might also be expected to be pyramidal in configuration. (However, in another study [27] of kinetic acidity of fluorinated hydrocarbons, it is concluded by the authors that $(CF_3)_3C^-$ is planar, largely as a result of a steric effect.) In the case of proton exchange of 9-trifluoromethylfluorene-9-*t* (9) with sodium methoxide in methanol the CF_3 group is again found to exert a normal inductive effect. Thus, it is observed that protodetritiation of 9-substituted fluorenes follows a linear log (rate) vs. σ^* (inductive parameter) relationship to which the CF_3 substituent-point adheres quite satisfactorily [28].

A more generalized plot has been constructed [29], of pK_a vs. $\Sigma\sigma^*$, which includes several bicyclic fluorinated hydrocarbons related to 8 as well as non-fluorinated hydrocarbons (Figure 3). The pK_a data are

---

*For a recent discussion of the conformational dependence of hyperconjugation, in positively and negatively charged species, see ref. 25.

derived from isotopic exchange rates and the values for the bicyclic carbon acids have been corrected (by 2.2 pK units) for differences in hybridization of the C—H bonds by means of the appropriate J(^{13}C—H) coupling constants. The $\Sigma\sigma^*$ values represent summations of inductive substituent constants, using $\sigma^*(CF_3) = 2.85$, $\sigma^*(\alpha$—F$) = 1.71$ and estimating the number of "effective" fluorines in the bicyclic molecules, while for other substituents standard inductive parameters [17, 18] are used. The degree of linearity of this plot is impressive considering that it extends over 30 pK_a units. Only the point for PhCH$_3$ deviates markedly from the line, which is in accord with an appreciable resonance stabilizing effect in the carbanion. This plot then provides strong evidence for the dominance of inductive effects in these carbanion-forming processes.

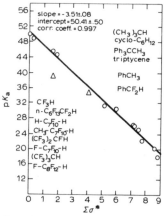

Figure 3. Correlation of the acidity of various compounds with inductive effects. Points marked $\triangle$ were omitted from the calculation of the least squares line [29]. (Reprinted with permission from D. Holtz, Prog. Phys. Org. Chem., 8 (1971) 1. Copyright by the American Chemical Society.)

In the above situations fluorine is found to exert a stabilizing effect on carbanion formation, though the mechanism of stabilization may be controversial. Now, under certain conditions a fluorine attached directly to a carbon bearing a negative charge can be *destabilizing*, when the carbanion is part of an extended conjugated system. Consider the case of 9-fluorofluorene which is observed to undergo proton exchange with methanolic methoxide at a *slower* rate, by a factor of 8, than fluorene itself does [30]. In contrast 9-chloro- and 9-bromofluorene are more reactive than fluorene itself, by factors of 4×10^2 and 7×10^2, respectively. The destabilizing effect of fluorine is apparently caused by interaction between an electron pair in a p-orbital of fluorine and the π-system of the fluorenyl carbanion. Essentially, we have here a repulsion of charge between the halogen p_π-electrons and the π-system, leading to destabilization. In the case of the planar delocalized fluorenyl carbanion the geometry of the system is ideally suited to the interaction with consequent destabilization. This type of electronic effect has been called a p–p lone-pair repulsion effect [29] and has also been described as a $+R$–p-orbital feedback mechanism in the destabilization of carbanions [27]. An α-fluorine atom is not the only substituent for which the effect will be operative, others being an α-oxygen (in the alkoxy function) and, to a much less extent, an α-chlorine [31]. This destabilizing mechanism is most effective in planar sp^2 hybridized carbanions.

Figure 4. Destabilizing interaction of α-fluorine with conjugated planar carbanions in the fluorenyl and nitromethane type systems.

The destabilizing effect of fluorine can also manifest itself on equilibrium acidity. Thus in pK_a measurements on substituted nitro-methanes, $XYCHNO_2$ where $Y = NO_2$, Cl, $CONH_2$ or COOEt, it is found that the acidity is uniformly *diminished* for $X = F$ compared with $X = H$, while for $X = Cl$ the acidity is increased in all but one case [32]. It is noteworthy that the carbanions formed on proton loss must be essentially planar due to conjugation with the nitro group, so that once again *alpha* fluorine has a destabilizing effect when attached to a conjugated sp^2 carbanion (Fig. 4).

It appears therefore that fluorine can interact with a carbanionic center in different ways, depending on the particular system. If the carbanion is pyramidal then an *alpha* fluorine is stabilizing while for conjugated, planar, carbanions fluorine is destabilizing. Fluorine attached to a *beta* carbon stabilizes carbanion formation and does so much more effectively than the other halogens.

What overall conclusion can we reach concerning the configuration of carbanions? The simplest carbanion, CH_3^-, is believed on theoretical grounds to be pyramidal, rather than planar, but with a low energy barrier to pyramidal inversion. The barrier to inversion can be increased significantly by multiple heteroatom substitution. One manifestation of this phenomenon is that the perhalo-substituted carbanion, ^-CFClBr, is asymmetric. The perfluoromethyl type carbanions such as CF_3^- and $(CF_3)_3C^-$ are probably also pyramidal, though the mode of action of fluorine in carbanionic systems is still controversial. However, the common carbanion-stabilizing groups, such as phenyl, nitro, cyano and keto, act primarily by their conjugative interaction with the carbanionic center and require coplanarity of the carbanionic system. It is important to note, however, that "purely pyramidal" or "purely planar" are extreme situations which will only rarely occur in practice and that actual carbanionic systems will generally, in the state of lowest energy, deviate somewhat in geometry from either extreme*.

A digression would appear to be in order before concluding this section. From the preceding discussion it is apparent that the effect of fluorine substituents on kinetic and equilibrium acidity can not be given a unique explanation, at least in a number of instances. However, it would be a mistake to assume that a considerably improved state of understanding has been reached with respect to the interpretation of other physical measurements for which fluorine substitution has a structural implication. We give one or two examples to illustrate this point.

Dipole moment measurements have allowed one to make useful deductions concerning the ground state electron distribution in molecules [19b]. Thus the fact that the dipole moment observed for *p*-dimethylaminonitrobenzene (6.89 D) is greater than the vector sum of the dipole moments for nitrobenzene (3.93 D) and for N,N-dimethyl-aniline (1.58 D) is generally interpreted as reflecting significant contribution from the canonical structure **10b**:

----

*The terms "purely pyramidal" and "purely planar" are somewhat loosely used in the context. Planarity is strictly an absolute term, i.e. trigonal bonds with bond angles of 120°. On the other hand the pyramidal configuration could encompass all other structural possibilities, although one might set a 90° bond angle as the limit in the case of the XY_3 species with equivalent bonds.

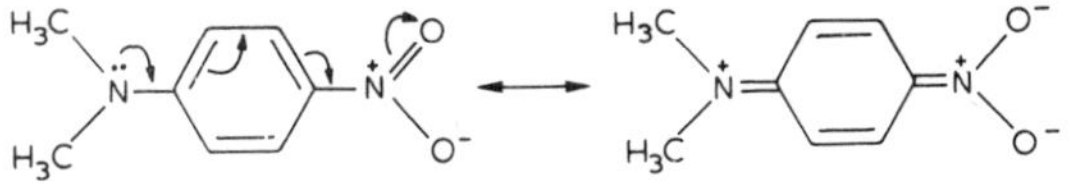

10 a   10 b

Now *p*-dimethylaminobenzotrifluoride also has an observed dipole moment (4.62 D) which is larger, by 0.43 D, than the sum of the component vector values. By analogy with **10b**, one may propose that structure **11b** also contributes significantly [33]:

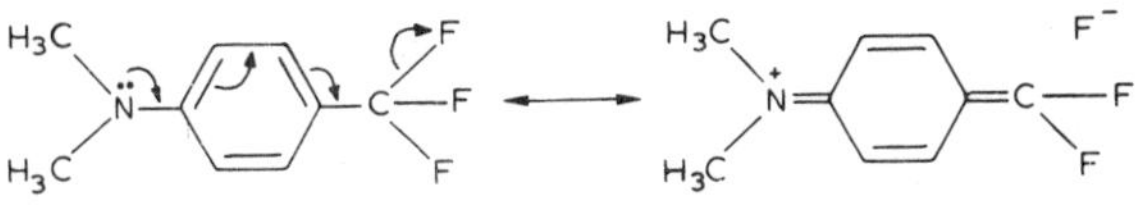

11 a   11 b

However, an alternative to the fluorine hyperconjugative mechanism of **11a** ↔ **11b** is the one shown in **11a** ↔ **11c** which has been termed the π-inductive mechanism [34]:

$(CH_3)_2N$—⟨ ⟩—CF_3 ⟶ $(CH_3)_2\overset{+}{N}$=⟨ ⟩—$\overset{+}{C}F_3$

11 a   11 c

Structure **11c** would represent a polarization of the π-electrons of the benzene ring which is induced by the powerful inductive electron withdrawal effect of the perfluoroalkyl group. It is difficult at present to make an enlightened choice between the alternative explanations. Similarly, it is difficult to determine whether the structures **12b** and/or **12c** make significant contribution to the electronic distribution in *p*-fluorobenzotrifluoride:

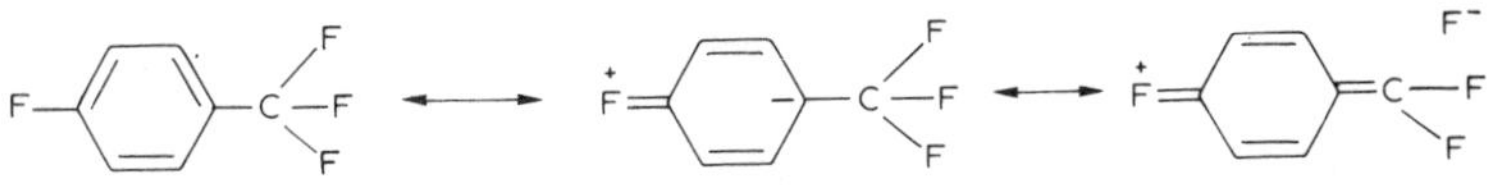

12 a   12 b   12 c

Much thought has already been given to these questions, and it can be hoped that the quantitative studies now underway in a number of

53

laboratories will provide definitive answers. The importance of organo-fluorine chemistry suggests that such studies are indeed well directed.

3. STEREOCHEMICAL RESULTS OF HYDROGEN-DEUTERIUM EXCHANGE IN REACTIONS INVOLVING SYMMETRICAL CARBANIONS

Hydrogen-deuterium exchange provides a powerful probe in the investigation of carbanions. When applied to a chiral substrate (cf. eq. (1)), proton exchange can in principle differentiate between a planar and a pyramidal carbanion, provided that the energy barrier to inversion is large enough for carbanion protonation to compete effective-ly with pyramidal inversion. Racemization as opposed to retention of configuration is the expected stereochemical result of exchange for the planar and the non-rapidly interconverting pyramidal systems.

The vast majority of carbanionic systems studied involve as carbanion-stabilizing substituents groups such as phenyl, keto, nitro or cyano, directly bonded to the carbanionic center. These carbanion-stabilizing groups act primarily through a conjugative interaction with the carbanionic center, and require coplanarity in the resulting carbanionic system (see for example structures in Chapter 1, Section 1). Since the resulting planar carbanions are no longer asymmetric, it follows that the stereochemical result of hydrogen-deuterium exchange should be one of complete racemization. If a kinetic study is carried out on such a system then the rate of carbanion formation will be given by the rate of racemization. This last statement, which forms our common notion in the isotopic study of carbanions, will now be critically examined taking account of the *actual total* carbanionic systems, which include the nature of the base and of the solvent medium.

There is a fundamental condition, not considered until now, which is necessary in order that the rate of racemization should equal the rate of carbanion formation. That is that the lifetime of the carbanion should be sufficiently long for the BH^+ species formed on abstraction of the proton by the base B to have the opportunity to diffuse away from the reaction site and thus allow the carbanion to be symmetrically solvated by solvent molecules acting as donors of the isotopic proton; only then would capture of a proton occur at equal rates from the backside and the frontside faces of the carbanion.

Experimentally a number of cases have been found in which the condition of the symmetrical environment does not hold. *In such cases the rate of deuterium exchange is not equal to the rate of racemization.* The following examples are illustrative of this phenomenon.

2-Phenylbutane-2-*d* (**13**) and 1-phenyl-1-methoxyethane-1-*d* (**14**) undergo hydrogen exchange in basic media (*t*-BuO⁻/*t*-BuOD/DMSO, etc.) [35] and 2-phenyl-1,1,1-trifluorobutane-2-*d* (**15**) reacts similarly [36].

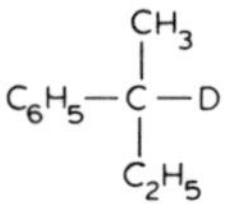

The stereochemical course of these carbanionic processes is found to vary *from racemization to net inversion, to net retention*, depending on the medium and base employed. Similarly, optically active 1-phenyl-ethane-1-*d* (**16**) undergoes hydrogen exchange with lithium cyclohexyl-amide in cyclohexylamine with predominant retention [37]. Thus, we have a variable stereochemical result with essentially planar carbanions. Let us now delve deeper into this facet.

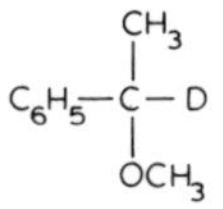

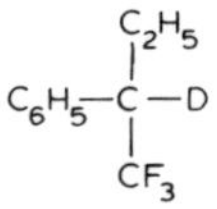

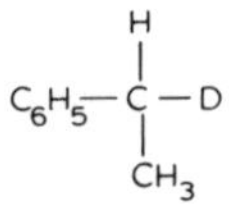

If we denote the rate constant for isotopic exchange by k_e and the rate constant for racemization by k_α then the various stereochemical results of isotopic exchange which have been recognized are characterized by the following limiting k_e/k_α ratios:

1. $k_e/k_\alpha = 1$; isotopic exchange occurs with net racemization.
2. $k_e/k_\alpha > 1$; isotopic exchange occurs with predominant retention.
3. $k_e/k_\alpha = 0.5$; isotopic exchange occurs with net inversion.
4. $k_e/k_\alpha < 0.5$; racemization is predominant over isotopic exchange.

A system which has provided extensive information on the stereo-chemical outcome of such isotopic exchange is 9-deuterio-(or protio)-9-methylfluorene, substituted unsymmetrically by the amide grouping (**17**) [38]. In principle, deuteron abstraction by base would yield the delocalized cyclopentadienide type anion, which is intrinsically planar and symmetrical, so that proton capture should lead to racemic material. *Racemization* is in fact the result (eq. (9)) when isotopic exchange is brought about by potassium phenoxide in *t*-butyl alcohol ($k_e/k_\alpha = 1.0$ at 27°):

(9)

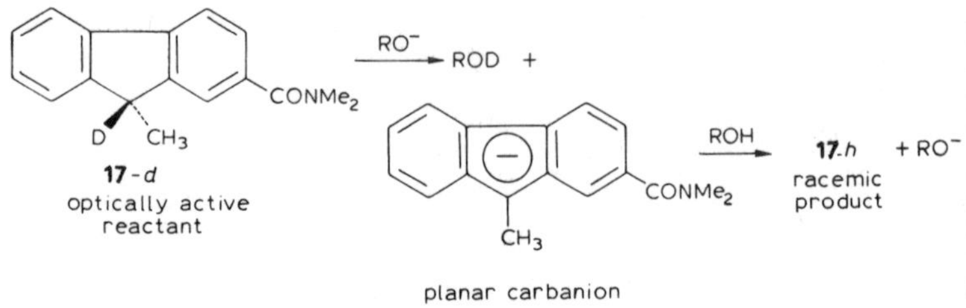

planar carbanion

Retention of configuration is the predominant result, however, when the base is changed to ammonia, while retaining the *t*-butyl alcohol medium ($k_e/k_\alpha > 50$ at $200°$). A plausible pathway for retention is shown in Figure 5. Proton abstraction occurs as previously but, prior to complete dissociation of the ammonium-carbanide ion pair, a rotation of the ammonium ion takes place, followed by recombination on the *same face* of the carbanion. The stereochemical result of this process is predominant retention. On the other hand with NH_3 as the base if the medium is changed to dimethyl sulfoxide, which is more dissociating than *t*-butyl alcohol, now the ion pairs dissociate at a faster rate than rotation can occur within the ammonium ion, so that a symmetrical solvation environment is reached, with racemization as the overall result ($k_e/k_\alpha = 1.0$ at $25°$).

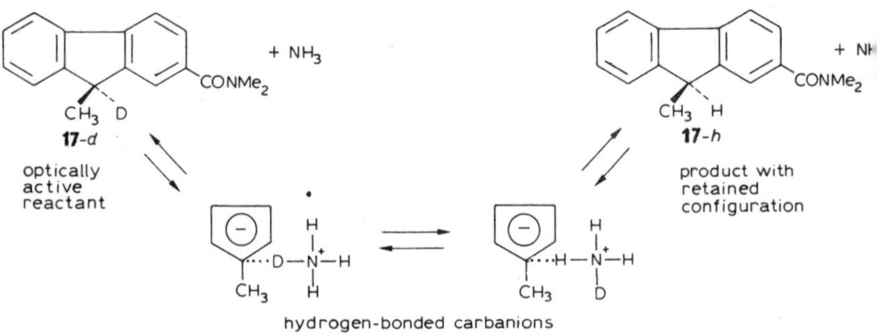

Figure 5. Retention mechanism for fluorenyl carbanion (abbreviated to show only the cyclopentadienyl moiety).

The *inversion* mechanism will operate in the situation that the carbanion is *asymmetrically solvated*, such that donation of the isotopic hydrogen will occur predominantly from the *opposite face* to the one

56

where proton abstraction occurred. This situation (Figure 6) was realized when optically active 17-d was treated with tripropylamine in methanol solvent: $k_e/k_\alpha = 0.65$ at 75°. A minor competing racemization process accounts for k_e/k_α being slightly greater than 0.50. The pathway for racemization would be the normal one, i.e. via a symmetrically solvated carbanion formed by displacement of the hydrogen-bonded DNPr_3^+ species by a molecule of solvent.

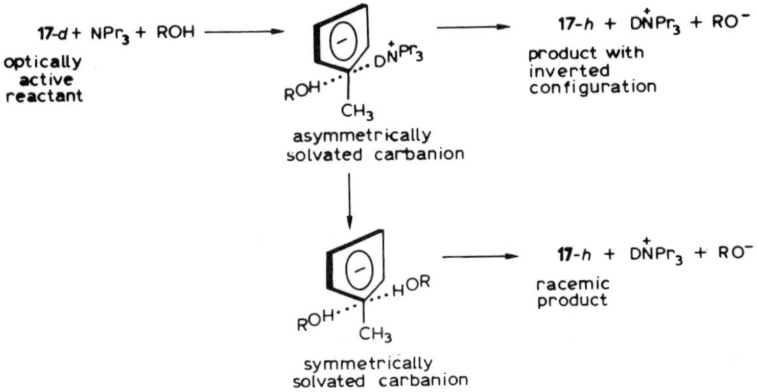

Figure 6. Inversion and racemization mechanisms for fluorenyl anion.

The last case type for the stereochemical outcome of isotopic exchange corresponds to $k_e/k_\alpha < 0.5$, that is racemization predominates over isotopic exchange. This process has been termed by Cram as *isoracemization*. Experimentally, a finding of $0.5 > k_e/k_\alpha > 0$ when the isotopic exchange-racemization experiment is conducted in the presence of an excess of the isotopic solvent pool indicates that isoracemization is a contributing pathway.

A system where isoracemization has been observed is the isotopic exchange-racemization of 2-phenylbutyronitrile, 18. In the basic system potassium t-butoxide/t-butyl alcohol this nitrile undergoes isotopic exchange with complete racemization, i.e. $k_e/k_\alpha = 1.0$ [39]. However, using tripropylamine as the base and the less dissociating medium tetra-hydrofuran/t-butyl alcohol (1.5 M), it is found that $k_e/k_\alpha = 0.05$ [40]. Thus racemization now far outweighs exchange. The reaction pathway proposed by Cram is shown in Figure 7; it is essentially an intramolecular

57

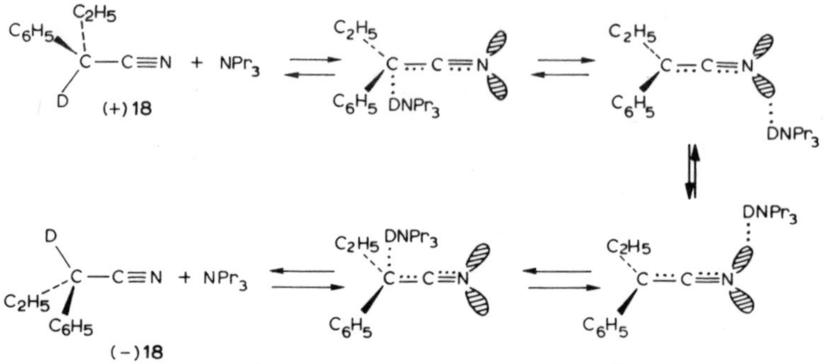

Figure 7. Isoracemization by the "conducted tour mechanism".

proton transfer-racemization process and was given the descriptive name "a conducted tour mechanism".

Essentially we have a series of planar, symmetrical reaction intermediates, but the protonated amine remains linked to the negatively charged residue through hydrogen-bonding. In this way the base takes the proton on a "conducted-tour" from one face of the molecule to the other; re-combination at the carbon site yields the enantiomeric nitrile in which the original hydrogen isotope is retained. This act of inversion without exchange has also been called *isoinversion* [41]. Of course, whenever dissociation of the ion pair occurs, then the $DNPr_3^+$ ion becomes exchanged for a proton-donating solvent molecule and isotopic exchange with the medium will hence be a competing process. The dissection of a given stereochemical result of isotopic exchange into the component mechanistic pathways has been achieved for several systems in which competing mechanisms operate simultaneously [41—46]. The dissection process is greatly aided by a powerful new technique called the *reresolution method* [44], which involves the resolution (fractional crystallization into racemate and optically pure material) and isotopic analysis of the component parts of a partially racemized—partially exchanged carbon acid.

We may conclude therefore that the observation of non-racemization as a stereochemical course in the deuterium exchange of 13 to 18 is the result of asymmetric solvation/ion-pairing effects superimposed on planar, symmetrical, carbanions. The principle of asymmetric solvation is of course well recognized in other areas also, as for example in carbonium ion processes [47].

58

4. THE CHIRALITY OF CARBANIONS ADJACENT TO SULFUR AND PHOSPHORUS

In the previous section we dealt with intrinsically planar carbanions, in the formation of which dissymmetry of the parent carbon acid has been destroyed, though the stereochemical course of reaction was governed by solvation factors. We now turn to carbanionic systems in which the stereochemical result of reaction (e.g. H/D exchange) is governed by the presence, adjacent to the carbanionic center, of certain sulfur or phosphorus functions and it is *these functions* which *are responsible for stereochemical control*. Thus, we shall now be considering carbanions which are *inherently dissymmetric*.

It has been known for some time that sulfur and phosphorus in certain valence states tend to promote hydrogen-deuterium exchange at adjacent carbon. For example, base-catalyzed hydrogen-deuterium exchange occurs quite readily in the trimethylsulfonium ion $(CH_3)_3S^+$ [48], in tetramethylphosphonium ion $(CH_3)_4P^+$ [48], in dimethyl-sulfone $CH_3SO_2CH_3$ [49], and less readily in dimethyl sulfoxide CH_3SOCH_3 [50]. A more complete listing of structural environments for sulfur and phosphorus adjacent to which base-catalyzed deuterium exchange is known to occur is given below (for some representative studies, in addition to those described in the following discussion, see refs. 51–54).

The observation of base-induced deuterium exchange in these systems shows that carbanions *alpha* to these functional groups are stabilized by

$$R_2\overset{+}{\underset{\cdot\cdot}{S}}-\overset{|}{\underset{|}{C}}-H \qquad R-\overset{\cdot\cdot}{\underset{\cdot\cdot}{S}}-\overset{|}{\underset{|}{C}}-H \qquad R-\overset{O}{\underset{\cdot\cdot}{\overset{\|}{S}}}-\overset{|}{\underset{|}{C}}-H \qquad R-\overset{O}{\underset{O}{\overset{\|}{\underset{\|}{S}}}}-\overset{|}{\underset{|}{C}}-H$$

| 19 | 20 | 21 | 22 |

$$RO-\overset{O}{\underset{O}{\overset{\|}{\underset{\|}{S}}}}-\overset{|}{\underset{|}{C}}-H \qquad R_2N-\overset{O}{\underset{O}{\overset{\|}{\underset{\|}{S}}}}-\overset{|}{\underset{|}{C}}-H \qquad M^+\,{}^-O-\overset{O}{\underset{O}{\overset{\|}{\underset{\|}{S}}}}-\overset{|}{\underset{|}{C}}-H$$

| 23 | 24 | 25 |

$$R_3\overset{+}{P}-\overset{|}{\underset{|}{C}}-H \qquad R_2\overset{O}{\overset{\|}{P}}-\overset{|}{\underset{|}{C}}-H \qquad (RO)_2\overset{O}{\overset{\|}{P}}-\overset{|}{\underset{|}{C}}-H \qquad R-\underset{M^+\,{}^-O}{\overset{O}{\overset{\|}{P}}}-\overset{|}{\underset{|}{C}}-H$$

| 26 | 27 | 28 | 29 |

59

some electronic effect(s) operating in these systems. Now it is well known that an inductively electron-withdrawing group tends to stabilize carbanion formation, but it appears that in the sulfur and phosphorus systems some other electronic effect must also be operative. A number of studies have been directed at elucidation of the origin of this effect. For example, one can make comparison of corresponding first row element vs. second row element in promoting carbanion formation in otherwise structurally analogous systems. The following data give results of this nature in the form of relative rates of base-catalyzed deuterium exchange [48, 55].

$(CH_3)_3\overset{+}{N}—CH_3$ vs. $(CH_3)_3\overset{+}{P}—CH_3$

Relative rate, 1   Relative rate, 2.4×10^6

Relative rate, 1   Relative rate, 10^6

We see that substitution of phosphorus for nitrogen, or sulfur for oxygen, enhances deuterium exchange by a factor of a million approximately, despite the greater electronegativity of the first-row heteroatom. If the inductive effect had been the chief factor operating then the relative rates would be expected to follow the reverse order.

The pairs phosphorus—nitrogen, and sulfur—oxygen, represent elements in the same vertical group in the Periodic Table, but phosphorus and sulfur are in the second row while nitrogen and oxygen belong to the first row of elements. (In this commonly used notation hydrogen and helium, which complete the first quantum level, are not reckoned as a "row". Hence elements in the "first row" are those for which the second quantum level is being filled and for elements in the "second row" it is the third quantum level which is filled.) Now elements in the second row of the Periodic Table have available 3d-orbitals of relatively low energy for bonding utilization; hence the stabilization of a carbanion *alpha* to phosphorus and sulfur would implicate the participation of such d-orbitals. The manner of participation could be through overlap of the carbon orbital (p or sp³) containing the unshared pair of electrons with a suitably oriented vacant 3d-orbital. A number of theoretical approaches to the problem of d-orbital participation have been made but

60

at present there appears to be no general agreement among various workers as to an explicit mechanism [56–60]. Indeed, some calculations suggest that d-orbitals do not participate at all in the stabilization of *alpha* carbanions adjacent to sulfur, but that specific electrostatic interactions are responsible for the observed effects [61, 62].

In the context of the chirality of carbanions of carbon acids and the effect of second row elements, we are concerned with a sulfur or phosphorus center having an adjacent chiral carbon atom with H (or D) as one of the substituents. Now of the functional sulfur and phosphorus compounds listed on p. 59 in which base-catalyzed hydrogen isotopic exchange has been studied, about half are known to lead to chiral carbanions, as judged by the criterion of isotopic exchange occurring with predominant *retention* of configuration ($k_e/k_\alpha \gg 1$) in solvent systems where asymmetric solvation or ion-pairing are *not* expected to be important.

Predominant retention in isotopic exchange under these conditions has been observed for sulfoxides (21) [63–66], sulfones (22) [67–70], sulfonates (23) [71], sulfonamides (24) [71] and for phosphinates (29) [71]. In the remaining functional classes of compounds listed on p. 59 the rate of isotopic exchange equals the rate of racemization, so that those carbanions do not retain their chirality. Some typical examples [68, 71] of k_e/k_α values are given below for two cases where the retention mechanism obtains and one case where complete racemization accompanies hydrogen-deuterium exchange.

$$k_e/k_\alpha = 59 \qquad k_e/k_\alpha = 34 \qquad k_e/k_\alpha = 1.0$$

How do we explain the observation that some carbanions adjacent to sulfur or phosphorus retain their chirality while others do not? A final answer to this question is not yet available at the present time. However, various proposals have been put forward to explain how retention of configuration might occur and this aspect of the problem will now be considered.

Our approach considers that proton abstraction from the chiral carbon acid can lead to either a pyramidal or a planar carbanion. We

61

then scrutinize the interactions between the various groups attached to the carbanionic center and the adjacent sulfur (or phosphorus) center, including also the unshared electron pairs which are regarded as coordinated ligands. In evaluation of the possible structures we use the principles of conformational analysis, but also valence and molecular orbital theory. Carbanions adjacent to SO_2 and SO have received the most consideration and our discussion here is limited to those cases.

Consider first the *alpha* sulfonyl carbanion in the pyramidal configuration. There are a number of possible conformations which the carbanion could assume, of which three are shown in Figure 8 as the Newman projections. Of these, conformer **30c** is the least likely energetically due to the presence of three eclipsing interactions (other eclipsed conformers have been omitted for that reason). Intuitively we expect conformer **30a** (or the alternative one obtained on counterclockwise rotation by $120°$) to be the most favoured energetically, since electrostatic repulsions between the electron pair in the sp^3 orbital and the partly negatively charged oxygens are expected to be minimized here [71]. Conformer **30b** provides another possibility, however, and theoretical calculations for the model system $^-CH_2-SO_2-H$ suggest that the configuration of lowest energy corresponds to **30b** ($R_1 = R_2 = R = H$) [72]. It is interesting that this structure, with the orbital containing the unshared pair on carbon being directed along the bisector of the OSO angle, *maximizes the gauche interactions* of the unshared electron pair with adjacent dipolar S–O electron pairs. This tendency has been referred to as the "gauche effect" [73]. In any case, whichever pyramidal conformation is the more stable, isotopic exchange will occur with retention provided the barrier to inversion is sufficiently high.

An alternative explanation for retention ascribes a *planar* (sp^2) structure to the carbanion, i.e. the electron pair on carbon now occupies a p-orbital. In the *alpha* sulfonyl carbanion case [74], one could have the asymmetric configuration **30d** with a high energy barrier to rotation which would convert **30d** to the symmetrical structure **30e**; protonation of **30d** preferentially from one face of the carbanion would result in retention.

The *alpha* sulfinyl carbanion can be considered analogously (Figure 9), with pyramidal (**31a** or **31b**) and planar (**31c**, **31d**) configurations. Note that since the sulfoxide grouping is itself asymmetric the *alpha* carbon is placed thereby in an asymmetric environment, even in the planar

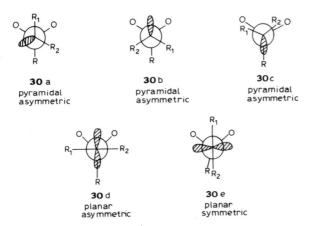

| 30 a | 30 b | 30 c |
| pyramidal asymmetric | pyramidal asymmetric | pyramidal asymmetric |

| 30 d | 30 e |
| planar asymmetric | planar symmetric |

Figure 8. Possible conformations for α-sulfonyl carbanion (RSO$_2$–CR$_1$R$_2^-$).

configuration **31d** (cf. **30e** for the sulfonyl analogue). This factor will have some effect on relative rates of protonation of **31d**, and unequal attack at either face would lead to incomplete racemization, though predominant retention as a result of this effect alone is unlikely. The argument for retention of configuration is analogous to the sulfonyl case, i.e. that there must be a sufficiently high energy barrier to inversion (via **31a** or **31b**), or to rotation (via **31c** or **31d**). It is interesting that theoretical calculations [75] for the model system $^-CH_2$–SO–H suggest that structure **31b** (R$_1$ = R$_2$ = R = H) is of lowest energy, consistent with the gauche effect.

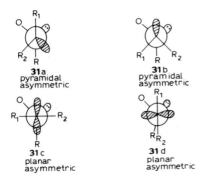

| 31a | 31b |
| pyramidal asymmetric | pyramidal asymmetric |

| 31c | 31d |
| planar asymmetric | planar asymmetric |

Figure 9. Possible conformations for α-sulfinyl carbanion (RSO–CR$_1$R$_2^-$).

63

To review the above discussion, though at present our knowledge as to why certain carbanions adjacent to sulfur (or phosphorus) retain their chirality while others do not is as yet incomplete, the general view is that the carbanions considered are pyramidal, i.e., that the electron pair on carbon occupies a sp^3 orbital. However, there is restricted rotation, or there exists a constraint on inversion of configuration within the carbanion, due to the presence of an electrostatic barrier. This barrier, which could operate as a result of an interaction between the electron-pair on carbon and those on sulfur (or phosphorus) and/or on the attached oxygen(s), results in proton capture by carbanion being a faster process than racemization.

Current developments, both experimental and theoretical, are expected to shed further light on these interesting problems. One experimental approach which yields valuable information on the stereochemistry of carbanions is to generate the carbanion by an independent route [4a], such as by formation of an alkali metal salt of the carbon acid, and then to study the stereochemical result of known processes, such as carbonation, or quenching with deuterium acids in carefully selected environments [76–78].

Another current approach, especially in the study of α-sulfinyl carbanions [79–81], involves the use of substrates in which a given configuration is imposed by conformational constraints. The stereochemistry of isotopic exchange and the geometrical preferences of the carbanions can be studied relatively unambiguously in such systems. For example [81], we have the *endo/exo* isomers **32, 33** and the biaryl sulfoxide **34**, for which assignment of the protons can be made unambiguously by NMR. One result which is already apparent from these studies is that solvent can have a striking effect on the stereochemical preference of isotopic exchange. This suggests that it will be difficult to reconcile experimental results with theoretical models based on molecular orbital calculations, owing to the computational limitations imposed on MO methods which hitherto have precluded rigorous consideration of solvation effects.

It may be noted that structures **32** and **33** possess a pair of diastereotopic protons adjacent to the sulfinyl group, while structure **34** has two such pairs of diastereotopic protons. These substrates, like benzyl methyl sulfoxide [63], characteristically exhibit a selectivity of isotopic exchange, i.e. one of the two diastereotopic hydrogens exchanges faster

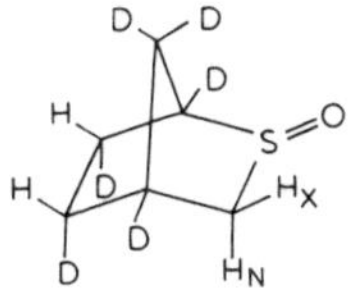

32

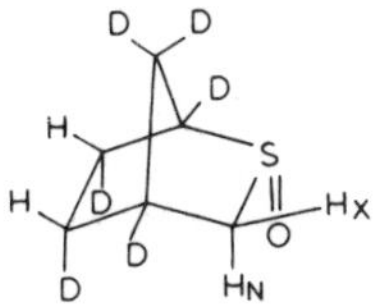

33

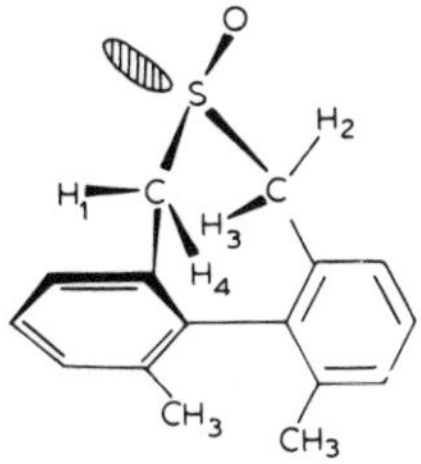

34

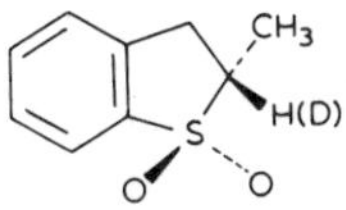

35

than the other (e.g. for **34** the relative rates are $H_1 : H_2 : H_3 : H_4$ = 1:1100:300:1300 in t-BuOD/t-BuOK at 30°, while for MeOD/MeONa the rates are in the order 1:200:7600:30). The stereoselectivity of exchange reflects the configurational dependence of the carbanion stabilizing effect of the sulfoxide group.

The last approach to the problem that we mention utilizes the reresolution method (see Section 3) for dissecting the stereochemical result of isotopic exchange. This method when applied to the cyclic sulfone **35** identifies the following exchange mechanisms to be occurring [82]: exchange with retention, exchange with inversion, and inversion without exchange. Moreover, the balance between these processes is solvent dependent. It thus appears that pyramidal inversion is an important component in the chemistry of α-sulfonyl carbanions. One can presume that an analogous situation will hold also for α-sulfinyl carbanions. Further studies of this nature with other carbon acids which lead to chiral carbanions should contribute greatly to our understanding of the stereochemistry of these processes.

65

REFERENCES

1   A. Rauk, L.C. Allen and K. Mislow, Angew. Chem. Internat. Ed., 9 (1970) 400.
2   J.M. Lehn, Fortschr. Chem. Forsch., 15 (1970) 311.
3   J.B. Lambert, Top. Stereochem., 6 (1971) 19.
4a D.J. Cram, "Fundamentals of Carbanion Chemistry", Academic Press, New York, 1966, Chapters 2–4.
4b D.J. Cram, J. Allinger and A. Langemann, Chem. Ind. (London), 919 (1955).
4c D.J. Cram, W.D. Nielsen and B. Rickborn, J. Amer. Chem. Soc., 82 (1960) 6415.
5a R.J. Gillespie and R.S. Nyholm, Quart. Rev. Chem. Soc. London, 11 (1957) 339; R.J. Gillespie, "Molecular Geometry", Van Nostrand Reinhold, New York, 1972.
5b S. Wolfe, L.M. Tel, W.J. Haines, M.A. Robb and I.G. Csizmadia, J. Amer. Chem. Soc., 95 (1973) 4863.
6a R.E. Kari and I.G. Csizmadia, J. Chem. Phys., 50 (1969) 1443.
6b P.H. Owens and A. Streitwieser, Jr., Tetrahedron, 27 (1971) 4471.
7   A. Rauk, L.C. Allen and E. Clementi, J. Chem. Phys., 52 (1970) 4133.
8   M.J.S. Dewar and M. Shanshal, J. Amer. Chem. Soc., 91 (1969) 3654.
9   A. Rauk, J.D. Andose, W.G. Frick, R. Tang and K. Mislow, J. Amer. Chem. Soc., 93 (1971) 6507.
10 G.W. Koeppl, D.S. Sagatys, G.S. Krishnamurthy and S.I. Miller, J. Amer. Chem. Soc., 89 (1967) 3396.
11 T.L. Brown, Accounts Chem. Res., 1 (1968) 23.
12 J.W. Henderson, Chem. Soc. Revs., 2 (1973) 397.
13 R.D. Baechler, J.D. Andose, J. Stackhouse and K. Mislow, J. Amer. Chem. Soc., 94 (1972) 8060.
14 S.I. Miller and W.G. Lee, J. Amer. Chem. Soc., 81 (1959) 6313.
15a H.M. Walborsky and L.M. Turner, J. Amer. Chem. Soc., 94 (1972) 2273.
15b J.F. Arnett and H.M. Walborsky, J. Org. Chem., 37 (1972) 3678.
15c H.M. Walborsky and J.M. Motes, J. Amer. Chem. Soc., 90 (1970) 2445.
16 G.A. Russell, J. Amer. Chem. Soc., 81 (1959) 2017.
17 R.W. Taft, Jr., "Separation of Polar, Steric, and Resonance Effects in Reactivity", in "Steric Effects in Organic Chemistry", edited by M.S. Newman, Wiley, New York, 1956.
18 J.E. Leffler and E. Grunwald, "Rates and Equilibria of Organic Reactions", Wiley, New York, 1963, Chapter 7.
19 C.K. Ingold, "Structure and Mechanism in Organic Chemistry", Cornell University Press, Ithaca, New York, 1969; (a) Chapter 16, (b) Chapter 3.
20 L.P. Hammett, "Physical Organic Chemistry", Second Edition, McGraw-Hill, New York, 1970, Chapter 11.
21 J. Hine, "Physical Organic Chemistry", McGraw-Hill, New York, 1962, p. 487.
22 M.K. Hargreaves and B. Modarai, Chem. Commun., 16 (1969); J. Chem. Soc. (C), 1013 (1971).
23 S. Andreades, J. Amer. Chem. Soc., 86 (1964) 2003.

24a L. Pauling, "The Nature of the Chemical Bond", Cornell University Press, Ithaca, New York, 1960, pp. 314–316.
24b J. Hine, J. Amer. Chem. Soc., 85 (1963) 3239.
25   R. Hoffmann, L. Radom, J.A. Pople, P.v.R. Schleyer, W.J. Hehre and L. Salem, J. Amer. Chem. Soc., 94 (1972) 6221.
26   A. Streitwieser, Jr. and D. Holtz, J. Amer. Chem. Soc., 89 (1967) 692.
27   K.J. Klabunde and D.J. Burton, J. Amer. Chem. Soc., 94 (1972) 5985.
28   A. Streitwieser, Jr., A.P. Marchand and A.H. Pudjaatmaka, J. Amer. Chem. Soc., 89 (1967) 693.
29   D. Holtz, Prog. Phys. Org. Chem., 8 (1971) 1.
30   A. Streitwieser, Jr. and F. Mares, J. Amer. Chem. Soc., 90 (1968) 2444.
31   J. Hine, L.G. Mahone and C.L. Liotta, J. Amer. Chem. Soc., 89 (1967) 5991; J. Hine and P.D. Dalsin, J. Amer. Chem. Soc., 94 (1972) 6998.
32   H.G. Adolph and M.J. Kamlet, J. Amer. Chem. Soc., 88 (1966) 4761.
33   J.D. Roberts, R.L. Webb and E.A. McElhill, J. Amer. Chem. Soc., 72 (1950) 408.
34   W.A. Sheppard, J. Amer. Chem. Soc., 87 (1965) 2410.
35   D.J. Cram, C.A. Kingsbury and B. Rickborn, J. Amer. Chem. Soc., 83 (1961) 3688.
36   D.J. Cram and A.S. Wingrove, J. Amer. Chem. Soc., 86 (1964) 5490.
37   A. Streitwieser, Jr., D.E. Van Sickle and L. Reif, J. Amer. Chem. Soc., 84 (1962) 258.
38   D.J. Cram and L. Gosser, J. Amer. Chem. Soc., 86 (1964) 5445.
39   D.J. Cram, B. Rickborn, C.A. Kingsbury and P. Haberfield, J. Amer. Chem. Soc., 83 (1961) 3678.
40   D.J. Cram and L. Gosser, J. Amer. Chem. Soc., 86 (1964) 5457.
41   W.T. Ford, E.W. Graham and D.J. Cram, J. Amer. Chem. Soc., 89 (1967) 4661.
42   D.J. Cram, W.T. Ford and L. Gosser, J. Amer. Chem. Soc., 90 (1968) 2598.
43   W.T. Ford and D.J. Cram, J. Amer. Chem. Soc., 90 (1968) 2612.
44   S.M. Wong, H.P. Fischer and D.J. Cram, J. Amer. Chem. Soc., 93 (1971) 2235.
45   K.C. Chu and D.J. Cram, J. Amer. Chem. Soc., 94 (1972) 3521.
46   J. Almy, D.H. Hoffman, K.C. Chu and D.J. Cram, J. Amer. Chem. Soc., 95 (1973) 1185.
47   D. Bethell and V. Gold, "Carbonium Ions. An Introduction". Academic Press, New York, 1967.
48   W. von E. Doering and A.K. Hoffmann, J. Amer. Chem. Soc., 77 (1955) 521.
49   J. Hochberg and K.F. Bonhoeffer, Z. Physik. Chem., A184 (1939) 419.
50   E. Buncel, E.A. Symons and A.W. Zabel, Chem. Commun., 173 (1965).
51a G. Barbarella, A. Garbesi and A. Fava, Helv. Chim. Acta, 54 (1971) 341, 2297.
51b A.A. Hartmann and E.L. Eliel, J. Amer. Chem. Soc., 93 (1971) 2572.
52a M. Fukunaga, K. Arai, H. Iwamura and M. Oki, Bull. Chem. Soc. Japan, 45 (1972) 302.
52b J.F. King, E.A. Luinstra and D.R.K. Harding, Chem. Commun., 1313 (1972).
53a F.G. Bordwell, D.D. Phillips and J.M. Williams, Jr., J. Amer. Chem. Soc., 90 (1968) 426.

53b L.A. Paquette, J.P. Freeman and M.J. Wyvratt, J. Amer. Chem. Soc., 93 (1971) 3216.

54 R. Kluger and P. Wasserstein, J. Amer. Chem. Soc., 95 (1973) 1071.

55 S. Oae, W. Tagaki and A. Ohno, Tetrahedron, 20 (1964) 417, 427.

56 C.C. Price and S. Oae, "Sulfur Bonding", Ronald Press, New York, 1962.

57 W.G. Salmond, Quart. Rev., 22 (1968) 253.

58 C.A. Coulson, Nature, 221 (1969) 1106.

59 D.L. Coffen, Rec. Chem. Prog., 30 (1969) 275.

60 K.A.R. Mitchell, Chem. Rev., 69 (1969) 157.

61 A. Rauk, S. Wolfe and I.G. Csizmadia, Can. J. Chem., 47 (1969) 113.

62 S. Wolfe, A. Rauk, L.M. Tel and I.G. Csizmadia, Chem. Commun., 96 (1970).

63 A. Rauk, E. Buncel, R.Y. Moir and S. Wolfe, J. Amer. Chem. Soc., 87 (1965) 5498.

64 E. Bullock, J.M.W. Scott and P.D. Golding, Chem. Commun., 168 (1967).

65 K. Nishihata and M. Nishio, Tetrahedron Lett., No. 47 (1972) 4839; M. Nishio, Chem. Commun., 562 (1968).

66 R.R. Fraser and F.J. Schuber, Chem. Commun., 397 (1969).

67 D.J. Cram, D.A. Scott and W.D. Nielsen, J. Amer. Chem. Soc., 83 (1961) 3696.

68 E.J. Corey and E.T. Kaiser, J. Amer. Chem. Soc., 83 (1961) 490; E.J. Corey, H. König and T.H. Lowry, Tetrahedron Lett., No. 12 (1962) 515.

69 H.L. Goering, D.L. Towns and B. Dittmar, J. Org. Chem., 27 (1962) 736.

70 M.D. Brown, M.J. Cook, B.J. Hutchinson and A.R. Katritzky, Tetrahedron, 27 (1971) 593.

71 D.J. Cram, R.D. Trepka and P. St. Janiak, J. Amer. Chem. Soc., 86 (1964) 2731; 88 (1966) 2749.

72 S. Wolfe, A. Rauk and I.G. Csizmadia, J. Amer. Chem. Soc., 91 (1969) 1567.

73 S. Wolfe, Accounts Chem. Res., 5 (1972) 102.

74 E.J. Corey and T.H. Lowry, Tetrahedron Lett., No. 13 (1965) 793, 803.

75 S. Wolfe, A. Rauk and I.G. Csizmadia, J. Amer. Chem. Soc., 89 (1967) 5710.

76 K.C. Bank and D.L. Coffen, Chem. Commun., 8 (1969).

77 R.T. Wragg, Tetrahedron Lett., No. 56 (1969) 4959.

78 T. Durst, R.R. Fraser, M.R. McClory, R.B. Swingle, R. Viau and Y.Y. Wigfield, Can. J. Chem., 48 (1970) 2148; R. Viau and T. Durst, J. Amer. Chem. Soc., 95 (1973) 1346.

79 B.J. Hutchinson, K.K. Anderson and A.R. Katritzky, J. Amer. Chem. Soc., 91 (1969) 3839.

80 R. Lett, S. Bory, B. Moreau and A. Marquet, Tetrahedron Lett., No. 35 (1971) 3255.

81 R.R. Fraser, F.J. Schuber and Y.Y. Wigfield, J. Amer. Chem. Soc., 94 (1972) 8795.

82 J.N. Roitman and D.J. Cram, J. Amer. Chem. Soc., 93 (1971) 2225.

Chapter 3

CARBANIONS AND TAUTOMERISM

In this chapter we consider equilibria between molecules in which the essential structural feature is a triad of atoms $X \cdot Y \cdot Z$ with a proton bonded to either of the terminal atomic centres:

$$(1) \quad \begin{matrix} X-Y=Z \\ | \\ H \end{matrix} \rightleftharpoons \begin{matrix} X=Y-Z \\ | \\ H \end{matrix}$$

Equilibria of this type are generally referred to as *tautomeric* equilibria, with the two isomers called *tautomers*. Alternatively, since the overall process by which the tautomers are interconverted is a 1,3-proton shift, one may also call this a *prototropic* equilibrium. This designation also serves to contrast prototropy from anionotropy, in which the fragment that migrates does so with its bonding pair of electrons.

In principle there is no limitation on the nature of the $X \cdot Y \cdot Z$ entity, though in the common cases, including those considered here, the triad is comprised of some combination of C, N, and O. In the case of keto-enol tautomerism the proton shift takes place between carbon and oxygen. The isomers constituting a given tautomeric system are generally interconvertible with relative ease, and often co-exist at equilibrium. This situation contrasts with isomerism in general, where in the typical case isomeric compounds are either not at all interconvertible or are only so under quite drastic conditions (see Chapter 6, however, for some exceptionally facile cases of isomerisation).

The interconversion of tautomers is generally brought about through the agency of a basic catalyst. This suggests that the reaction mechanism is one involving an anionic intermediate species:

$$(2) \quad B + \begin{matrix} X-Y=Z \\ | \\ H \end{matrix} \rightleftharpoons BH^+ + [X \cdots Y \cdots Z]^- \rightleftharpoons \begin{matrix} X=Y-Z \\ | \\ H \end{matrix} + B$$

One would expect the function of B in proton abstraction to be performed by any base in the Brönsted sense. It is also conceivable, however, that proton abstraction by B could occur *simultaneously* with the donation of a proton to Z:

(3)  $X{-}Y{=}Z$ $\rightleftharpoons$ $X{=}Y{-}Z$
        B:H   H—A   B—H   H:A

In this case the function of HA as the proton donating species should be taken by any general acid, or simply by a molecule of solvent. The distinction between the carbanion intermediate mechanism of prototropy and the concerted proton transfer mechanism may be quite subtle in practice, particularly so in the case in which the solvent donates the proton, since the observed kinetic order with respect to reactants would be the same in both cases. A third possibility, that of *bifunctional catalysis*, arises when the basic and acidic components of the catalytic species are constituents of the same molecule. Finally, the carbonyl or other (XYZ) compound may itself contain an acidic and/or basic center elsewhere within the molecule, leading to the possibility of *intramolecular catalysis*.

Elucidation of the phenomenon of tautomerism has been one of the interesting episodes in the development of the theories of organic chemistry. Isotopic studies have played an important role in this elucidation and, as in other situations, they have been used most successfully in conjunction with other criteria — in the present case the parallel measurement of the kinetics of isotopic exchange, racemization, and halogenation. Since proton transfer is a key step in the interconversion of tautomers, the application of kinetic isotope effects would be expected to be especially informative.

1. KETO-ENOL TAUTOMERISM

(a) Investigation of keto-enol equilibria

The keto-enol tautomeric system is of special interest historically and, moreover, research work into its various facets continues unabated at the present time. Since the principles of tautomerism are well represented in the keto-enol system we consider this case in detail. Some of the subject matter discussed will be concerned with matters outside strictly carbanionic chemistry, but is included to provide a fully balanced account.

The discovery of ethyl acetoacetate by Geuther in 1863 marked the first instance of a compound which could simultaneously exhibit the properties normally associated with two different substances [1]. On the one hand ethyl acetoacetate exhibited the typical properties of a methylene *ketone*, $-CO-CH_2-$, e.g. it formed adducts with cyanide ion and with hydroxylamine and could undergo alkylation at the methylene carbon atom *alpha* to the carbonyl group. On the other hand, ethyl acetoacetate exhibited the properties of an *enol*, $-C(OH)=CH-$, e.g. in its rapid reaction with bromine and in the formation of an alcoholate salt, RO^-Na^+. It was some 50 years after this paradox had been made apparent before the two isomeric forms of ethyl acetoacetate were separately isolated and made available in pure form for independent examination [2]. The keto-isomer, which has a m.p. of $-39°$, could be crystallized out at $-78°$ from a solution of the mixture in petroleum ether. The enol–isomer was obtained in pure form by acidification of the sodium salt with hydrogen chloride at $-78°$; it melts below $-78°$. At low temperature the two forms could be kept in the pure state for prolonged periods; however, in the presence of acids or bases, even in trace amounts, an equilibrium mixture was obtained consisting predominantly of the keto form.

(4) $CH_3-\overset{O}{\overset{\|}{C}}-CH_2-\overset{O}{\overset{\|}{C}}-OC_2H_5 \rightleftharpoons CH_3-\overset{OH}{\overset{|}{C}}=CH-\overset{O}{\overset{\|}{C}}-OC_2H_5$

keto form — of ethyl acetoacetate   enol form — of ethyl acetoacetate

At room temperature the equilibrium mixture of ethyl acetoacetate contains 92.5% of the keto tautomer and 7.5% of the enol tautomer; this value was easily estimated experimentally by volumetric determination of the amount of bromine consumed, since only the enol form reacts rapidly in the bromination reaction (see, however, below). The enol content of ethyl acetoacetate varies in different media, increasing from 0.4% in water to 46.4% in hexane, indicating increased enol content as the polarity of the medium is decreased.

The position of the keto-enol equilibrium in other systems was determined subsequently by various adaptations of the halogenation method [3–8]. The *direct* method, involving titration with a standard solution of bromine, may often be inadequate since the keto-enol equilibrium may itself be affected by the analytical procedure. Thus hydrogen bromide, a product of the reaction, is a catalyst for the keto-enol interconversion, which often leads to an unsteady end point and high enol values. Most of the recent determinations employ a potentiometric method for estimating the bromine concentration [4, 7, 8]. One of these uses a redox electrode of platinum in combination with a standard glass electrode; the method is capable of measuring bromine concentrations down to 10^{-8} M, and is hence especially applicable to substrates of low enol content [8]. For cyclopentanone and cyclohexanone the reaction between enol and bromine was complete in 5 sec, while the keto-enol equilibrium was re-established in 10 min. The values for the enol content so obtained are considerably smaller than those from previous measurements.

Spectroscopic methods of determination of keto-enol equilibria are coming into increasing use as alternatives to the halogenation method; they have the advantage that the tautomeric equilibrium is unperturbed by the process of measurement. Ultraviolet [9], infrared [10] and Raman [11] spectroscopy have been used and, latterly, nuclear magnetic resonance spectroscopy [12–16]. The use of time-averaging computers should considerably enhance the value of these methods by making them applicable in the region of very small enol (or keto) content. It may be noted that not only is proton resonance applicable for this purpose but also carbon-13 [15] and oxygen-17 [16] magnetic resonance.

The relaxation technique has been applied with much effect to the study of keto-enol equilibria. By use of the temperature-jump method, and monitoring of the rate of change of optical density due to the enolate ion, the constituent rate constants for the following equilibria can be measured [17]:

$$(5) \quad EH + OH^- \; \underset{k_{21}}{\overset{k_{12}}{\rightleftharpoons}} \; E^- + H_2O \; \underset{k_{32}}{\overset{k_{23}}{\rightleftharpoons}} \; CH + OH^-$$

In this equation EH, CH, and E^- represent the enol, keto, and enolate forms, respectively. Thus the kinetic measurements allow determination of the position of the tautomeric equilibrium as well as the acidities (pK_{EH} and pK_{CH}) of the keto and enol forms.

In Table 1 are presented selected data on keto-enol equilibria for some representative cases (for comprehensive tabulations see refs. 18–21). For mono-functional compounds, such as acetone, the enol

72

Table 1

EQUILIBRIUM ENOL CONTENT IN TAUTOMERIC KETO-ENOL SYSTEMS
AS A FUNCTION OF STRUCTURE

Compound	% enol	Medium	Reference
Monoketones			
Acetone	2.0×10^{-4}	Liquid	6
	2.5×10^{-6}	Water	4
	0.9×10^{-6}	Water	8
Ethyl methyl ketone	1.2×10^{-3}	Liquid	6
Butyl methyl ketone	1.1×10^{-3}	Liquid	6
Acetophenone	3.5×10^{-4}	Liquid	6
Cyclopentanone	4.8×10^{-3}	Liquid	6
	1.3×10^{-5}	Water	8
Cyclohexanone	2.0×10^{-4}	Water	4
	4.1×10^{-6}	Water	8
Cyclooctanone	1.1×10^{-1}	Liquid	6
	$<5 \times 10^{-3}$	Methanol	9
β-Ketoesters and 1,3-diketones			
Ethyl acetoacetate	7.5	Liquid	3
Ethyl cyclohexanone-			
2-carboxylate	74	Liquid	5
Acetylacetone	78	Liquid	5
	90	Cyclohexane	13
1,1-Dimethylcyclohexane-			
3,5-dione	95	Water	4

content in the liquid state is quite small, but it increases markedly for
β-ketoesters and for 1,3-diketones. Other trends are perhaps slightly less
obvious, such as increased enol content in cyclic compounds compared
with corresponding acyclic compounds. Extensive discussions of the
factors influencing keto-enol equilibria have been given [18–21]. A
particularly challenging problem is to rationalize the effect of substituents
on keto-enol equilibria. A number of approaches, including application
of *ab initio* molecular orbital theory [22], have been used to this end.
However, at the present time many aspects of the problem are still
incompletely understood.

While simple aliphatic enols are thermodynamically unstable relative
to the keto tautomers, it may nevertheless be possible to *observe* the
enol as a fleeting species, at a considerably *greater concentration* than

when equilibrium is reached. A striking example of how this situation may be achieved has been reported recently [23].

The heterocyclic dioxolane derivative 1, which is stable in dry dimethylformamide (DMF) for extended periods, is acted upon by a catalytic amount of benzoic acid to yield the enol 2 by loss of DMF. The enol is then gradually transformed into the keto tautomer 3. Starting with a 2 M concentration of 1 and 0.003 M benzoic acid, the concentration of 2 reaches a maximum of 0.5 M after 15 h at $25°$ and then begins to decline. Simultaneously, the concentration of the keto form increases, and the transformation is essentially complete after a further 15 h. (In dimethyl sulfoxide the enol is formed in even greater concentration and its decay is slower.) The enol was characterized by its NMR and IR spectra, which were fully in accord with the structure 2. On treatment with D_2O in CCl_4 the enolic proton could be exchanged rapidly and quantitatively; the deuterioenol (2-d) so produced subsequently rearranged to the deuterioketone (3-d) [23].

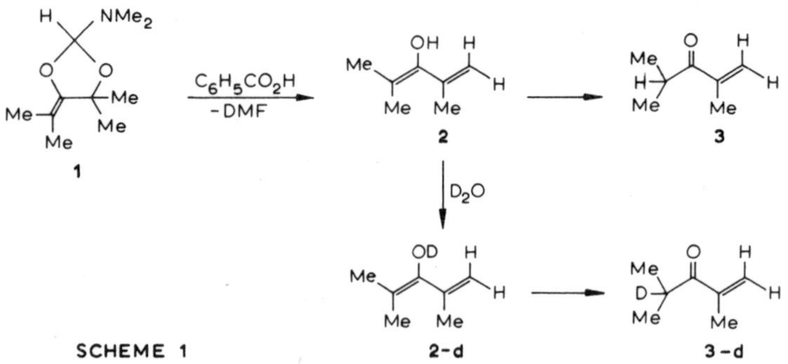

SCHEME 1

It should be pointed out that several factors in this system favour stabilization of the enol form; alkyl substitution and delocalization by conjugation are obvious factors. Perhaps even more important is the point that the enol has been generated, under mild conditions, in a polar aprotic solvent which is capable of hydrogen bonding with the enol form but does not provide protons to catalyze the rearrangement. This work should provide an impetus for further studies of this kind.

(b) Halogenation studies

Study of the kinetics of halogenation of ketones has provided a key to our understanding of keto-enol tautomerism. Lapworth in 1904 made the highly interesting and pertinent discovery that the rate of bromination of acetone in acidic aqueous solution was directly proportional to the concentration of acetone and of acid but *independent of the concentration of bromine* [24]. Somewhat later, Dawson observed that for the iodination of acetone in basic aqueous solution the rate of iodination was likewise proportional to the concentration of substrate and of base and independent of the concentration of halogen [25]. These workers concluded therefore that *halogenation is a fast process* in each case, following rate-determining enolization, as shown in eq. (6) and (7):

$$(6) \quad CH_3-\overset{O}{\overset{\|}{C}}-CH_3 \quad \underset{slow}{\overset{H^+}{\longrightarrow}} \quad CH_3-\overset{OH}{\overset{|}{C}}=CH_2 \quad \underset{fast}{\overset{X_2}{\longrightarrow}} \quad CH_3-\overset{O}{\overset{\|}{C}}-CH_2X$$

$$(7) \quad CH_3-\overset{O}{\overset{\|}{C}}-CH_3 \quad \underset{slow}{\overset{OH^-}{\longrightarrow}} \quad CH_3-\overset{O^-}{\overset{|}{C}}=CH_2 \quad \underset{fast}{\overset{X_2}{\longrightarrow}} \quad CH_3-\overset{O}{\overset{\|}{C}}-CH_2X$$

Additional evidence for the two-stage mechanism of halogenation of acetone, and of other ketones and keto-esters, was provided by the finding that the rate of chlorination, bromination, and iodination, were all equal under given conditions of acidity or basicity, solvent and temperature [25–27]. Hence the rate-determining step in all these halogenations is simply the rate of enolization.

The above conclusions concerning the mechanism of halogenation, which were arrived at on the basis of studies worthy in every sense of being called classical, need some qualification under the special conditions in which the halogen concentration is made extremely small (ca. 10^{-7}–10^{-5} M) while the concentration of ketone and acid are maintained at "normal" values (ca. 10^{-2}–1 M). With these extremely small chlorine or bromine concentrations *the halogenation step becomes rate-controlling*, since the rate of attack of halogen on the enol or enolate ion has been reduced to a value which is below that of the enolization step [28, 29]. The results of these studies, performed with ketones as well as with keto esters, illustrate the importance of experimental design, and serve to confirm rather than contradict the previous hypothesis [24, 25] concerning the mechanism of halogenation. Thus the two-stage nature of the reaction is established beyond reasonable doubt through the demonstration that either stage can be made to be rate-controlling when the concentrations of reagents are appropriately chosen. The studies substantiate the conclusion that in ketone halogenation under normal conditions the rate-controlling step is enolization, and that halogenation takes place as a subsequent rapid reaction.

This section on halogenation is concluded on a more practical note, with a brief account of the well known haloform reaction. The iodoform reaction, for example, involves a sequence of enolization–halogenation processes terminated by a cleavage at the carbonyl group:

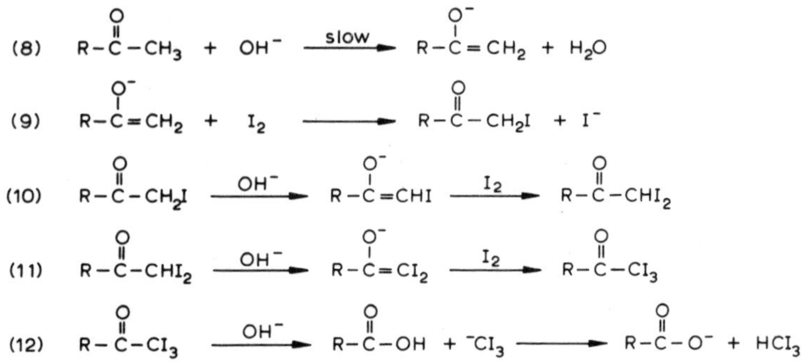

$$(8) \quad R-\overset{O}{\overset{\|}{C}}-CH_3 + OH^- \xrightarrow{\text{slow}} R-\overset{O^-}{\overset{|}{C}}=CH_2 + H_2O$$

$$(9) \quad R-\overset{O^-}{\overset{|}{C}}=CH_2 + I_2 \longrightarrow R-\overset{O}{\overset{\|}{C}}-CH_2I + I^-$$

$$(10) \quad R-\overset{O}{\overset{\|}{C}}-CH_2I \xrightarrow{OH^-} R-\overset{O^-}{\overset{|}{C}}=CHI \xrightarrow{I_2} R-\overset{O}{\overset{\|}{C}}-CHI_2$$

$$(11) \quad R-\overset{O}{\overset{\|}{C}}-CHI_2 \xrightarrow{OH^-} R-\overset{O^-}{\overset{|}{C}}=CI_2 \xrightarrow{I_2} R-\overset{O}{\overset{\|}{C}}-CI_3$$

$$(12) \quad R-\overset{O}{\overset{\|}{C}}-CI_3 \xrightarrow{OH^-} R-\overset{O}{\overset{\|}{C}}-OH + {}^-CI_3 \longrightarrow R-\overset{O}{\overset{\|}{C}}-O^- + HCI_3$$

The second enolization (eq. (10)) proceeds more rapidly than the first (eq. (8)) since the proton to be abstracted from $RCOCH_2I$ is more acidic than the one abstracted from $RCOCH_3$; the enolization of $RCOCHI_2$ (eq. (11)) proceeds even more rapidly. Thus the above sequence of reactions has a single rate-controlling step, the initial enolization (eq. (8)). We should note, however, that the halogenating species is not restricted to molecular iodine; other active halogenating agents are the tri-iodide ion, I_3^-, and the hypoiodite ion, IO^-.

The above scheme for the haloform reaction needs some qualification when applied to more complex carbonyl compounds and when quantitative reactivity data are desired. Thus a competing reaction which can often become important is replacement of halogen in the various halogeno-ketone intermediates by hydroxide ion (or by other nucleophilic species which may be present). This competing reaction has only moderate effect with acetone but it can have major importance with other substrates. In such cases reaction leads to a variety of products

76

depending on the relative rates of halogenation and of substitution. These complications will be taken up further in Section (d).

(c) Mechanism of enolization

An important indication of the mechanism of enolization is given by the nature of the catalysis. Thus it is known that in addition to catalysis by hydronium ion and hydroxide ion, other Brönsted acids and bases are also effective catalytic species, according to their acid or base strengths. In other words, the enolization is subject to *general acid and general base catalysis* [30–33].

According to current theory there are two distinct stages in the acid catalyzed reaction. First there is a pre-equilibrium step in which the carbonyl oxygen is protonated by the acid (HA) to give an oxonium ion. Then follows the rate-determining attack on the *alpha*-hydrogen by the conjugate base A, with formation of enol. The initial protonation makes the *alpha*-hydrogen more susceptible to attack by the weakly basic species A (generally H_2O). Taking the case of acetone as an example, we have:

$$(13) \quad HA + CH_3-\overset{\overset{\ddot{O}:}{\|}}{C}-CH_3 \ \rightleftharpoons \ CH_3-\overset{\overset{+OH}{\|}}{C}-CH_3 + A^- \ \xrightarrow{\text{slow}} \ CH_3-\overset{\overset{OH}{|}}{C}=CH_2 + HA$$

Bromination of the enol takes place by the usual electrophilic mechanism for alkene halogenation; the reagent can be regarded as a source of a formally positive species, e.g. Br^+ in the case of bromination. The direction of addition of Br^+, which attaches to the terminal carbon, can be explained in terms of the resonance stabilization associated with delocalization of charge in the bromo-carbonium-oxonium ion:

$$(14) \quad CH_3-\overset{\overset{OH}{|}}{C}=CH_2 \ \xrightarrow[\;(-Br^-)\;]{Br-Br} \ [CH_3-\overset{\overset{OH}{|}}{\underset{+}{C}}-CH_2Br \ \longleftrightarrow \ CH_3-\overset{\overset{+OH}{\|}}{C}-CH_2Br] \ \rightleftharpoons \ CH_3-\overset{\overset{O}{\|}}{C}-CH_2Br + H^+$$

The reaction of a ketone with base is analogous in principle, though somewhat simpler in that the enolate ion is formed directly through proton abstraction:

$$(15) \quad CH_3-\overset{\overset{O}{\|}}{C}-CH_3 + B \ \xrightarrow{\text{slow}} \ [CH_3-\overset{\overset{O}{\|}}{C}-CH_2^- \ \longleftrightarrow \ CH_3-\overset{\overset{O^-}{|}}{C}=CH_2] + BH^+$$

The resulting delocalized anion has the charge residing predominantly on the electronegative oxygen; hence the term "enolate" is more appropriate than merely "carbanion", although enolate ions can act as both carbon and oxygen nucleophiles [34—37]. It is due to this delocalization that proton abstraction from a carbon *alpha* to the carbonyl is so much easier (in terms of energy requirement) than it would be from a carbon further removed from the carbonyl group, as with the *beta* hydrogens of $RCOCH_2CH_3$. However, we should also contrast proton abstraction from acetone, which is a slow process, with that from a typical hydroxy compound such as methanol, which occurs very rapidly. In the latter case the residual negative charge in CH_3O^- is restricted to the atom from which the proton departed, whereas in the former case an extensive electronic reorganization is necessary in the covalency changes which accompany proton abstraction. Another factor, which undoubtedly plays an important role, is solvation of the ionic species that are produced. Solvation effects are often quite complex [38], and subject to uncertainty in interpretation, but the greater solvation of methoxide ion than of the enolate ion in a hydroxylic solvent will clearly stabilize the former ionic species relative to the latter*.

There are also other possible pathways for achieving catalysis of enolization. One possibility is that the acid HA and base B act in a *concerted* manner, as illustrated in eq. (16):

$$(16) \quad HA + CH_3\text{-}\overset{\displaystyle O}{\overset{\|}{C}}\text{-}CH_3 + B \longrightarrow [CH_3\text{-}\overset{\displaystyle O\text{---}H\text{---}A}{\overset{\|}{C}}\text{:=}CH_2\cdots H\cdots B] \longrightarrow A^- + CH_3\text{-}\overset{\displaystyle OH}{\overset{|}{C}}\text{=}CH_2 + BH^+$$

This mechanism, which has sometimes been called a "push-pull" mechanism, would require a termolecular rate law for enolization. Several studies have in fact reported the findings that a termolecular term contributes to a minor extent, representing $\leqslant 20\%$ of the overall enolization pathways in the systems investigated [40].

It might well have been expected that the concerted mechanism would become more favourable if both the acidic and basic centers were parts of one molecule, as for example in amino-carboxylic acids. This

---

*As has been pointed out by Eigen [39], another factor favouring proton transfer from oxygen, relative to carbon, is the greater facility of hydrogen bridge formation requisite in proton transfer processes for OH relative to CH.

situation would be one of *bifunctional catalysis*. However, glycine was not found to be a more effective catalyst in the enolization of acetone than simple amines [41], indicating that in this case there was no assistance from the internal carboxylate ion. Possibly carboxylate ion does not compete effectively with water acting as a general base, or some unfavourable orientation in this system precludes bifunctional catalysis from being effective. As yet there are very few instances where bifunctional catalysis has been demonstrated to occur [42].

It appears that orientation factors can be more favourable in *intramolecular* processes than in intermolecular processes and it is known in fact that intramolecular catalysis of enolization can provide an effective reaction pathway, at least in certain situations. For example, the keto carboxylic acids 4, n = 2–5, give evidence of intramolecular general acid catalysis of enolization [43]. However, pyruvic acid itself (n = 0) apparently does not take part in such a process since its rate of iodination is equal to that of the methyl and ethyl esters [44]. In the more rigid system given by 5 (R = H, Me) there is conclusive evidence of intramolecular catalysis by the carboxylate group [45, 46]. Similarly, the amino ketone 6 presents evidence of intramolecular base catalysis, through proton abstraction by the amino group [47]. However, ketone 7, which contains an acidic as well as a basic centre, showed catalytic behaviour similar to that of the monoamino ketone 6, indicating that intramolecular general base and general acid catalysis was not occurring concertedly [47].

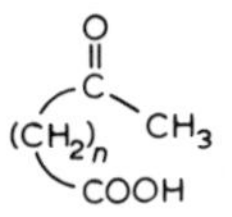

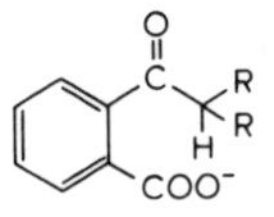

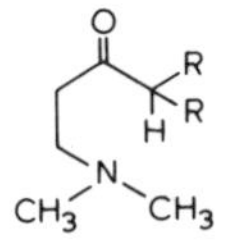

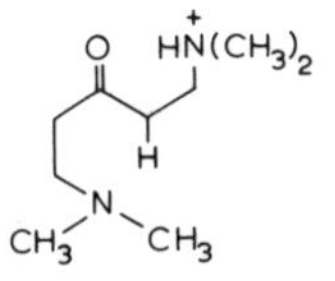

Studies of intramolecular catalysis, and the search for bifunctional catalysis in these and in related systems, are of considerable interest as models for enzyme-catalyzed reactions, for which concerted processes of this type are thought to be the rule rather than the exception [48–50].

(d) Isotopic studies

Isotopic studies are particularly suited to help elucidate reactions in which proton-transfer is an important constituent process, as is the case in keto-enol isomerism. It is immediately apparent that enolization provides a pathway for the exchange of isotopic hydrogen between the substrate and the reaction medium, and hence the rate of hydrogen isotope exchange should give a direct measure of the rate of enolization.

Acid-catalyzed hydrogen-deuterium exchange in acetone, for example, occurs through enol formation (cf. eq. (6) and (13)):

$$(17) \quad CH_3-\overset{O}{\overset{\|}{C}}-CH_3 \xrightarrow[\text{fast}]{D_3O^+} CH_3-\overset{^+OD}{\overset{\|}{C}}-CH_3 \xrightarrow[\text{slow}]{D_2O} CH_3-\overset{OD}{\overset{|}{C}}=CH_2 \xrightarrow[\text{fast}]{D_2O} CH_3-\overset{O}{\overset{\|}{C}}-CH_2D$$

In the overall reaction one molecule of D_2O becomes exchanged into HOD, which then dilutes the enrichment of the isotopic pool. Since the overall exchange process is reversible, this isotopic dilution may result in a slowing up of the observed exchange rate as the reaction proceeds. In practice this complication is largely, but not completely, avoided by using a considerable excess of heavy solvent (at least a 20:1 ratio in terms of exchangeable $D:H$). Even then, a correction factor should strictly be applied when accurate exchange rate constants are desired.

Base-catalyzed hydrogen-deuterium exchange in acetone also occurs via the enolate ion (cf. eq. (7) and (15)):

$$(18) \quad CH_3-\overset{O}{\overset{\|}{C}}-CH_3 \xrightarrow[\text{slow}]{OD^-} CH_3-\overset{O^-}{\overset{|}{C}}=CH_2 \xrightarrow[\text{fast}]{D_2O} CH_3-\overset{O}{\overset{\|}{C}}-CH_2D$$

We see that the rate-determining step in either case is enolization and that the act of deuteration is a secondary fast step. It follows then that *the rate of deuteration should be equal to the rate of halogenation*. This prediction was first shown to hold in 1937 by Reitz for the case of acid-catalyzed bromination and hydrogen-deuterium exchange of acetone [51].

Active interest in the mechanistic aspects of enolization has since continued unabated [52–61] and though the majority of studies have confirmed Reitz's initial finding, it has been inferred on occasion [54] that the rate-determining step for halogenation and deuterium exchange

80

are *not* the same. Such conclusions have generally been drawn from estimates of halogenation rates, as determined from analysis of products of halogenation. However, it has been shown that halogenated ketones may undergo substitution reactions, which can lead to misleading conclusions if not allowed for in a quantitative manner [60, 61]. For example, the bromination of 2-butanone in aqueous alkaline medium gives initially two products, $BrCH_2COCH_2CH_3$ and $CH_3COCHBrCH_3$. Now of these the former tends to be rapidly brominated and cleaved, leading to $CHBr_3$ and $CH_3CH_2CO_2^-$. The latter, in contrast, is preferentially hydrolyzed to $CH_3COCH(OH)CH_3$, which is then halogenated and cleaved to $CHBr_3$ and $CH_3CH(OH)CO_2^-$ (further bromination in $CH_3COCHBrCH_3$ is suppressed as a result of steric hindrance). Moreover, $CHBr_3$ is slowly brominated to CBr_4 under the conditions. It is only by thoroughly investigating the fate of each intermediate under the reaction conditions that meaningful conclusions concerning enolization rates at the two sites can be drawn, and comparison made with deuterium exchange at the corresponding sites. Measurement of halogenation rates by following the rate of consumption of halogenating agent needs also to be approached with caution since the rate law may not always be straightforward and simple [59, 61].

The methods employed in isotopic exchange studies also need critical examination. For deuterium exchange, in the usual experimental approach one starts with ketone which is isotopically normal and a medium which is deuterated (though the reverse situation is equally valid in principle) and one measures the rate of uptake of deuterium into the substrate under acidic or basic conditions. In the early work deuterium uptake by the substrate had to be measured rather tediously by its combustion followed by isotopic analysis of the resulting water involving careful measurement of its density or refractive index. Nowadays the deuterium content of the substrate is commonly determined directly by infrared, NMR or mass spectrometry without the time-consuming combustion. For tritium exchange, experiments are carried out in tritiated water and the rate of uptake of radioactivity by the ketone is determined. Alternatively the tritiated ketone can be used and the rate of loss of radioactivity in an isotopically normal medium determined. Results obtained by these experimental methods have generally been accepted as unambiguous. However, recently there has been a report of an anomalous isotopic result, since it has been found that the base-catalyzed proton transfer from diisopropyl ketone shows a deuterium isotope effect but not a tritium isotope effect [62].

A particularly challenging aspect of recent work on enolization in ketones is concerned with the effect of substituents on rates of enolization. Let us consider, for example, the case of 2-butanone, $CH_3COCH_2CH_3$, for which the NMR method allows the measurement of isotopic exchange to be followed at both *alpha* carbon sites simultaneously. Would the two exchange rates (as calculated per hydrogen atom, i.e. after allowing for a statistical factor) be expected to be equal? In fact they are generally not, path (a) being preferred over path (b) with *p*-nitrophenoxide or acetate ion as catalyst, while with deuteroxide ion the two sites are about equally reactive [63]. Under acidic conditions the preference for exchange at the methylene carbon is again observed [64, 65].

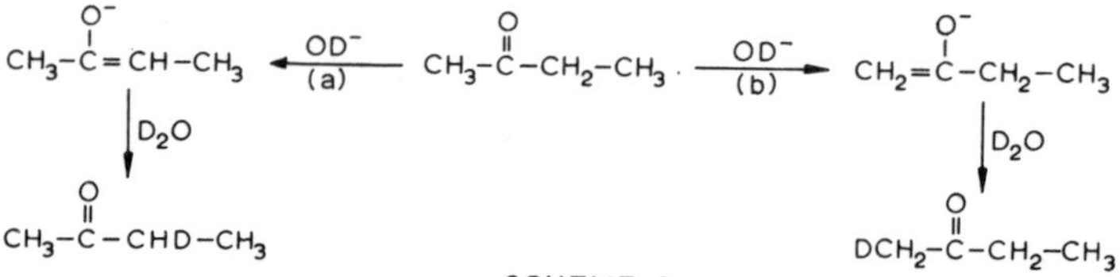

SCHEME 2

One would have expected the transition states in the acid and base processes to carry opposite charges, in which case a methyl substituent, through its electronic effect, should inhibit the base-catalyzed process at the methylene site and facilitate the acid-catalyzed process at this site. The intriguing suggestion has been made [63] that even in base catalyzed enolization the transition state can resemble enol in structure as a result of attack by a solvent molecule on the carbonyl oxygen being concerted with proton removal by base:

(19)

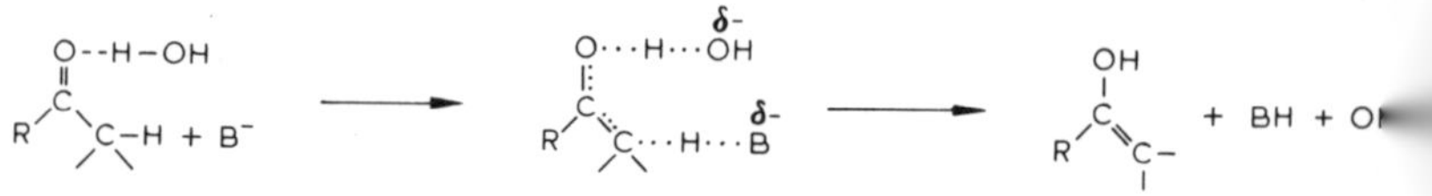

In an unusual isotopic tracer study of enolization, radioactive iodine was allowed to react with 2-butanone and exchange rates at the methyl and at the methylene group were determined from the activity of the iodoform and iodopropionic acid produced [66]. The results in this case indicated preferential enolization towards the methylene group in acid media and toward the methyl group in basic media.

On the other hand, for exchange in CD_3COCD_2X, with $X = OCH_3$, deuterium exchange at the methyl site occurs more rapidly than at the methylene site for phenoxide ion and amine catalysis in aqueous solution; however, the relative rates are reversed for hydroxide ion catalysis [56a]. The effect of halogen substituents on enolization sites is also complex [67]. It would appear that substituents affect enolization rates through both steric and electronic effects and that the balance of these factors for different situations may be difficult to predict. A theoretical (CNDO/2) treatment of the problem has been presented [68].

An important aspect of the isotopic exchange process has not yet been dealt with explicitly. How is deuterium actually incorporated onto the enolate ion; should one not expect protonation to occur on the electronegative oxygen, rather than on carbon? The answer to this question brings forth the principle of kinetic control versus thermodynamic control in protonation of the high energy enolate ion intermediate:

(20)
$$R-\underset{\substack{\text{less stable}\\\text{thermodynamically}}}{\overset{\overset{\displaystyle OD}{|}}{C}}=CH_2 \xrightarrow[\text{fast}]{D_2O} R-\underset{}{\overset{\overset{\displaystyle O^-}{|}}{C}}=CH_2 \xrightarrow[\text{slow}]{D_2O} \underset{\substack{\text{more stable}\\\text{thermodynamically}}}{R-\overset{\overset{\displaystyle O}{\|}}{C}-CH_2D}$$

A lesser electronic reorganization, and hence a smaller activation energy, is needed in conversion of the enolate ion to the thermodynamically less stable enol tautomer [69—71]. As a result of these relationships, though protonation of the enolate ion on oxygen occurs the more rapidly[*], the resulting enol will re-dissociate and go over into

---

[*]Evidence from kinetic measurements of fast proton transfer processes [39] indicates that in *acidic* media the protonation of simple enolates (such as the enolate of acetone) at either carbon or at oxygen is a diffusion-controlled process so that there is no kinetic preference for attack at one of the two sites. (Protonation of the enolate anion of 4-*t*-butylcyclohexanone has been shown to occur at carbon and at oxygen simultaneously, on the basis of stereochemical evidence [34].) On the other hand, anions derived from relatively acidic carbon acids (such as 1,3-diketones, enones, or nitroalkanes, Section 2a), do appear to exhibit a kinetic preference for protonation at oxygen rather than carbon.

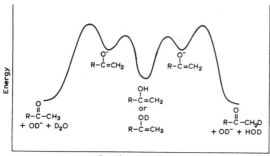

Figure 1. Intermediates in the deuteroxide-catalyzed hydrogen-deuterium exchange of ketone $RCOCH_3$ as illustrated by means of activation energy-reaction coordinate profile.

the more stable keto tautomer*. In the case of acetone, for which the enol tautomer is a much higher energy species than the keto tautomer, the proportion of molecules which at a given instant are in the enol (or enolate) form will be quite small.

These points are illustrated by means of the potential energy-reaction coordinate diagram for the $RCOCH_3$ system undergoing deuteroxide ion-catalyzed hydrogen-deuterium exchange, shown in Figure 1.

We note the symmetry of the energy profile, as is required for isotopic exchange processes since isotopic substitution cannot affect the nature of the potential energy surface. An isotopic exchange process in which only a single reaction intermediate is involved would show a single minimum on the energy profile.

---

*Alternative pathways to the conversion of an initially formed enol to the keto tautomer are the following:

(a) $\displaystyle \underset{}{\ce{C=C}} \text{—}OH \xrightarrow{H^+} \text{—}\overset{}{\underset{}{C}}\text{—}\overset{+}{C}\text{—}OH \longleftrightarrow \text{—}C\text{—}C\overset{+}{\underset{}{OH}} \xrightarrow[-H_3O^+]{H_2O} \text{—}C\text{—}C\text{=}O$

(b) $\displaystyle \underset{}{\ce{C=C}}\text{—}O\text{—}H \xrightarrow[H_2O]{H^+} \ce{C=C} \text{—} O\text{—}H \xrightarrow{-H_3O^+} \text{—}C\text{—}C\text{=}O$

84

Acid catalyzed hydrogen-deuterium exchange in the case of a ketone will be characterized by a somewhat similar energy diagram (Figure 2). Starting with ketone plus acid, the first intermediate formed is the conjugate acid of the ketone and the energy barrier for its formation is relatively low. Deuteriation of the enol occurs by the addition of D^+ (solvated) to the double bond, analogous to the electrophilic addition of Br^+ (eq. (14); cf. footnote on p. 84).

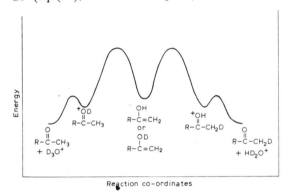

Figure 2. Intermediates in the acid-catalyzed hydrogen-deuterium exchange of ketone $RCOCH_3$ on energy-coordinate profile.

The kinetic isotope effect may also be applied as a mechanistic probe in the study of enolization. Since the rate-determining step in enolization is proton abstraction from *alpha*-carbon, it follows that the rates of enolization of $RCOCH_3$, $RCOCD_3$ and $RCOCT_3$ should be subject to primary isotope effects [72–75], namely $k_H/k_D \sim 7$–10, $k_H/k_T \sim$ 15–20 and $k_D/k_T \sim 2.5$–3. (These values of the isotope effect are near to maximal and apply to transition states in which the proton is about half transferred from the carbon acid to the base. In the case of unsymmetrical transition states the kinetic isotope effect can take on considerably smaller values – see also Chapter 1, Section 5.) This criterion was used in one of the classical studies on the mechanism of enolization [76]. It was found that in the acetate ion-catalyzed bromination of acetone $k_H/k_D = 7.0$, while in the hydronium ion-catalyzed bromination $k_H/k_D = 7.7$. Since that time measurements of isotope effects have been used in related enolization studies [77–82]; in some instances [77, 81, 82] proton tunneling [83] has been found to occur.

85

Also of interest is the application of solvent [84, 85] and of secondary [86] isotope effects in enolization processes.

(e) Stereochemical consequences of keto-enol tautomerism

Another approach to the problem of keto-enol tautomerism makes use of the stereochemical consequences which are implicit in the inter-conversion of the two isomers. Consider a ketone RCOCHRR' which is optically active by virtue of a chiral carbon atom from which the hydrogen is abstracted in enolization. Then in the formation of the planar enol, RC(OH)=CRR', or enolate ion, RC(O⁻)=CRR', the chirality will have been destroyed. When the more stable keto tautomer is regenerated, the two optically active antipodes will in principle be produced in equal amounts, to give a racemic mixture. Therefore the rate of enolization of an optically active ketone should be measurable by following the rate of racemization, which can be done by observing the rate of loss of optical activity by means of a polarimeter.

We see that racemization of an optically active ketone provides another criterion of mechanism, in addition to halogenation and isotopic exchange. In some ingeneously devised studies, racemization has been used *in conjunction* with one of the other criteria of enolization. For the keto-acid **8** it was shown that the rate of racemization was equal to the rate of bromination in acid medium* [87], while for ketone **9** the rate of racemization was found to be equal to the rate of iodination [89]. Base-catalyzed hydrogen-deuterium exchange was determined for ketone **8** and was found to be equal to the rate of racemization [90].

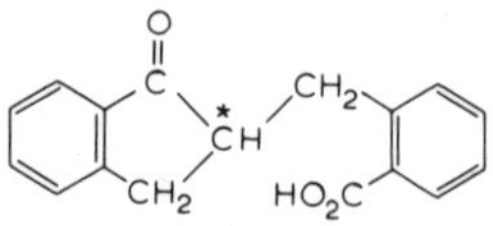

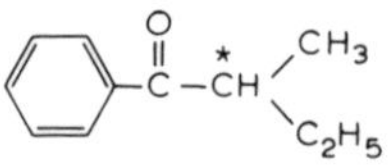

It thus appears that the rate of enolization can be determined by measuring the rate of racemization, halogenation, or hydrogen isotope exchange.

---

*Under basic conditions there is evidence of intramolecular catalysis by the carboxylate anion in 8 [88].

However, the equality of the rates of racemization and of hydrogen isotope exchange is subject to a restriction which has not been mentioned in this chapter. This is that the enolate ion must be a free entity, completely separated from the proton which was abstracted by the basic species, in order that reprotonation by a molecule from the bulk medium should take place. In those circumstances isotopic exchange will be a true measure of racemization and of enolization.

Next we may ask whether it is conceivable that the enolate ion could not be a free entity. That can, indeed, be made possible, by hydrogen-bonding of the type $-C=C-O^-\ \cdots H-B$, where the abstracted hydrogen, though covalently bonded to base B, is electrostatically bound to the electronegative oxygen. Though the incidence of cases of hydrogen-bonding of the type $C^-\cdots H-B$ is not uncommon in the chemistry of carbanions (see ref. 91, and Chapter 2, Section 3; also several instances are described later in this chapter), there are only very few reports of this phenomenon with enolate ions. The following study provides an example.

It was found that when N-carbobenzoxy-S-benzyl-L-cysteine penta-chlorophenyl ester was allowed to interact with triethylamine in chloroform solution containing CH_3OD for a period sufficient for almost complete racemization, the extent of deuteration on the *alpha* carbon was only 20% [92]. The proposed mechanism for racemization involves proton abstraction and recombination within hydrogen-bonded inter-mediates:

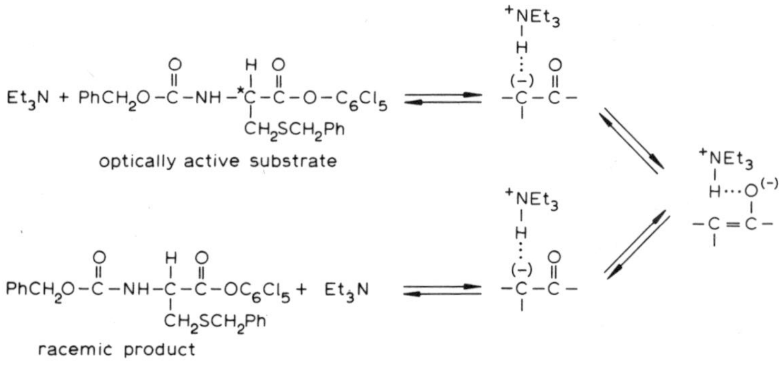

Figure 3. A "conducted tour" mechanism for racemization in the absence of isotopic exchange.

In the cases in which the hydrogen-bond is ruptured, the Et_3NH^+ species will be displaced by a molecule of CH_3OD, resulting in isotopic exchange. Presumably the choice of the non-polar reaction medium, in which complete separation of ions is less likely, is the main factor responsible for the observation of base catalyzed racemization without complete isotopic exchange in this system. A qualitatively similar result was obtained with N-carbobenzoxy-L-phenylalanine pentachlorophenyl ester, indicating that the process is not specific for cysteine derivatives [93].

There is an important stereochemical implication of the requirement for delocalization in the enolate ion in providing stabilization, viz. that the four atomic centers bonded to the carbon—carbon double bond in the enolate ion must be coplanar. It follows, therefore, that proton abstraction from the *alpha* carbon of a ketone should only be practicable when this coplanarity criterion can be satisfied, for it is only then that resonance stabilization in the anion can be effective. An early illustration of this principle was provided by the observation that the hydrogen at the bridgehead-carbon of the bicyclic β-diketone **10** was much less acidic than the hydrogens flanked by two keto groups in the monocyclic analogue **11** [94].

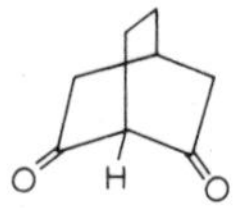

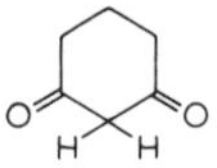

We note that formation of the enolate ion from **10** would require introduction of a double bond at the bridgehead, thus leading to a strained ring system (Bredt's rule [95] — for a recent evaluation see ref. 96). The strain energy involved in such a process would make enolization in the bicyclic system unfavourable. Now the strain energy factor has an inverse dependence on ring size; thus bicyclic alkenes in which the bridgehead double bonds are endocyclic in rings of eight carbons or more are generally sufficiently stable to allow their synthesis. On this basis the observation of bridgehead exchange in ketone **12** under relatively mild conditions (CH_3ONa/CH_3OD at $100°$) can readily be rationalized [97]. The bridgehead alkene **13** has been isolated, and its strain energy determined to be 12 kcal/mole [98].

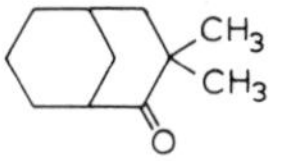

12 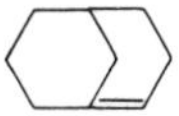   **13**

The bicyclooctenes **14** and **15** have only been obtained as transient species (characterized as the Diels–Alder adducts) [99]. Evidence has also been put forth for the generation of 1-norbornene (**16**) as a transient species; this molecule is expected to be highly strained, and may possibly exist in a triplet ground state [100].

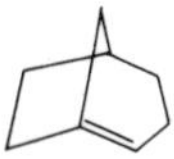

14 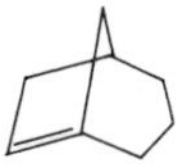   **15** 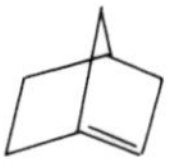   **16**

In light of the above considerations, the hydrogen-deuterium exchange studies [101] in 7-ketonorbornane (**17**) and in nortricyclanone (**18**) take on special significance. The bridgehead hydrogen (H1) in **17** becomes labile under stringent conditions; on treatment with potassium t-butoxide in *t*-butyl alcohol-O-*d*, at 200° for 48 hours, the extent of exchange at C1 is 25%. Exchange in **18** proceeds more readily and at several centers; in *t*-BuOK/*t*-BuOD at 200° and 22 h, there is 73% exchange at C1, 70% exchange at C2 and 68% exchange at C4.

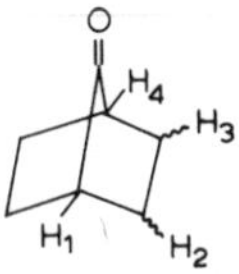

17 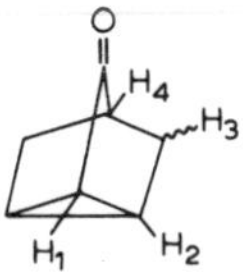   **18** 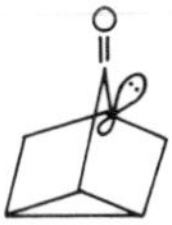   **19**

How is one to explain the occurrence of bridgehead exchange in **17** and **18**? It may be presumed that enolate ion formation is energetically prohibitive due to the excessive strain requisite in introducing a double bond at the bridgehead in either molecule. Thus the carbanionic center which is formed on proton abstraction will be essentially in the sp^3 hybridization state. The degree of overlap between the electron pair in the sp^3 orbital and the p-π orbitals of the carbonyl group is zero, since the projected angle between the interacting orbitals is 90°. It would appear, therefore, that the carbanion formed on proton abstraction is

89

stabilized solely through an inductive effect of the carbonyl function.
This is an important finding since it implies that, in the other systems
that we have considered, some part of the carbanion-stabilizing capability
of the carbonyl group should be apportioned to the inductive effect; a
quantitative separation of the two effects — inductive and resonance —
would be desirable. Also of note is the increased exchange rate at the
bridgehead (C4) in **18**, compared with **17**. This is probably due to interaction of the sp^3 orbital (**19**) with the cyclopropyl ring, whose carbon—
carbon bonds have enhanced p-character. The relatively facile exchange
of H1 and H2 in **18** is in accord with the high degree of s-character in
the cyclopropyl C—H bonds (see also Chapter 1, Section 1).

The cyclopropane ring system as a substituent *alpha* to the carbonyl
group provides an interesting instance of the effect of steric strain on
enolization and on isotopic exchange rates. The introduction of a
double bond exocyclic to a cyclopropane ring is well known to be an
energetically unfavourable process, since the deviation from the normal
bond angle value (with respect to acyclic systems) is increased thereby
from 49° in the initial state to 60° in the final state. Hence proton
abstraction from cyclopropane activated by an *alpha* carbonyl should be
suppressed because of steric inhibition of delocalization in the transition
state for enolization. The results of several hydrogen-deuterium exchange
studies [52, 53, 102, 103] are in agreement with this prediction, e.g. [102],

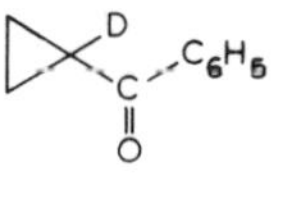

Relative rate, 1 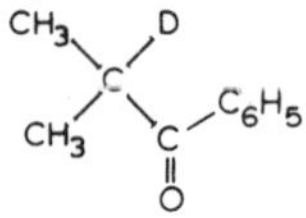   Relative rate, 170

In the case of the optically active cyclopropyl ketone, 1-benzoyl-2,2-
diphenylcyclopropane, it was found that hydrogen-deuterium exchange
with methanolic sodium methoxide proceeded with partial (27%)
retention of configuration [104a]. This suggests that there is a delicate
balance between the delocalization energy for enolization, ring strain
energy, and hydrogen bonding effects in this system (cf. [104b]).

2. SOME FURTHER TAUTOMERIC SYSTEMS

Tautomerism is not restricted to the keto-enol system, though this is probably the most widely investigated case. There are numerous known tautomeric systems, of which only a few representative types will be discussed here. Our discussion will, in the main, highlight the use of isotopic studies in the elucidation of the characteristics governing those particular systems.

Closely related to the tautomerism shown by ketones is that exhibited by the related functional classes of compounds containing the carbonyl grouping, such as amides and esters of carboxylic acids. Though the $-NR_2$ and $-OR$ functions in $R'CONR_2$ and $R'COOR$ will affect the tautomeric equilibria of amides and esters in comparison with ketones, the basic principle that the carbonyl group must act to delocalize the charge in the *alpha* carbanions remains unchanged. The delocalized carbanions are throughout planar and symmetrical, as shown by concurrent racemization and hydrogen-deuterium exchange studies with optically active amides and esters. Thus amide **20** and ester **21** undergo proton exchange in basic media (e.g. t-BuO$^-$/t-BuOD) with complete racemization [105].

$$C_6H_5CH_2-\underset{\underset{H}{|}}{\overset{\overset{CH_3}{|}}{C^*}}-\underset{\underset{O}{\|}}{C}-N(C_2H_5)_2 \qquad\qquad C_6H_5-\underset{\underset{H}{|}}{\overset{\overset{CH_3}{|}}{C^*}}-\underset{\underset{O}{\|}}{C}-OC(CH_3)_3$$

20   **21**

It is noteworthy that enolization studies with aldehydes and with carboxylic acids are relatively rare. A contributing factor in the case of aldehydes is their great tendency to undergo hydration in aqueous media:

$$(21)\qquad \overset{R_1}{\underset{R_2}{>}}CH-\overset{\overset{O}{\|}}{C}-H \;\rightleftharpoons\; \overset{R_1}{\underset{R_2}{>}}CH-CH\overset{OH}{\underset{OH}{<}}$$

This reaction is catalyzed by acids and by bases [106]. Another complication is presented by the facility with which aldehydes undergo the aldol condensation [35, 107]. However, in highly crowded molecules (e.g., in eq. (21), R_1 = phenyl, R_2 = mesityl), in which steric hindrance to hydration or to intermolecular condensation is large, the keto-enol tautomeric equilibrium can be observed [108]. Isobutyraldehyde-2-*d* undergoes amine-catalyzed isotopic exchange, but the reaction is thought to proceed via the intermediate iminium ion [109]:

(22) $\quad Me_2CDCHO + RNH_3^+ \rightleftharpoons Me_2CDCH=NHR^+ + H_2O$

(23) $\quad Me_2CDCH=NHR^+ + B \longrightarrow Me_2C=CHNHR + BD^+$

α-Anions of carboxylic acids have been prepared by the action of strong base on the carboxylic acids in aprotic media:

(24)

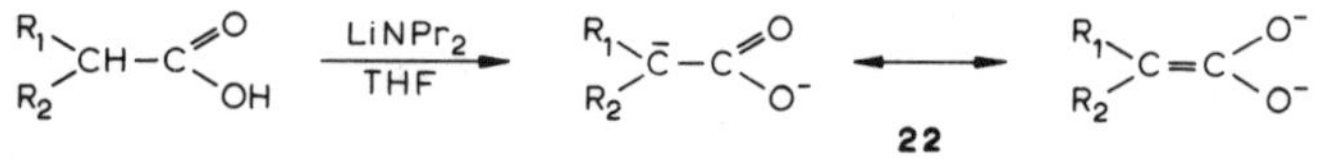

The dianion 22 can be deuterated, carboxylated, or alkylated, at the α-carbon centre [110]. A small equilibrium concentration of the α-anion can be established in protic media, as demonstrated by hydrogen-deuterium exchange of substituted phenylacetic acids with $NaOD/D_2O$ [111] or with Et_3N/D_2O [112]. Tautomerism in acid anhydrides has also been observed and 2-phenylbutyric acid anhydride has been shown, by NMR and IR spectroscopy, to exist as an equilibrium mixture of the diketo, keto-enol and dienol forms [113]:

(25)

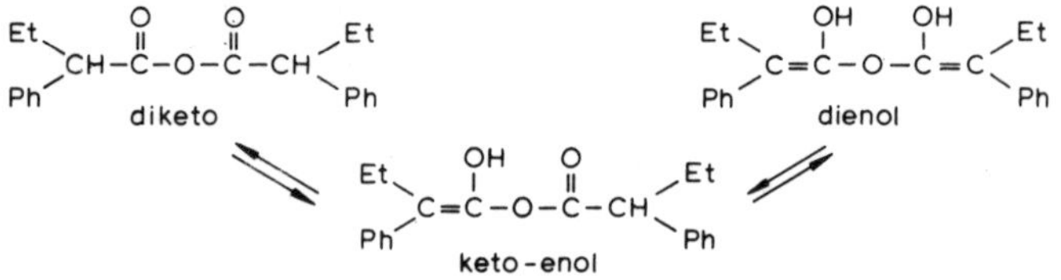

It is interesting that primary and secondary amides (as opposed to tertiary amides such as 20) also exhibit another type of tautomeric equilibrium:

(26)

Though the structure on the left ("lactam" form) predominates, diverse evidence indicates the existence of the tautomeric structure on the right ("lactim" form). For instance, alkylation of amides can give

92

rise to both O- and N-alkylated products. Of course, lactam–lactim tautomerism is fundamentally related to keto-enol tautomerism in that the activating carbonyl is common to both, but with the former the proton is removed from an *alpha* nitrogen while in the latter case it is removed from an *alpha* carbon.

A special case of keto-enol tautomerism is that of aromatic hydroxy compounds. It is seen that the Kekulé valence bond structure for phenol is tautomeric with the keto form, 2,4-cyclohexadienone:

(27)

The equilibrium concentration of the keto tautomer is of course extremely small (the ratio [enol]/[keto] has been estimated [114] as 10^{14}), due to the special stability associated with the closed shell π-system of the benzenoid ring. Thus phenols are at the other end of the scale from simple ketones as far as the position of equilibrium between keto and enol forms is concerned. The Hückel molecular orbital method has been applied to a number of aromatic hydroxy compounds and the total delocalization energies of the enol and keto tautomers have been evaluated [21].

Phenoxide ions are formally analogous to enolate ions and once again the negative charge may be borne by oxygen or by carbon. However, since the carbon skeleton of the ring forms a conjugated system, the charge can be located on the *ortho* and the *para* carbons:

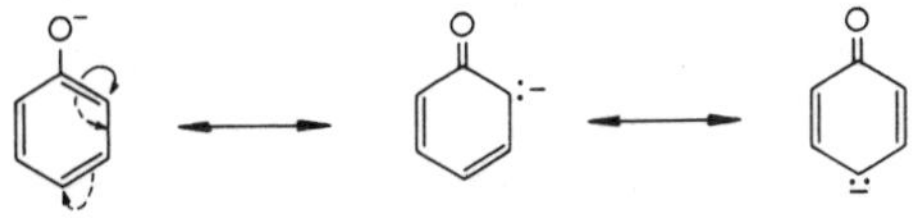

Experimental evidence supports this formulation. Thus alkylation of phenoxide ion, though commonly occurring on oxygen, may also take place on the *ortho* and *para* carbons under certain reaction conditions [115]. Similarly the Kolbe and Reimer-Tiemann reactions involve carbon as the nucleophilic centre rather than oxygen. Evidence from hydrogen-exchange studies [116–119] is also in accord with charge delocalization in the phenoxide ion. Thus phenol in a basic medium of

93

sodium deuteroxide in D_2O undergoes isotopic exchange at the *ortho* and *para* carbons. This is illustrated for *para* exchange in eq. (28):

(28)

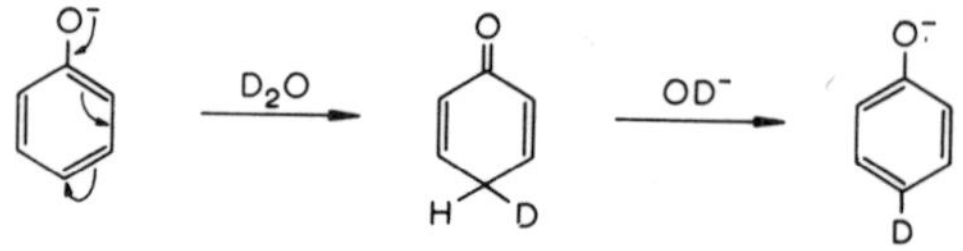

(a) Nitro-aci-nitro tautomerism

After the keto-enol system, the most extensively studied case of tautomerism is probably that of aliphatic nitro compounds. In this case, also, the proton shift is between carbon and oxygen, but the central atom of the $X \cdot Y \cdot Z$ triad is now nitrogen. The generalized tautomeric equilibrium in this system is as follows, with the isomer on the right hand side termed an *aci*-nitro compound (or a nitronic acid):

(29)

     nitro tautomer   *aci*-nitro tautomer

Hantzsch was the first to examine this nitro-*aci*-nitro tautomeric system, and in this context he coined the term "pseudo-acid" to describe the measurable rate of ionization of a carbon acid [120]. Phenylnitromethane, $C_6H_5CH_2NO_2$, for instance, reacts with sodium hydroxide at a measurable rate to form a sodium salt; the rate of this transformation can readily be followed conductometrically. On completion of the reaction, if the salt is carefully acidified with one equivalent of acid the *aci* form is obtained as a crystalline solid. Phenylnitromethane, a pale yellow liquid, contains only ca. 0.1% of *aci* form at equilibrium. The reverse reaction, from *aci*-nitro to the nitro isomer, occurs spontaneously in aqueous solution. Since the *aci*-nitro isomer is more highly ionized than the nitro isomer, the rate of the reverse reaction can conveniently be followed by the concurrent decrease in conductivity [121]. A recent spectroscopic study [122] (NMR, IR and UV) of phenylnitromethane, its *aci*-form, and the sodium salt, confirms the structural relationships which had been deduced by Hantzsch.

The position of equilibrium in the nitro-*aci*-nitro system is markedly dependent on structure, as with the keto-enol system. The proportion of the *aci*-nitro isomer increases from $1 \times 10^{-5}\%$ with nitromethane to $9 \times 10^{-3}\%$ with nitroethane, to 0.3% with 2-nitropropane. *p*-Nitro-phenylnitromethane contains 0.8% of the *aci* form in aqueous methanol and 16% in pyridine.

An interesting structural situation arises with benzoylnitromethane, $C_6H_5COCH_2NO_2$, since the possibility arises here of either keto-enol or of nitro-*aci*-nitro tautomerism, as shown below. Both tautomers provide opportunity for strong internal hydrogen bonding.

enol tautomer
of benzoylnitromethane

aci-nitro-tautomer
of benzoylnitromethane

The observation that methylation with diazomethane leads to an enolic ether (23) as well as a nitronic ether (24) suggests that both tautomers are present in the equilibrium mixture:

23

24

The anion formed on proton abstraction from a nitroalkane $R_1R_2CHNO_2$ is a resonance hybrid, with the following contributing resonance structures:

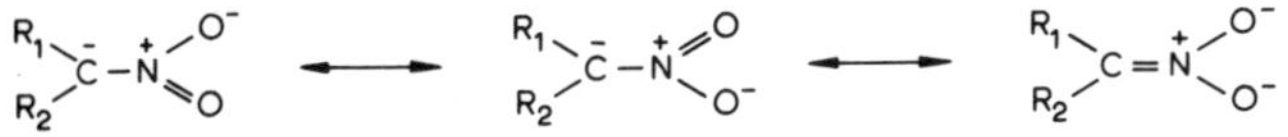

25 a   25 b   25 c

Structure 25c with the negative charge borne by oxygen is expected to contribute much more heavily than 25a or 25b in which carbon bears this charge (see also Chapter 1, Section 1). A test of this expectation should be provided by the stereochemical criterion since the planarity of 25c implies that an optically active nitro compound $R_1R_2CHNO_2$ should give rise to a racemic salt $[R_1R_2CNO_2]^-Na^+$. This is, in fact, the case [123]. For synthetic uses of nitroanions, see refs. 35, 107, 124.

Proton abstraction from nitro-compounds has been studied by hydrogen-deuterium exchange, by conductometric measurement

95

of ionization, and by halogenation [125], and consistent results
are obtained by the three methods. The proton abstraction process is
subject to general base catalysis [126–131] and shows a primary
hydrogen/deuterium isotope effect [126, 130–132]. For example, for
the ionization of nitroethane at $30°$ with piperidine as the base [130],

$$\frac{k(CH_3CH_2NO_2)}{k(CH_3CD_2NO_2)} = 8.1$$

With hydroxide ion as base the isotope effect has a value of 9.3 [130].
These values fall into the range of "normal" hydrogen/deuterium isotope
effects for rate-determining proton abstraction [72–75]. On the other
hand, for the ionization of 2-nitropropane much larger k_H/k_D values
are obtained with certain bases, e.g., with 2,4,6-trimethylpyridine as
base [126],

$$\frac{k(Me_2CHNO_2)}{k(Me_2CDNO_2)} = 24.3$$

This "abnormally" large value of the kinetic isotope effect is indicative
of proton tunneling [83].

Proton transfer from nitro-substituted toluenes and some structurally
related compounds is governed by similar principles [133]. For
example, the 2,4,6-trinitrobenzyl anion is obtained on treatment of
2,4,6-trinitrotoluene (TNT) with ethoxide ion in ethanol [134]:

(30)

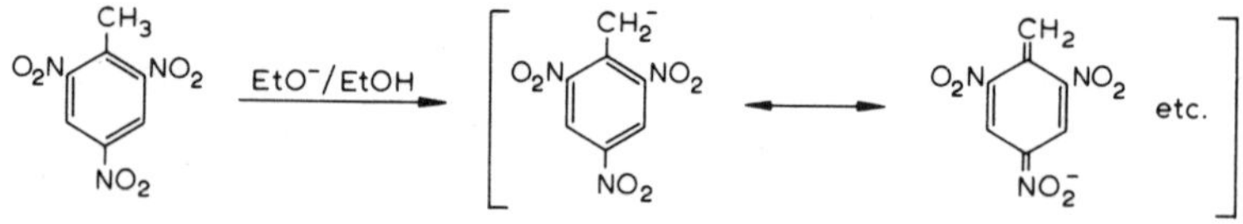

26

The resonance-delocalized species **26** is purple-colored. The proton
abstraction is subject to a primary isotope effect of 7.0 at $25°$, as
measured by comparison of the rates of reaction of TNT and TND-d_3
[135]. However, a complicating factor in TNT-alkoxide ion systems is
the formation of a σ-adduct (a "Meisenheimer complex") by addition
of alkoxide ion to aromatic carbon [136]. Proton transfer has been
observed in analogous fashion with tri-(4-nitrophenyl)methane [137]
and di-(4-nitrophenyl)methane [138].

96

(b) Tautomerism of nitriles

Nitriles give evidence of an equilibrium between a cyano form and an imino form, with the former predominating under most circumstances:

(31) $R_1R_2CH-C\equiv N \rightleftharpoons R_1R_2C=C=NH$

Evidence for this tautomeric equilibrium is provided in studies of concurrent racemization and hydrogen-deuterium exchange with optically active nitriles. Thus it is found that 2-phenylbutyronitrile (27) undergoes hydrogen-deuterium exchange in t-butyl alcohol containing potassium t-butoxide with *complete racemization* [139]. This is in accord with formation of an intermediate in which chirality has been destroyed. The mechanism of isotopic exchange is analogous to that in ketones, depicted in Figure 1.

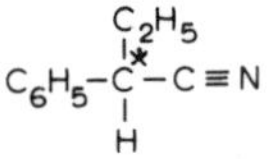

27 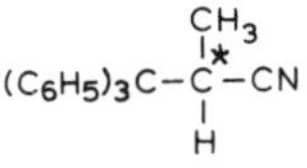   28

The nitrile system also provides one of the classical examples in which, as a result of hydrogen-bonding between the imine tautomer and the protonated base BH, the proton can be conducted by B to the opposite face of the erstwhile chiral carbon, leading to racemization without isotopic exchange. This "conducted tour mechanism" (cf. Figure 3) was observed [139] with 2-phenylbutyronitrile and tripropylamine as base in the less dissociating tetrahydrofuran/t-butyl alcohol (1.5 M) medium (see also Chapter 2, Section 3). Nitrile 28 also shows a medium dependence in the racemization vs. isotopic exchange relationship [140].

Proton abstraction from malononitrile and from t-butylmalononitrile shows general base catalysis [141], but is subject to an unusually small kinetic isotope effect [142]. For example, for t-butylmalononitrile $k_H/k_T = 1.74$ (acetate ion catalysis, at $25°$), from which one may calculate [143] $k_H/k_D = 1.47$. This small value of the isotope effect indicates a transition state structure in which the proton is almost completely transferred to the base, an interpretation which is consistent with the near unit value of the Brönsted exponent β obtained for catalysis by carboxylate anions in this system [142] (cf. Chapt. 1, Sect. 3).

97

(c) Tautomerism of propenes

The tautomeric system consisting of a 3-carbon triad is involved in the important area of olefin isomerisation [144]:

(32) $R_1R_2CH-CR_3=CR_4R_5 \rightleftharpoons R_1R_2C=CR_3-CHR_4R_5$

This isomerisation occurs under strongly basic conditions, suggesting that a carbanionic mechanism is operative. A number of kinetic studies have been reported [145–149] (e.g. the isomerisation of allylbenzene to propenylbenzene), and are in general agreement with such a mechanism. However, studies involving deuterated compounds have shown that in a number of systems proton shift leading to isomerisation occurs, at least in part, *intramolecularly*. For example, the isomerisation of deuterated 1-pentene in dimethyl sulfoxide containing *t*-butyl alcohol and potassium *t*-butoxide proceeds 16 times more rapidly than does isotopic exchange with the medium [150]. The predominant intramolecularity of the process is explained by the following scheme:

(33) $D_2C=CD-CD_2-C_2D_5 + RO^-$ ⟶

hydrogen-bonded
allylic anion

⟶ $CD_3-CD=CD-C_2D_5 + RO^-$

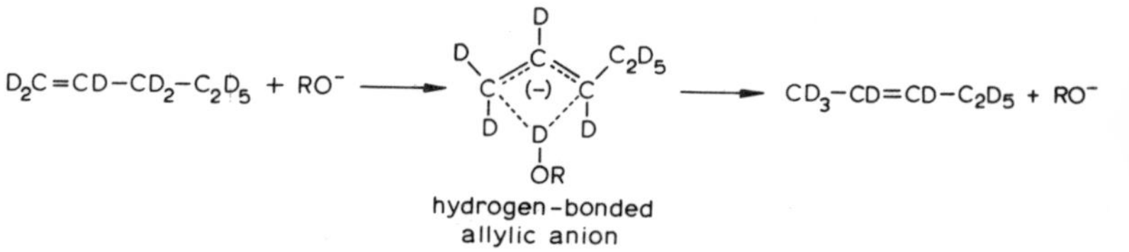

The central feature of this mechanism is that deuteron abstraction by RO⁻ yields an allylic ion in which the ROD molecule is hydrogen-bonded to the two terminal centers, so that collapse within this ion-pair leads to isomerisation without isotopic exchange. Other olefin isomerisations have also been found to occur with partial intramolecularity [149–156].

(d) Tautomerism of α,β- and β,γ- unsaturated derivatives

Propenes containing a strongly electronegative group X on one of the terminal carbons also form a well defined tautomeric system. The two tautomers are generally referred to as the allyl (β,γ-) and vinyl (α,β-) isomeric derivatives. Among the functional groups X, those commonly studied are CN, OR, SR, C(O)R, S(O)R and SO₂R [157, 158].

(34) $R_1R_2C=CR_3-\overset{\overset{\displaystyle H}{|}}{C}R_4-X$ ⇌ $R_1R_2\overset{\overset{\displaystyle H}{|}}{C}-CR_3=CR_4-X$

allyl (β,γ-) tautomer   vinyl (α,β-) tautomer

Of the factors which may influence such equilibria, the electronic effects of substituents on the stabilities of carbon—carbon double bonds must be of prime importance. It has been found, however, that the interaction of substituents across the double bond must also be allowed for in correlating the tautomeric equilibrium constants [158].

A classical study of this type of tautomeric system was performed with X = CN on the cyclohexyl derivatives 29 and 30 [159]. The conjugated vinyl cyanide 30 is the more stable form, as expected. Here both isomers were examined with respect to rate of isotopic exchange and isomerisation (29 → 30) in basic medium (EtO⁻/EtOD). The observed rate relationships, viz., H/D exchange of 29 ≫ isomerisation (29 → 30) ≫ H/D exchange of 30, are in accord with the energy diagram for hydrogen exchange and tautomerisation in the keto-enol system (Figure 1).

(35)

        29   30

The less stable tautomer 29 (the equivalent of the enol tautomer) undergoes isotopic exchange and isomerisation via the delocalized allylic carbanion (the equivalent of the enolate anion) to the more stable tautomer 30 (the equivalent of the keto form). The parallelism in the two systems is that the high energy intermediate protonates more rapidly to form the less stable isomer in both cases.

On the other hand, in the unsaturated ether system, 31 ⇌ 32 (i.e. X = OR), a different relationship obtains between proton exchange and tautomerisation [160]. The dodecyl allyl ether 31 is isomerised by t-BuOK/t-BuOD to the *cis* propenyl ether 32, with incorporation of deuterium at the terminal carbon. Under the same conditions 32 undergoes isotopic exchange at very much slower rate and there is no detectable isomerisation to 31.

(36) $CH_2=CH-CH_2-OC_{12}H_{25}$ $\xrightarrow{t\text{-BuOK}/t\text{-BuOD}}$ $DCH_2-CH=CH-OC_{12}H_{25}$

                    31   32

It follows then that in this system *the anionic intermediate protonates more rapidly to form the more stable isomer*. The energy diagram applicable to this case is given in Figure 4 (see also refs. 39 and 91).

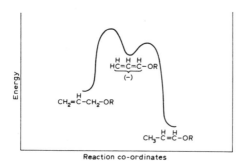

Figure 4. Activation energy-reaction coordinates diagram for isomerisation in the unsaturated ether system.

(e) The methyleneazomethine (imine) tautomeric system

The triad of atoms $C \cdot N \cdot C$ provide the structural unit for the methyleneazomethine tautomeric equilibrium, which we will call simply the imine equilibrium. This is illustrated with an actual example in eq. (37):

$$(37) \quad C_6H_5-\underset{\underset{H}{|}}{\overset{\overset{CH_3}{|}}{C}}-N=\overset{\overset{C_6H_5}{|}}{C}-C_6H_4-Cl \ (p) \ \rightleftharpoons \ C_6H_5-\overset{\overset{CH_3}{|}}{C}=N-\underset{\underset{H}{|}}{\overset{\overset{C_6H_5}{|}}{C}}-C_6H_4-Cl \ (p)$$

In contrast to tautomeric systems such as those of ketones and nitriles, in which the position of equilibrium is normally greatly in favour of one of the tautomers, the imine system is one in which neither tautomer is predominant. This is in accord with the symmetry of the system, provided that the substituents on the terminal carbon centres are not greatly dissimilar. One may call this a thermodynamically balanced system.

An interesting application of the isotopic method to evaluate the imine tautomeric equilibrium has been reported [161]. The labelled compound N-benzylidinebenzylamine (methine-^{14}C) was prepared by

100

condensation of $C_6H_5-{}^{14}CHO$ with $C_6H_5CH_2NH_2$, and equilibration was effected by means of sodium ethoxide (1.5 M) in ethanol:

(38) $C_6H_5-{}^{14}CH=N-CH_2-C_6H_5 \rightleftharpoons C_6H_5-{}^{14}CH_2-N=CH-C_6H_5$

Analysis of the equilibrium mixture was performed by acid quenching, formation of the 2,4-dinitrophenylhydrazone of benzaldehyde, and monitoring its radioactivity. It was found in this way that equilibrium is reached in 36 h at 40° and 0.5 h at 82° and the equilibrium constants at these temperatures are 0.983 ± 0.002 and 0.986 ± 0.002, respectively. Thus the equilibrium favours carbon-14 at the sp^2-position, which is in accord with theoretical considerations [162].

An extensive investigation [163] of the imine equilibrium which included measurement of rates of isomerisation, racemization, and deuterium uptake in ethanol-O-*d* containing sodium ethoxide, suggested that tautomerisation proceeds by a *concerted* mechanism (i.e. that a free carbanion is not present):

(39)

$$R_2-\underset{\underset{RO^{\curvearrowright}H}{|}}{\overset{\overset{R_1}{|}}{C}}-N=\underset{\underset{D-OR}{|}}{\overset{\overset{R_3}{|}}{C}}-R_4 \rightleftharpoons R_2-\overset{\overset{R_1}{|}}{C}=N-\underset{\underset{D^{\curvearrowleft}OR}{|}}{\overset{\overset{R_3}{|}}{C}}-R_4$$

However, another investigation [164] of the imine tautomeric system, with structurally different methyleneazomethines and using another base, gave results which indicated that a free carbanion does in fact intervene:

(40) $RO^- + R_1R_2\underset{\underset{H}{|}}{C}-N=CR_3R_4 \rightleftharpoons ROH + [R_1R_2C \overset{\cdots}{=} N \overset{\cdots}{=} CR_3R_4]^- \rightleftharpoons R_1R_2C=N-\underset{\underset{H}{|}}{C}R_3R_4 + RO^-$

Debate on these alternative mechanisms is still continuing [165, 166] and it would be premature to come to a firm conclusion at the present time. We may recall that for the keto-enol system, also, there was some evidence that enolization can follow a concerted pathway. The distinction is an important one, since the concerted pathway may well be expected to provide a valid alternative to the carbanionic mechanism. At the present time our knowledge as to the factors which would favour the alternative pathways is incomplete.

It is fitting perhaps to conclude this chapter by pointing out that even with a relatively simple overall process such as the prototropic equilibrium the detailed mechanism is not yet fully understood. The challenge remains with us, to devise new experiments and novel approaches, so that we may gain further insight into the manner of transformation between molecular species.

REFERENCES

1   A. Guether, Arch. Pharm., 106 (1863) 97; Jahresber., (1863) 323.
2   L. Knorr, O. Rothe and H. Averbeck, Ber., 44 (1911) 1138.
3   K.H. Meyer, Ann., 380 (1911) 212; Ber., 47 (1914) 826.
4   G. Schwarzenbach and C. Witwer, Helv. Chim. Acta, 30 (1947) 659, 663, 669.
5   R. Schreck, J. Amer. Chem. Soc., 71 (1949) 1881.
6   A. Gero, J. Org. Chem., 19 (1954) 469, 1960.
7   W. Walisch and J.E. Dubois, Ber., 92 (1959) 1028.
8   R.P. Bell and P.W. Smith, J. Chem. Soc. (B), 241 (1966).
9   M. Bergon and J.-P. Calmon, Bull. Soc. Chim. France, 1819 (1972).
10 R. Mecke, in "Hydrogen Bonding", edited by D. Hadži and H.W. Thompson, Pergamon Press, London, 1959.
11 D.N. Shigorin, Zhur. Fiz. Khim., 27 (1953) 689.
12 N.L. Allinger, L.W. Chow and R.A. Ford, J. Org. Chem., 32 (1967) 1994.
13 D.W. Thompson and A.L. Allred, J. Phys. Chem., 75 (1971) 433.
14 A.I. Koltsov and G.M. Kheifets, Russ. Chem. Rev., 40 (1971) 773; Uspekhi Khim., 40 (1971) 1646.
15 J.H. Billman, S.A. Sojka and P.R. Taylor, J.C.S. Perkin II, 2034 (1972).
16 M. Gorodetsky, Z. Luz and Y. Mazur, J. Amer. Chem. Soc., 89 (1967) 1183.
17 P. Alcais and R. Brouillard, J.C.S. Perkin II, 1214 (1972).
18 G.S. Hammond, in "Steric Effects in Organic Chemistry", edited by M.S. Newman, Wiley, New York, 1956.
19 G.W. Wheland, "Advanced Organic Chemistry", 3rd Ed., Wiley, New York, 1960, Chapter 14.
20 C.K. Ingold, "Structure and Mechanism in Organic Chemistry", 2nd Ed., Cornell University Press, Ithaca, N.Y., 1969, Chapter 11.
21 S. Forsén and M. Nilsson, in "The Chemistry of the Carbonyl Group", Vol. 2, edited by J. Zabicky, Interscience Publishers, New York, 1970.
22 W.J. Hehre and W.A. Lathan, Chem. Commun., 771 (1972).
23 H.M.R. Hoffmann and E.A. Schmidt, J. Amer. Chem. Soc., 94 (1972) 1373; E.A. Schmidt and H.M.R. Hoffmann, J. Amer. Chem. Soc., 94 (1972) 7832.
24 A. Lapworth, J. Chem. Soc., 85 (1904) 30.
25 H.M. Dawson and J.S. Carter, J. Chem. Soc., 2282 (1926); H.M. Dawson and C.R. Hoskins, ibid., 3166 (1926).
26 P.D. Bartlett, J. Amer. Chem. Soc., 56 (1934) 967.
27 R.P. Bell and H.C. Longuet-Higgins, J. Chem. Soc., 636 (1946).
28 R.P. Bell and K. Yates, J. Chem. Soc., 1927 (1962); K. Yates and W.V. Wright, Can. J. Chem., 41 (1963) 2882.

29   J.E. Dubois and J. Toullec, J. Chim. Phys., 65 (1968) 2166; Chem. Commun., 292 (1969); Tetrahedron Lett., 3373, 3377 (1971).
30   R.P. Bell, "Acid-Base Catalysis", Oxford University Press, 1941.
31   C.G. Swain, A.J. Di Milo and J.P. Cordner, J. Amer. Chem. Soc., 80 (1958) 5983.
32   J.A. Feather and V. Gold, J. Chem. Soc., 1752 (1965).
33   G.E. Lienhard and T.-C. Wang, J. Amer. Chem. Soc., 91 (1969) 1146.
34   H.O. House, B.A. Tefertiller and H.D. Olmstead, J. Org. Chem., 33 (1968) 935; H.O. House, R.A. Auerbach, M. Gall and N.P. Peet, ibid., 38 (1973) 514.
35   H.O. House, "Modern Synthetic Reactions", 2nd Ed., Benjamin, Menlo Park, 1972.
36   H.D. Zook and J.A. Miller, J. Org. Chem., 36 (1971) 1112.
37   A.L. Kurts, N.K. Genkina, A. Macias, I.P. Beletskaya and O.A. Reutov, Tetrahedron, 27 (1971) 4757.
38   E.S. Amis, "Solvent Effects on Reaction Rates and Mechanisms", Academic Press, New York, 1966; J.F. Coetzee and C.D. Ritchie (editors), "Solute-Solvent Interactions", M. Dekker, New York, 1969.
39   M. Eigen, Angew. Chem. Intern. Ed., 3 (1964) 1.
40   R.P. Bell, "The Proton in Chemistry", Cornell University Press, Ithaca, N.Y., 1959, Chapter 9.
41   M.L. Bender and A. Williams, J. Amer. Chem. Soc., 88 (1966) 2502.
42   J. Hine, M.S. Cholod and J.H. Jensen, J. Amer. Chem. Soc., 93 (1971) 2321; J. Hine, J.L. Lynne, Jr., J.H. Jensen and F.C. Schmalstieg, J. Amer. Chem. Soc., 95 (1973) 1577.
43   R.P. Bell and M.A.D. Fluendy, Trans. Far. Soc., 59 (1963) 1623.
44   M. Hegazi and J.E. Meany, J. Phys. Chem., 76 (1972) 3121.
45   E.T. Harper and M.L. Bender, J. Amer. Chem. Soc., 87 (1965) 5625.
46   R.P. Bell, B.G. Cox and J.B. Henshall, J.C.S. Perkin II, 1232 (1972).
47   J.K. Coward and T.C. Bruice, J. Amer. Chem. Soc., 91 (1969) 5339.
48   T.C. Bruice and S.J. Benkovic, "Bioorganic Mechanisms", Benjamin, New York, 1966.
49   W.P. Jencks, "Catalysis in Chemistry and Enzymology", McGraw-Hill, New York, 1969.
50   M.L. Bender, "Mechanisms of Homogenous Catalysis from Protons to Proteins", Wiley, New York, 1971.
51   O. Reitz, Z. Physik. Chem., 179 (1937) 119.
52   H. Schechter, M.J. Collis, R. Dessy, Y. Okuzumi and A. Chen, J. Amer. Chem. Soc., 84 (1962) 2905; H.W. Amburn, K.C. Kauffman and H. Schechter, ibid., 91 (1969) 530.
53   E.F. Ullman and E. Buncel, J. Amer. Chem. Soc., 85 (1963) 2106.
54   C. Rappe, Acta Chem. Scand., 21 (1967) 1823; 22 (1968) 219.
55   M.S. Sytilin, Zhur. Fiz. Khim., 41 (1967) 1200; Russ. J. Phys. Chem., 41 (1967) 640.
56a J. Hine, K.G. Hampton and B.C. Menon, J. Amer. Chem. Soc., 89 (1967) 2664.
56b B.C. Menon and E. Kiehlmann, Can. J. Chem., 49 (1971) 3648.

57   J.R. Jones and R. Stewart, J. Chem. Soc. (B), 1173 (1967); J.R. Jones,
      R.E. Marks and S.C. Subba Rao, Trans. Far. Soc., 63 (1967) 111.
58   J.E. Dubois, P. Alcais, R. Brouillard and J. Toullec, J. Org. Chem., 36 (1971)
      4129.
59   R.-R. Lii and S.I. Miller, J. Chem. Soc. (B), 2269 (1971).
60   C.G. Swain and R.P. Dunlap, J. Amer. Chem. Soc., 94 (1972) 7204.
61   J.W. Thorpe and J. Warkentin, Can. J. Chem., 50 (1972) 3229; R.A. Cox and
      J. Warkentin, Can. J. Chem., 50 (1972) 3233; 50 (1972) 3242.
62   R.A. Lynch, S.P. Vincenti, Y.T. Lin, L.D. Smucker and S.C. Subba Rao,
      J. Amer. Chem. Soc., 94 (1972) 8351.
63   J. Warkentin and O.S. Tee, J. Amer. Chem. Soc., 88 (1966) 5540; R.A. Cox,
      J.W. Thorpe and J. Warkentin, Can. J. Chem., 50 (1972) 3239.
64   A.A. Bothner-By and C. Sun, J. Org. Chem., 32 (1966) 492.
65   W.H. Sachs and C. Rappe, Acta Chem. Scand., 22 (1968) 2031.
66   A. Schellenberger and G. Hübner, Chem. Ber., 98 (1965) 1938.
67   J. Jullien and Nguyen Thoi-Lai, Bull. Soc. Chim. France, 4669 (1968);
      M. Chevalier, J. Jullien and Nguyen Thoi-Lai, ibid., 3332 (1969).
68   J.R. Grunwell and J.F. Sebastian, Tetrahedron, 27 (1971) 4387.
69   A.G. Catchpole, E.D. Hughes and C.K. Ingold, J. Chem. Soc., 11 (1948).
70   G.S. Hammond, J. Amer. Chem. Soc., 77 (1955) 334.
71   J. Hine, J. Org. Chem., 31 (1966) 1236.
72   L. Melander, "Isotope Effects on Reaction Rates", The Ronald Press, New
      York, 1960.
73   W.A. Van Hook, in "Isotope Effects in Chemical Reactions", edited by
      C.J. Collins and N.S. Bowman, ACS Monograph 167, Van Nostrand Reinhold,
      New York, 1970.
74   R.A. More O'Ferrall, J. Chem. Soc. (B), 785 (1970).
75   M. Wolfsberg, Accounts Chem. Res., 5 (1972) 225.
76   O. Reitz and J. Kopp, Z. Physikal. Chem., 184A (1939) 429.
77   R.P. Bell, J.A. Fendley and J.R. Hulett, Proc. Roy. Soc. A, 235 (1956) 453.
78   C.G. Swain, E.C. Stivers, J.F. Reuwer, Jr. and L.J. Schaad, J. Amer. Chem.
      Soc., 80 (1958) 5885.
79   Y. Pocker, Chem. Ind., 1383 (1959).
80   F.A. Long and D. Watson, J. Chem. Soc., 2019 (1958); T. Riley and
      F.A. Long, J. Amer. Chem. Soc., 84 (1962) 522.
81   J.R. Jones, R.E. Marks and S.C. Subba Rao, Trans. Far. Soc., 63 (1967) 993;
      D.W. Earls, J.R. Jones and T.G. Rumney, J.C.S. Faraday I, 68 (1972) 925.
82   J.P. Calmon, M. Calmon and V. Gold, J. Chem. Soc. (B), 659 (1969).
83   E.F. Caldin, Chem. Rev., 69 (1969) 135.
84   C.G. Swain and A.S. Rosenberg, J. Amer. Chem. Soc., 83 (1961) 2154.
85   V. Gold and S. Grist, J. Chem. Soc. (B), 2282 (1971).
86   G. Lamaty, A. Roques and L. Fonzes, Compt. Rend. Acad. Sci. Serial C,
      273 (1971) 521.
87   C.K. Ingold and C.L. Wilson, J. Chem. Soc., 773 (1934).
88   C. Rappe and H. Bergander, Acta Chem. Scand., 23 (1969) 214.

89   P.D. Bartlett and C.H. Stauffer, J. Amer. Chem. Soc., 37 (1935) 2580.
90   S.K. Hsü, C.K. Ingold and C.L. Wilson, J. Chem. Soc., 78 (1938).
91   D.J. Cram, "Fundamentals of Carbanion Chemistry", Academic Press, New York, 1965, Chapters 3 and 5.
92   G.L. Mayers and J. Kovacs, Chem. Commun., 1145 (1970).
93   J. Kovacs, H. Cortegiano, R.E. Cover and G.L. Mayers, J. Amer. Chem. Soc., 93 (1971) 1541.
94   P.D. Bartlett and G.F. Woods, J. Amer. Chem. Soc., 62 (1940) 2933.
95   J. Bredt, Ann., 437 (1924) 1.
96   G.L. Buchanan and G. Jamieson, Tetrahedron, 28 (1972) 1123, 1129; G. Köbrich, Angew. Chem. Intern. Ed., 12 (1973) 464.
97   E.N. Marvell, G.J. Gleicher, D. Sturmer and K. Salisbury, J. Org. Chem., 33 (1968) 3393.
98   P.M. Lesko and R.B. Turner, J. Amer. Chem. Soc., 90 (1968) 6888.
99   J.A. Chong and J.R. Wiseman, J. Amer. Chem. Soc., 94 (1972) 8627.
100 R. Keese and E.-P. Krebs, Angew. Chem. Int. Ed., 10 (1971) 262.
101 P.G. Gassman and F.V. Zalar, J. Amer. Chem. Soc., 88 (1966) 3070.
102 W.Th. Van Wijnen, H. Steinberg and Th.J. De Boer, Rec. Trav. Chim. Pays-Bas, 87 (1968) 844; Tetrahedron, 28 (1972) 5423.
103 C. Rappe and W.H. Sachs, Tetrahedron, 24 (1968) 6287.
104a J.M. Motes and H.M. Walborsky, J. Amer. Chem. Soc., 92 (1970) 3697.
104b J.-O. Levin and C. Rappe, Chemica Scripta, 1 (1971) 233.
105 D.J. Cram, B. Rickborn, C.A. Kingsbury and P. Haberfield, J. Amer. Chem. Soc., 83 (1961) 3678.
106 R.P. Bell, Adv. Phys. Org. Chem., 4 (1966) 1.
107 D.C. Ayres, "Carbanions in Synthesis", Oldbourne Press, London, 1966, Chapter 4.
108 R.C. Fuson and T.-L. Tan, J. Amer. Chem. Soc., 70 (1948) 602.
109 J. Hine, J.G. Houston, J.H. Jensen and J. Mulders, J. Amer. Chem. Soc., 87 (1965) 5050.
110 P.E. Pfeffer, L.S. Silbert and J.M. Chirinko, Jr., J. Org. Chem., 37 (1972) 451.
111 P. Belanger, J.G. Atkinson and R.S. Stuart, Chem. Commun., 1067 (1969).
112 D.J. Barnes and J.M.W. Scott, Can. J. Chem., 51 (1973) 411.
113 J.E. Hendon, A.W. Gordon and M. Gordon, J. Org. Chem., 37 (1972) 3184.
114 J.B. Conant and G.B. Kistiakowsky, Chem. Rev., 20 (1937) 181.
115 N. Kornblum, P.J. Berrigan and W.J. Le Noble, J. Amer. Chem. Soc., 85 (1963) 1141.
116 C.K. Ingold, C.G. Raisin and C.L. Wilson, J. Chem. Soc., 1637 (1936).
117 G.W. Kirby and L. Ogunkoya, J. Chem. Soc., 6914 (1965).
118 J. Massicot, Bull. Soc. Chim. France, 2204 (1967).
119 V. Gold, J.R. Lee and A. Gitter, J. Chem. Soc. (B), 32 (1971).
120 A. Hantzsch and O.W. Schultze, Ber., 29 (1896) 699, 2251.
121 A.T. Nielsen, in "The Chemistry of the Nitro and Nitroso Groups", Part 1, ed. H. Feuer, Interscience, New York, 1969.

122   S.S. Novikov, V.A. Shlyapochnikov, G.I. Oleneva and V.G. Osipov, Bull. Acad. Sci. U.S.S.R., 151 (1971); Izvest. Acad. Nauk SSSR, Ser. Khim., 171 (1971).

123   N. Kornblum, J.T. Patton and J.B. Nordmann, J. Amer. Chem. Soc., 70 (1948) 746.

124a H.H. Baer and L. Urbas, in "The Chemistry of the Nitro and Nitroso Groups", Part 2, ed. H. Feuer, Interscience, New York, 1970.

124b L.A. Kaplan, in ref. 124a.

125   R.G. Pearson and R.L. Dillon, J. Amer. Chem. Soc., 75 (1953) 2439.

126   E.S. Lewis and L.H. Funderburk, J. Amer. Chem. Soc., 89 (1967) 2322.

127   M.J. Gregory and T.C. Bruice, J. Amer. Chem. Soc., 89 (1967) 2327.

128   F.G. Bordwell, W.J. Boyle, Jr., J.A. Hautala and K.C. Yee, J. Amer. Chem. Soc., 91 (1969) 4002; F.G. Bordwell and W.J. Boyle, Jr., J. Amer. Chem. Soc., 93 (1971) 511; 94 (1972) 3907.

129   M. Fukuyama, P.W.K. Flanagan, F.T. Williams, Jr., L. Frainier, S.A. Miller and H. Schechter, J. Amer. Chem. Soc., 92 (1970) 4689.

130   J.E. Dixon and T.C. Bruice, J. Amer. Chem. Soc., 92 (1970) 905.

131   F.G. Bordwell and W.J. Boyle, Jr., J. Amer. Chem. Soc., 93 (1971) 513.

132   P.W.K. Flanagan, H.W. Amburn, H.W. Stone, J.G. Traynham and H. Schechter, J. Amer. Chem. Soc., 91 (1969) 2797.

133   E. Buncel, A.R. Norris and K.E. Russell, Quart. Rev., 22 (1968) 123.

134   E. Buncel, A.R. Norris, K.E. Russell and R. Tucker, J. Amer. Chem. Soc., 94 (1972) 1646.

135   E. Buncel, K.E. Russell and J. Wood, Chem. Commun., 252 (1968).

136   E. Buncel, A.R. Norris, K.E. Russell and H. Wilson, unpublished results.

137   E.F. Caldin, A. Jarczewski and K.T. Leffek, Trans. Far. Soc., 67 (1971) 110.

138   A. Jarczewski and K.T. Leffek, Can. J. Chem., 50 (1972) 24.

139   D.J. Cram and L. Gosser, J. Amer. Chem. Soc., 86 (1964) 5457.

140   S.M. Wong, H.P. Fischer and D.J. Cram, J. Amer. Chem. Soc., 93 (1971) 2235.

141   F. Hibbert, F.A. Long and E.A. Walters, J. Amer. Chem. Soc., 93 (1971) 2829.

142   F. Hibbert and F.A. Long, J. Amer. Chem. Soc., 93 (1971) 2836.

143   C.G. Swain, E.C. Stivers, J.F. Reuwer, Jr. and J.L. Schaad, J. Amer. Chem. Soc., 80 (1958) 5885.

144   A. Schriesheim, Trans. New York Acad. Sci., Ser. II, 31 (1969) 97.

145   S.W. Ela and D.J. Cram, J. Amer. Chem. Soc., 88 (1966) 5791.

146   N.J. Van Hoboken and H. Steinberg, Rec. Trav. Chim. Pays-Bas, 91 (1972) 153.

147   S. Bank, J. Org. Chem., 37 (1972) 114.

148   K. Bowden and R.S. Cook, J.C.S. Perkin II, 1407 (1972).

149   J.M. Figuera, J.M. Gamboa and J. Santos, J.C.S. Perkin II, 1434 (1972).

150   S. Bank, C.A. Rowe, Jr. and A. Schriesheim, J. Amer. Chem. Soc., 85 (1963) 2115.

151   W. von E. Doering and P.P. Gaspar, J. Amer. Chem. Soc., 85 (1963) 3043.

152 R.B. Bates, R.H. Carnighan and C.E. Staples, J. Amer. Chem. Soc., 85 (1963) 3032.
153 P. Ahlberg and F. Ladhar, Chemica Scripta, 3 (1973) 31.
154 D.J. Cram, F. Willey, H.P. Fischer, H.M. Relles and D.A. Scott, J. Amer. Chem. Soc., 88 (1966) 2759.
155 S. McLean, C.J. Webster and R.J.D. Rutherford, Can. J. Chem., 47 (1969) 1555.
156 J. Klein and S. Brenner, Chem. Commun., 1020 (1969).
157 C.D. Broaddus, Accounts Chem. Res., 1 (1968) 231.
158 J. Hine and N.W. Flachskam, J. Amer. Chem. Soc., 95 (1973) 1179.
159 C.K. Ingold, E. de Salas and C.L. Wilson, J. Chem. Soc., 1328 (1936).
160 C.D. Broaddus, J. Amer. Chem. Soc., 87 (1965) 3706.
161 Y. Yukawa, T. Ando and T. Otsubo, Bull. Chem. Soc. Japan, 45 (1972) 2645.
162 J. Bigeleisen and M.G. Mayer, J. Chem. Phys., 15 (1947) 261.
163 S.K. Hsü, C.K. Ingold and C.L. Wilson, J. Chem. Soc., 1774 (1935); R.P. Ossorio and E.D. Hughes, ibid., 426 (1952); R.P. Ossorio and V.S. Del Olmo, Anales Real Soc. Espan. Fis. Quim. (Madrid), 56B (1960) 915.
164 D.J. Cram and R.D. Guthrie, J. Amer. Chem. Soc., 87 (1965) 397; 88 (1966) 5760.
165 C.K. Ingold, "Structure and Mechanism in Organic Chemistry", 2nd Ed., Cornell University Press., Ithaca, N.Y., 1969, p. 837.
166 R.D. Guthrie, D.A. Jaeger, W. Meister and D.J. Cram, J. Amer. Chem. Soc., 93 (1971) 5137.

Chapter 4

CLASSICAL AND NONCLASSICAL CARBANIONS

1. INTRODUCTION

The subject matter considered in Chapter 1, formation of carbanions by ionization of carbon acids, was concerned essentially with structure-stability relationships among classical carbanions. In the present discussion the term "classical" can be taken as referring largely to chemical bonding, the typical "classical" carbanion being adequately described by a conventional bonding structure. Carbanions illustrative of such classical bonding are the methyl, phenyl, and benzyl carbanions, of which the first two are charge-localized species while the third is a charge-delocalized species involving p-π orbital overlap. Charge delocalization in arylmethyl carbanions leads to increased thermodynamic stability, and thus to a higher acid dissociation constant for the parent hydrocarbon. (The pK_a of toluene is 41, while methane has a pK_a of ~48.)

Central to carbanion chemistry is a logically constructed acidity scale. Now if our appreciation of the acidity scale is to extend to carbon acids of a variety of structural types, other than those mentioned above, we must consider also some further carbanion-stabilizing characteristics. Chief among these is the presence in the molecule of electronegative substituents (e.g. pK_a of CHF_3 is 31) and of functional groups such as carbonyl or cyano, suitably located so as to act as the seat for negative charge through delocalization. The latter factor accounts for the considerable acidity of compounds such as acetone (pK_a 20) and nitromethane (pK_a 11), as well as for the even greater acidity of compounds such as tricyanomethane (pK_a ~ 0).

The factors considered above do not suffice to explain the unusually large acidity of the hydrocarbon cyclopentadiene (pK_a 15) and of structurally related compounds such as fluorene (pK_a 23) or fluoradene (pK_a 11). The stability of the cyclopentadienyl anion structure is interpreted in terms of its closed-shell aromatic system involving 6 π-electrons, which falls under the Hückel rule applicable to planar monocyclic conjugated systems containing 4n + 2 π-electrons [1]. In this sense the

109

cyclopentadienide carbanion may also be called a classical aromatic system as it contains a completed molecular orbital configuration. The cyclooctatetraenyl dianion with 10 π-electrons likewise falls into this category (see Chapter 1, Section 1, for other examples).

Nonclassical ions characteristically exhibit the properties associated with charge delocalization, but under structural circumstances in which the usual principles of orthodox bonding and conjugation appear not to apply. Historically, the concept of nonclassical ions arose in the chemistry of carbonium ions, and so that the significance of nonclassical carbanions be more fully placed in context, a brief review of this aspect of carbonium ion chemistry is presented, with emphasis on basic principles.

2. NONCLASSICAL CARBONIUM IONS: SOME BASIC PRINCIPLES

Nonclassical carbonium ions were first proposed as intermediates in certain solvolytic ionization processes, as a result of some unexpected discoveries. The observations included the finding of retention of configuration, of unexpectedly high rates of solvolysis, and the formation of characteristic rearranged products.

The most striking discoveries were made during studies of the solvolysis of cholesteryl derivatives [2, 3]. It was found that when cholesteryl chloride (1, X = Cl) undergoes solvolysis in acetic acid containing potassium acetate, the reaction product is cholesteryl acetate (1, X = OAc) with *retained configuration* at C3. Normally solvolytic processes lead to racemization or partial inversion of configuration. In the solvolysis of cholesteryl tosylate (1, X = OTs) in methanol containing acetate ions the reaction product is the *rearranged ether* 2 (Y = OMe). A kinetic study showed that cholesteryl tosylate undergoes acetolysis with a *rate enhancement* of 100 compared to that of cyclohexyl tosylate [3]. Similarly the rate of ethanolysis of cholesteryl tosylate was observed to proceed 40 times faster than that of cholestanyl tosylate, the saturated derivative corresponding to 1.

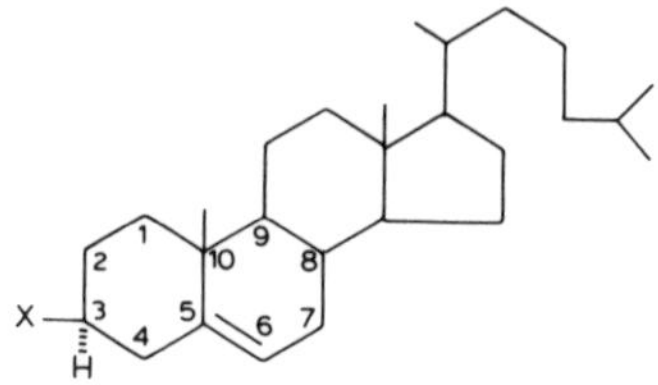

1

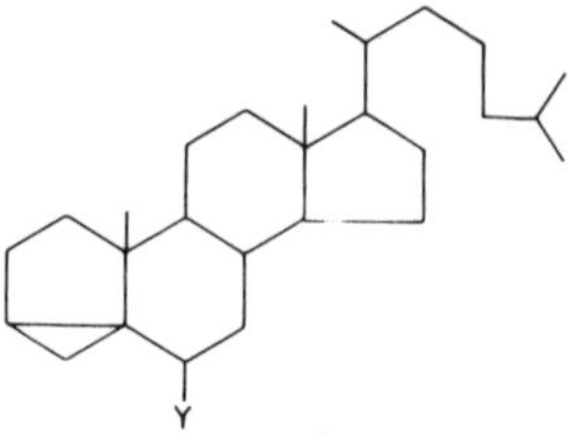

2

These observations were interpreted on the basis that *ionization of* 1 *gives directly the bridged nonclassical carbonium ion* 3. The dotted lines in that structure represent partial or non-orthodox bonds of bond-order between zero and one. An alternative possibility, that ionization of 1 initially yields the carbonium ion **4a** and that the latter subsequently rearranges to **4b**, could be discounted, since there should then be no unusual rate acceleration in the solvolytic ionization process. However, the enhanced rate can be fully accounted for by participation of the double bond π-electrons in the ionization process, thus providing the driving force for the reaction. This type of participation may be referred to as "anchimeric assistance" or "synartesis".

| 3 | 4 a | 4 b |

The observed retention of configuration at C3 in the solvolysis of cholesteryl chloride and tosylate (1) is now recognized as one of the criteria of participation involving bridged intermediates [4]. Molecular models show that nucleophilic attack at C3 in the bridged species must occur from a rearward direction to the bridge. It is also pertinent that solvolysis of 2 (Y = Cl), which proceeds much more rapidly than of 1 (X = Cl), also occurs stereospecifically. This reverse rearrangement apparently proceeds through the same intermediate 3 [5].

The relationship between the pathway involving the bridged ion 3 and that involving the separate classical open ions **4a, b** can usefully be illustrated by means of a potential energy-reaction coordinate diagram (Figure 1). It is noted that the bridged, delocalized, nonclassical ion 3 is a lower energy species than either **4a** or **4b** and, equally important, that the transition states leading to the formation of 3 are lower in energy than those leading to **4a** (from 1) or **4b** (from 2). However, probably the most significant distinction between the mechanisms represented by Figures 1a and 1b is that in the former there is a single energy minimum on the energy profile, corresponding to the nonclassical structure (3), while in the latter there are two energy minima, corresponding to the equilibrating classical structures (**4a** and **4b**).

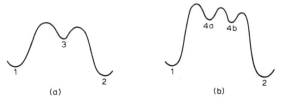

Figure 1. Potential energy-reaction coordinate representations for solvolysis of cholesteryl derivative (1), contrasting the energy profiles for the two reaction pathways: (a) a single, nonclassical, reaction intermediate (3), and (b) two equilibrating classical ions (4a and 4b).

The ability of a suitably placed β-olefinic group to participate with a developing carbonium ion centre and thus assist in the ionization process, as exemplified in the cholesteryl system, has been extended and generalized as the phenomenon of "homoallylic resonance" or "homo-conjugation" [6, 7]. The analogy with the allyl system is seen more clearly in the following representations:

allylic resonance:

5a   **5b**

homoallylic resonance:

6a   **6b**

The orbital representation **6c** indicates that, assuming the proper spatial orientations of the carbon atoms, the interposed methylene group does not insulate the carbonium ion center from the double bond, and that appreciable overlap between the π-orbitals and the vacant p-orbital is possible. Molecular orbital calculations [6] support this pictorial representation.

6c

112

A fundamentally different type of nonclassical interaction is proposed in the solvolysis of 2-norbornyl derivatives. A variety of evidence indicates that a symmetrical bridged carbonium ion (7) is formed in the ionization process:

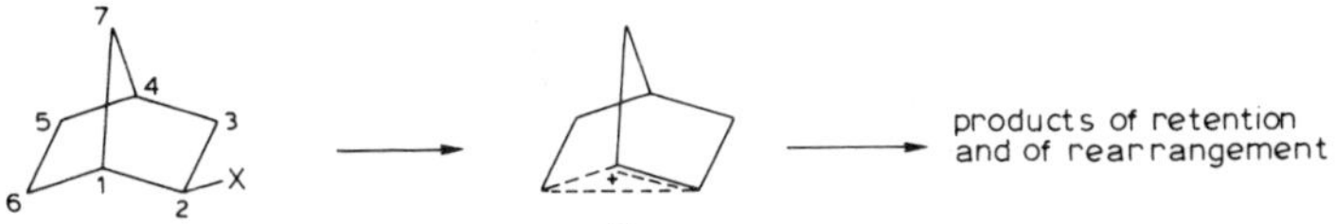

In this ionization process anchimeric assistance is provided by the electron-pair of a σ-bond (C1–C6), in contrast to the cases considered hitherto where an ethylenic linkage participated in ionization. Thus whereas with the homoallylic ions 3 and 6 the bridging cyclopropane ring could formally be represented in the conventional way with three single covalent bonds (4b and 6b), this is not possible in the case of the bridged ion 7. In that sense 7 is an electron-deficient bridged ion, and recalls the stabilization associated with the Hückel 2 π-electron system of the cyclopropenium ion. Just as the valence bond structures 4a and 4b are the limiting representations of 3, so the limiting classical carbonium ion structures by which 7 may be represented are given by 8a, b, c:

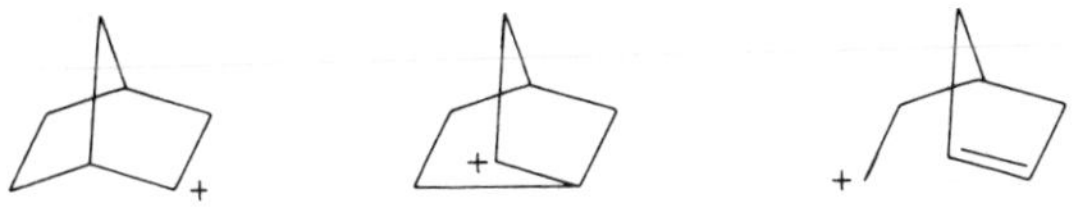

The norbornyl nonclassical carbonium ion represents probably the most extensively investigated, though still controversial [8], system. Several reviews are available [7–12], which deal with the problem in depth.

We conclude this section with discussion of the concept of homo-aromaticity [7, 13]. As the terminology suggests, the central feature here is the resonance stabilization characteristic of cyclic 4n + 2 π-electron systems. The nonclassical interaction which is invoked involves the homoallylic type as well as the kind of interaction present in the 2-norbornyl cation system. Several examples will illustrate this concept.

Ionization of pentamethylcyclobutenyl chloride is observed to give rise to a stable solution of cation **9** [14, 15]:

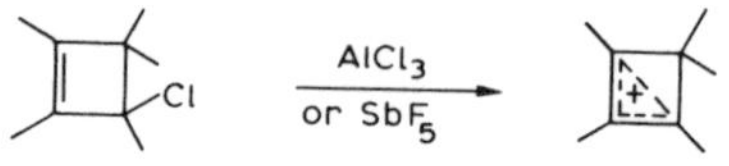

The NMR (proton and carbon-13) and ultraviolet spectra indicate that the structure of the cation has characteristics intermediate between an allylic and a cyclopropenyl cation. It is apparent that the geometry of the cyclobutenyl system allows appreciable 1,3-π-interaction; in fact the spectral data point to a 1,3-resonance integral of 0.33 β_0 where β_0 is the normal 1,2-π-resonance integral. The analogy between **9** and the cyclopropenium ion is that both involve in essence the 2 π-electron cyclically delocalized aromatic system. Thus **9** is an example of a homo-aromatic system and may be designated as a *monohomocyclopropenyl* cation since one bond of the cyclopropyl σ-bond network has been interrupted. Similarly we can have a *bishomocyclopropenyl* cation and a *trishomocyclopropenyl* cation. The nonclassical ions **10** [16] and **11** [13] provide examples of such homoaromatic systems.

The homotropylium ion, $C_8H_9^+$, provides a case of a homoaromatic cationic system containing 6 π-electrons. The ion **12** has been fully characterized [17–20] and is obtained, for example, on protonation of cyclooctatetraene in concentrated sulfuric acid. Similarly, 1,3- and 1,4-bishomotropylium ions have been observed in several laboratories [21–23].

114

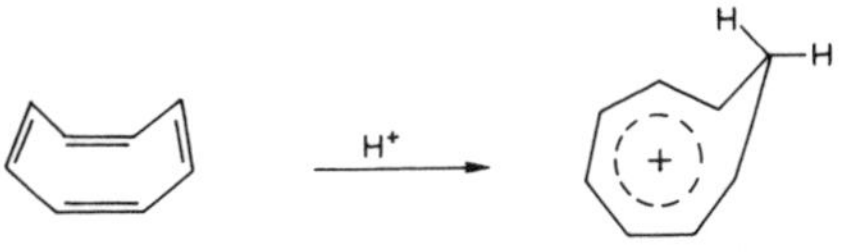

12

3. HOMOCONJUGATION IN UNCHARGED SYSTEMS

The concept of homoconjugation and homoaromaticity applies also to systems of other charge types. The application to anionic systems is the subject matter of the remaining sections of this chapter. It seems appropriate however to include here a brief consideration of homoconjugation in systems of overall electrical neutrality. A pertinent case is that of spiro[2.4]hepta-4,6-diene (13), which is represented, on the basis of evidence derived from NMR [24] and photoelectron [25a] spectroscopy, as well as theoretical considerations [26a], as a charge delocalized species 13a, 13b. (See, however, ref. 26b for a dissenting viewpoint, which is based on electron diffraction measurements with 13.)

13a   13b

Similarly NMR and electronic spectral characteristics of benzo[a]spiro-[2,5]octa-1,4-diene-3-one (14) have been interpreted in terms of appreciable aromatic character in the ground state as a result of cyclopropyl conjugation [27], via 14b:

14a   14b

Electron delocalization in 13 and 14 can be viewed essentially as an intramolecular charge-transfer interaction, which is particularly effective in these systems since the donor and acceptor functions thereby achieve aromatic character.

115

The type of delocalization represented in **13** and **14**, which has been termed *spiroconjugation*, should be differentiated from *cross-conjugation* of cyclic π-systems as exemplified by the resonance contributing structures **15a** and **15b** for calicene [28].

The homobenzenes provide a most interesting structural possibility for homoaromaticity. HMO calculations [29] on *cis,cis,cis*-cyclonona-1,4,7-triene, **16**, which may be considered a homoaromatic analog of benzene with three insulating methylene groups, indicated a small delocalization energy, ca. 8% of the value in benzene; this would be difficult to determine unambiguously by conventional measurements. Some evidence of homoaromaticity in this molecule was detected by photoelectron spectroscopy [25b].

The problem of homoconjugation in neutral compounds is receiving considerable current attention, from the theoretical as well as experimental approaches [30, 31]. A difficulty that has yet to be surmounted is the lack of definitive criteria which would be independent of a reference model and the necessity for assumptions concerning cancellation of extraneous steric and electronic effects. This difficulty need also be borne in mind in the ensuing discussion on homoconjugation in anionic species.

4. HOMOENOLATE IONS

The first studies aimed at exploring nonclassical interactions in carbanion chemistry, analogous to the interactions discussed for carbonium ions, were concerned with carbanions stabilized by the carbonyl group. The interaction between carbonyl and a negatively charged *alpha* carbon, resulting in the resonance stabilized enolate ion,

116

is responsible for the relative ease of proton abstraction from carbonyl compounds such as acetone:

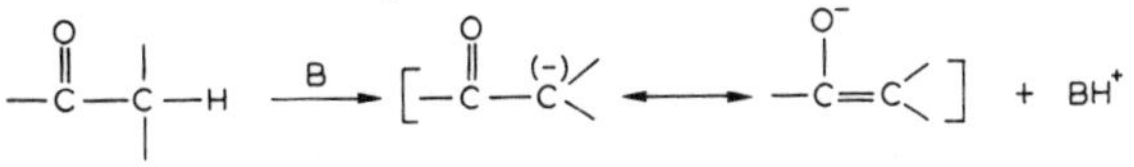

Now by analogy with homoallylic resonance, we can conceive of an interaction involving carbonyl as taking place across an intervening methylene group, to yield a homoenolate ion:

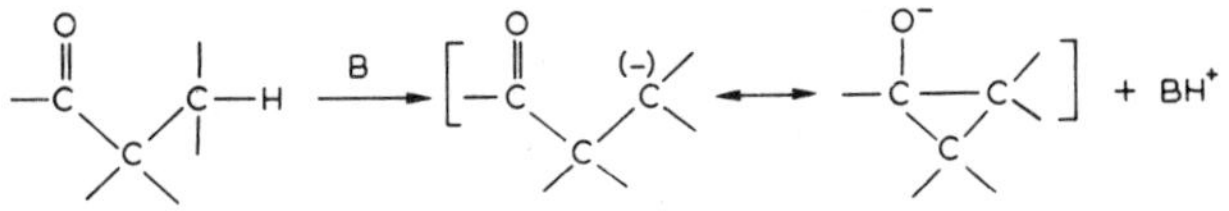

Charge delocalization in the enolate and homoenolate ions may be visualized through diagrams showing overlap between the relevant atomic orbitals:

(a)

(b)

Figure 2. Orbital representations to show delocalization in the enolate (a) and homoenolate (b) ions.

It is seen that the intervening methylene in the homoenolate species may not act as a perfect insulator under favourable circumstances, i.e. with a correct orientation of the atomic centres. In principle, more than one methylene could intervene between the carbanionic centre and carbonyl, though orientation factors could well become critical then if an adequate overlap is to be feasible. These considerations are analogous to the ones governing homoallylic interactions in carbonium ions, discussed previously.

The most favourable structural environment for orbital overlap between noncontiguous atomic centres should be found in ring systems, since the orientation of atoms relative to one another is then largely fixed. Thus just as cyclic compounds have provided a fertile ground for observations of nonclassical interactions in carbonium ions, so in the case of carbanion chemistry much of the evidence has been derived from studies with appropriately chosen cyclic systems.

The bicyclic ketone camphenilone (**17**) is almost ideally suited for testing the possibility of a homoenolate type of interaction [32]. There is only a single *alpha* C–H, and since this hydrogen is situated on a bridgehead carbon and thus especially difficult to abstract in a normal enolization process (cf. Chapter 3, Section 1e), this leaves the hydrogens on C6 as the most likely potential candidates for proton abstraction in a homoenolization process. The hydrogens on C7, and the methyl hydrogens, should be less likely possibilities on stereoelectronic grounds.

When optically active camphenilone (**17**) is heated with potassium *t*-butoxide in *t*-butyl alcohol at 165°–185° a gradual racemization occurs [32]. The loss in optical activity is consistent with formation of the bridged nonclassical ion **18** (limiting classical structures **20a, b, c**) on proton abstraction from C6.

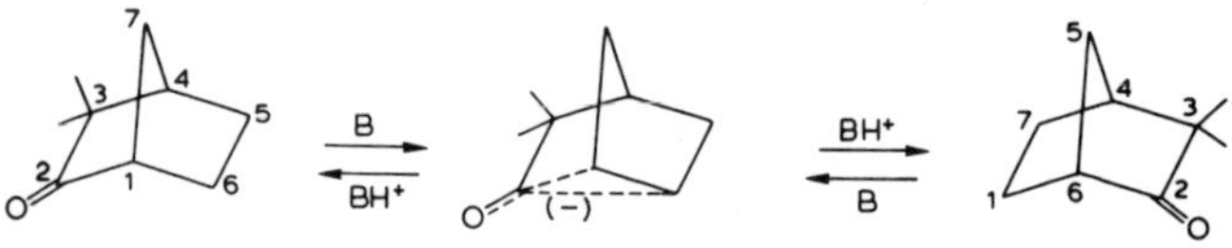

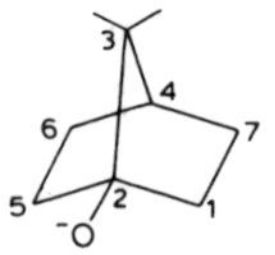

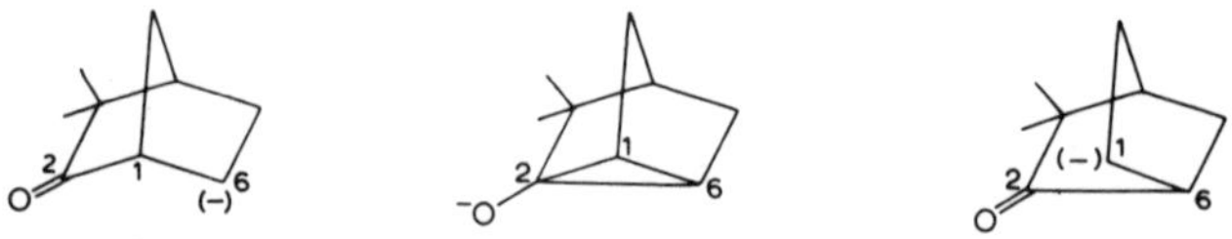

Since structure **20b** has a plane of symmetry (as shown more clearly in **20b'**) it follows that the involvement of this configuration in the reaction pathway as a transient intermediate will lead to racemization;

118

thus protonation will regenerate **17** and its enantiomer **19**. It will be recalled that racemization of a ketone under base catalysis was one of the criteria of enolization, so the observation of racemization in the camphenilone system provides strong evidence for a homoenolization process.

Homoenolization should also provide a pathway for deuterium exchange. It was found, in fact, that with *t*-butoxide in *t*-butyl alcohol-O-*d*, under the conditions of racemization, up to three atoms of deuterium were incorporated into camphenilone recovered from the base treatment, as estimated by mass spectrometry or by the more conventional combustion analysis [32]. (More recently ^{13}C nmr has been used to measure the extent of deuterium incorporation [33]; this method promises to have wide applicability.) The fact that three atoms of hydrogen are exchangeable (rather than two) is fully consistent with formation of the homoenolate ion, since through the process of homo-enolization C6 of structure **17** becomes equivalent to C1 of structure **19**. Moreover, the extent of deuterium incorporation was found to be equal to that of racemization. Such equivalence of racemization and of deuterium exchange is normal in enolization of ketones, as we have seen [34].

Further information on homoenolizable sites in camphenilone was obtained by conducting the exchange experiments at considerably higher temperatures and for prolonged reaction times [35]. Under these conditions exchange of the methyl hydrogens also takes place but not of the C4, C7, or C5 protons. These observations are consistent with predictions based on the geometric requirements of homoenolization.

In a related study it was found that hydrogen exchange, with sodium deuteroxide in aqueous dioxane at $25°$, occurs much faster with the diketo compound **22** than with **21** [36a]. This large rate enhancement* is in accord with formation of the homodienolate ion **23** on proton abstraction from **22**. It is interesting that the data indicate a partial stereoselectivity for exchange in both **21** and **22**, a result also found for exchange in other bicyclic ketones [37]. Both steric and electronic

---

*A detailed breakdown has been given of the factors which may contribute to the overall observed rate enhancement, in terms of a hybridization effect, torsional effect, conformational effect, strain effect, inductive effect and homoconjugative effect [36b].

119

factors probably contribute to the observed stereoselectivity of exchange [38, 39].

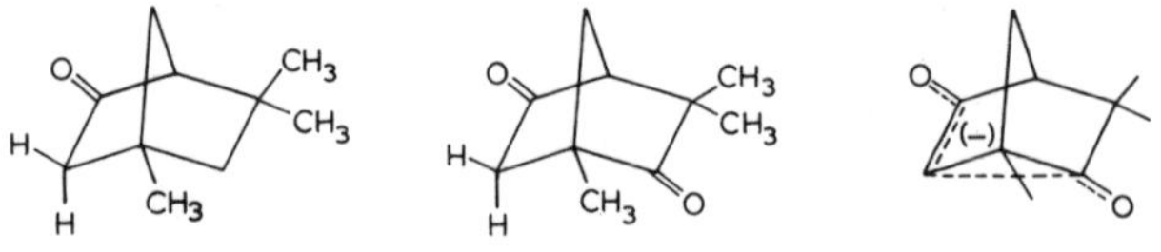

| relative rates | endo | 1 | 1800 |
| of H/D exchange | exo | 380 | 80000 |

| **21** | **22** | **23** |

Specific deuteration of the camphor skeleton has been achieved utilizing the principle of homoenolization [40]. It is noteworthy that longicamphor, **24**, when treated with t-BuOK/t-BuOD at 185° for 48 hours, incorporates upto three atoms of deuterium one of which proved to be at the bridge-head position [41]. Enolization at bridgehead carbon and the apparent implied "violation" of Bredt's rule is considered further in Chapter 3, Section 1e.

24   **24-d_3**

A remarkable case of homoallylic resonance is observed in a carbanionic process which involves the abstraction of positive halogen by trialkyl phosphine [42]. The rate of abstraction of Cl^+ from 4-trichloromethyl-cyclohexadienone (**25**) with tris(dibutylamino)phosphine in ethanol solution, to yield the dichloromethyl derivative **28**, proceeds ca. 10^6 times faster than reaction of CH_3CCl_3 or of $CHCl_3$, and is comparable in rate to that of $C_6H_5CCl_3$. Thus, the cyclohexadienone structure has a carbanion stabilizing effect comparable to that of phenyl. It is proposed therefore that abstraction of Cl^+ from **25** yields the resonance stabilized homodienolate ion **26**. An alternative nonclassical interaction, leading to **27**, is less likely since the strain energy associated with **27** should be greater than that associated with **26**.

120

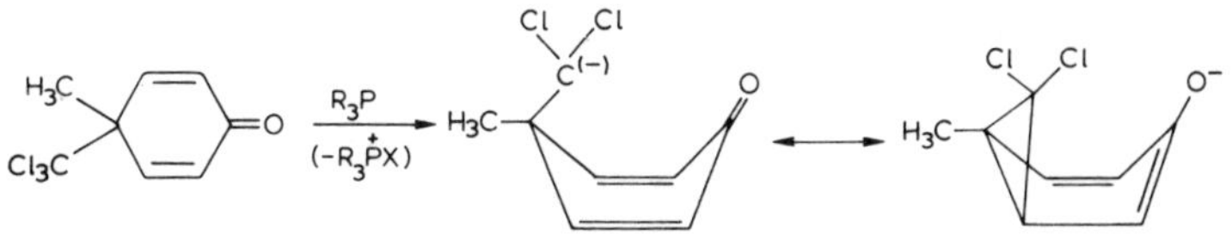

25 → **26** 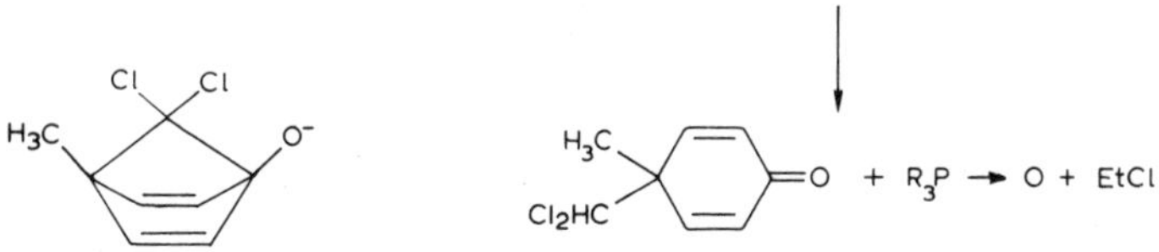 → **27** / **28**

Homoenolate ion formation in acyclic systems has also been proposed. For example, the possible formation of homoenolate ion **30** was suggested for the reaction of the bromo-ketone **29** with lithium, in order to account for formation of the cyclopropyl acetate **31** on treatment with acetic anhydride [43].

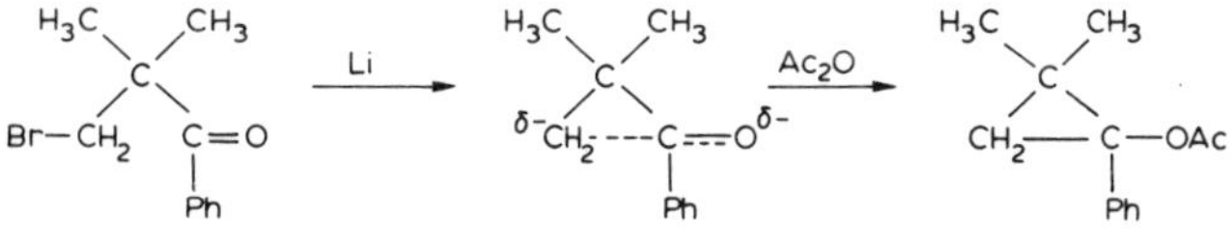

29 → **30** → **31**

32 → **33**

However, it is important to consider whether in this system the carbonyl actually participates in the rate-determining step in carbanion formation, or whether reaction occurs sequentially, $29 \rightarrow 32 \rightarrow 33 \rightarrow 31$, in which case a true homoenolate mechanism is not operative. Differentiation between these alternatives is fundamental to the problem of participation involving nonclassical structures. In practice this may be a very difficult task; generally one seeks unambiguous evidence on the question through rate studies, isotopic labelling, stereochemical investigations, etc.

121

In other work on carbonyl-anion interactions in acyclic systems detailed structural evidence was obtained and was indicative of the intermediacy of nonclassical homoenolate species, in preference to equilibrating classical enolate ion intermediates [44, 45]. Homoenolization processes are considered further in Chapter 5 under the heading "Rearrangements of Homoenolate Anions".

5. HOMOCONJUGATED AND HOMOAROMATIC CARBANIONS

Having established the effectiveness of a long range interaction between the carbonyl group and an anionic centre on a noncontiguous carbon, we now turn to examine the evidence for carbanion stabilization by the homoallylic or homoconjugative mechanism in systems which do not contain heteroatom functions capable of acting as carbanion stabilizing sites.

In the first instance we consider the possibility of a 1,3-homoallylic interaction between a carbanion and an olefinic group, as the (homo) analog of the delocalized allylic anion:

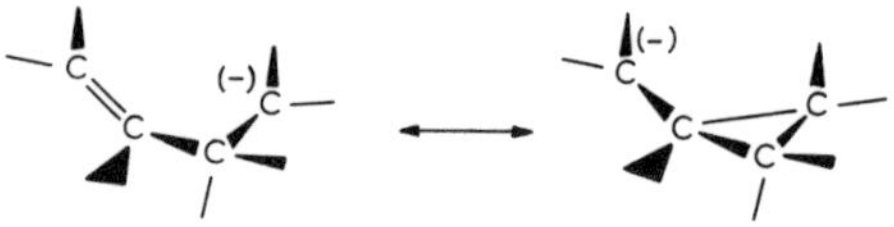

34a   34b

Up to the present time only few studies have been reported in which the importance of this type of interaction could be evaluated. A relevant investigation is concerned with hydrogen-deuterium exchange, catalyzed by $NaOD/D_2O$, in the bicyclo-enone 35 and comparison with the rates of exchange of 21 and 22 [36].

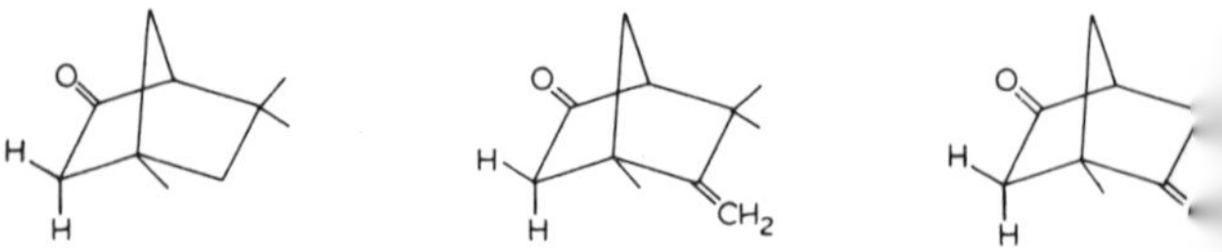

		21	35	22
relative rates	endo	1	3	1800
of H/D exchange	exo	380	2700	80,000

122

The results provide an answer to the question of whether the carbanion formed *alpha* to carbonyl would be further stabilized by the more distant double bond. It is seen in fact that the olefinic group in **35** causes only a moderate increase (3 fold for the *endo* hydrogen, 7 fold for *exo*) in the rate of hydrogen exchange over that in **21**. This contrasts with the situation in **22**, where the homoenolate interaction is clearly indicated by the very large rate enhancement of hydrogen exchange. Since a Coulombic interaction with the double bond can account for part of the observed rate increase, it is clear that the homoallylic stabilization of the carbanion generated in this system must be small.

The norbornenyl system, which has yielded valuable information concerning nonclassical interactions in carbonium ion processes, is now subjected to scrutiny as to the possibility of homoallylic carbanion stabilization. Entry into norbornenyl carbanionic species has been gained through alkaline oxidative cleavage of the norbornenylhydrazine derivative **36** and the nortricyclyl derivative **37** [46]. It is found that the reaction of either **36** or **37** in water or in *t*-butyl alcohol yields a 43:57 mixture of norbornene (**38**) to nortricyclene (**39**):

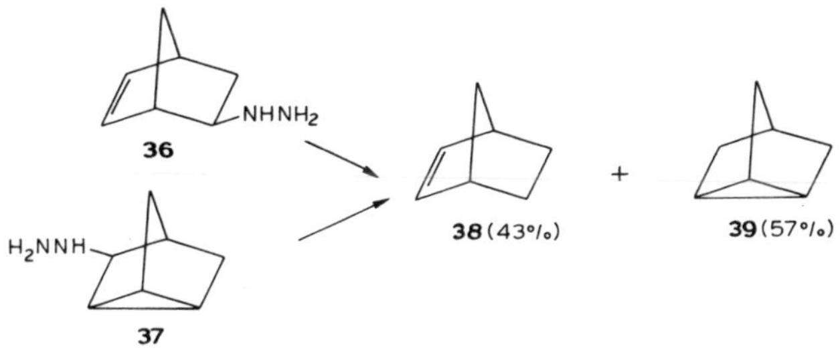

This result may be explained *either* by the postulate that there is an equilibration between the classical ions **40a** and **40b**, and that this process is rapid compared to proton capture from solvent, *or* that a non-classical ion such as **41** or **42** intervenes (note that of the latter two structures only **42** is symmetrical and is favoured on other grounds) [46].

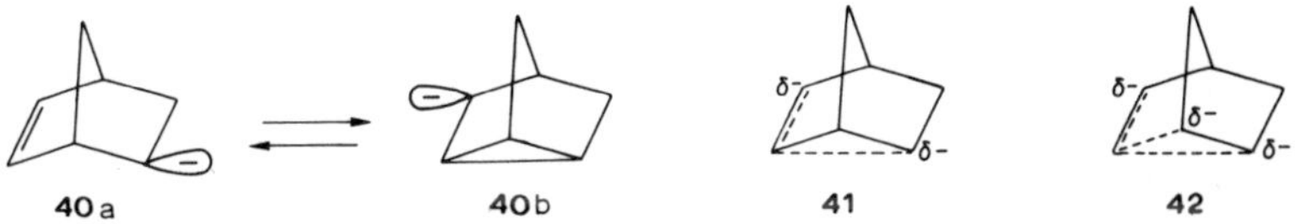

This dual explanation is reminiscent of the discussion of carbonium ion processes given earlier in this chapter, and illustrated in Figure 1 for the cholesteryl system. In that case kinetic and stereochemical evidence pointed convincingly to the nonclassical alternative. One may look forward to corresponding studies in the norbornenyl anion system.

An interesting situation arises when one considers the possibility of stabilization due to charge delocalization in the interaction of an allylic carbanion with an olefinic double bond. For such a case, represented by **43**, HMO calculations indicate a substantial delocalization energy even with small values of the 1,3-overlap integral between the non-contiguous atoms, as seen in the following results:

	Delocalization energy (β-units) as function of overlap			
$\beta_{1,3}/\beta_0$	0	0.3	0.5	1.0
DE	0.828	1.032	1.346	2.472

43

In essence one is considering in structure **43** a delocalization involving six π-electrons over a pseudo five-membered ring system. Thus **43** is a homoaromatic analog of the cyclopentadienide anion, with two sides of the σ-bond network interrupted. In terms of the concept of homoaromaticity, **43** may be called a *bishomocyclopentadienide* nonclassical system [7].

The first experimental substantiation of this type of nonclassical interaction was obtained in hydrogen exchange studies with the bicyclooctadiene **45** [47]. It was observed that **45** undergoes hydrogen-deuterium exchange in strongly basic medium (t-BuOK/CD$_3$SOCD$_3$) at the allylic 4-position at a rate which is 30,000 times greater than that of the bicyclooctene **44**. The latter exchanges at a rate comparable with that of cyclohexene. The observed rate enhancement, which is far greater than could be expected on the basis of a field effect, was rationalized in

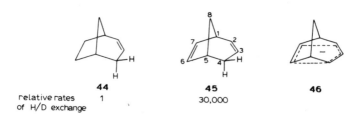

44	**45**	**46**
relative rates of H/D exchange 1	30,000	

124

terms of participation, or anchimeric assistance, by the C6,7 double
bond and the incipient allylic carbanion, with formation of the nonclassical species **46**. It should be noted that the symmetry of **46** requires
that positions 2 and 4 become equivalent; in accord with this requirement a total of three hydrogens are found to undergo exchange in **45**
on prolonged reaction in a basic medium [48].

Direct observation of carbanion **46** by NMR spectroscopy confirms
the conclusions derived from the isotopic exchange experiments. Thus
a solution of **46**, which is orange in color, is obtained when 4-methoxy-
bicyclooctadiene is treated with sodium-potassium alloy in tetrahydrofuran or in dimethoxyethane [49, 50]. The NMR spectrum, which is
stable for prolonged periods, clearly shows the delocalization of negative
charge to C6 and C7, and indicates also that an aromatic ring current is
operative. This provides strong evidence for the bishomocyclopentadienide formulation.

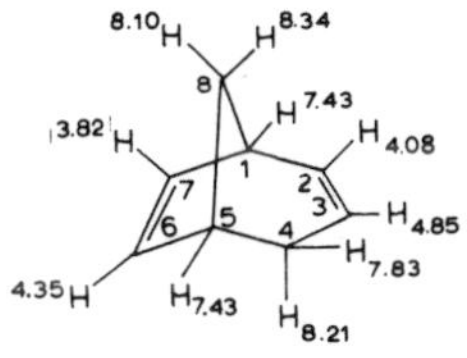

45a   **46a**

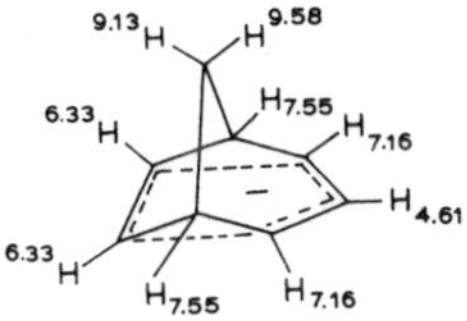

The actual NMR parameters of the bicyclooctadiene precursor and
of the carbanion produced (in perdeuteriotetrahydrofuran) are shown
under **45a** and **46a**, respectively. A particularly noteworthy feature in
46a is the upfield shift, relative to **45a**, of 2.3 ppm (average) sustained
by the C-6,7 protons (roughly two-thirds as large as for the C-2
proton), which is indicative of extensive delocalization of charge to C6
and C7. This situation is in contrast with the C-1,5 hydrogens which
experience only slight change. The C-2,4 protons in **46a** appear as a well-
defined triplet and so does the C-3 proton, while the signal for the C-6,7
protons is a singlet. For the bridgehead C-1,5 hydrogens a skewed triplet
is observed; for the hydrogens on C-8 the signal at 9.58 is a doublet and
that at 9.13 a multiplet. The slight downfield shift of the C-3 proton
resonance in **46a**, relative to **45a**, indicates that the deshielding due to
the aromatic ring current more than compensates the shielding effect of

125

the negative charge at C3. Calculations of charge densities in **46** by the HMO method show that charge will be highest at C2 and C4 (0.426), smaller at C6 and C7 (0.064) and least at C3 (0.021).

Hydrogen exchange and NMR studies with the benzobicyclooctadiene **47** show that the aromatic nucleus is also capable of providing anchimeric assistance to the allylic carbanion [51]. However, the rate of hydrogen exchange of **47** is only 1/8 that of **45** under comparable conditions, indicating a somewhat smaller anchimeric assistance by the arene. Thus, the necessary perturbation of the benzenoid nucleus leads to a reduced capacity for homoaromatic stabilization; this result is similar to that obtained with nonclassical carbonium ions. The nonclassical carbanion structure **48** is fully substantiated by NMR measurements, with results and arguments similar to those given for **46**.

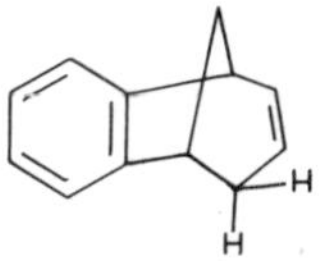

47 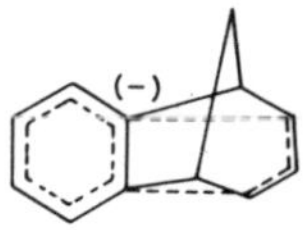   **48**

With the nonclassical carbanions discussed above the question of a possible stereoselectivity of proton abstraction and donation is an interesting one, and there is analogy to the situation in nonclassical carbonium ion chemistry where stereoselectivity is the generally observed rule. Thus, it is found that the *exo* proton at C4 in **45** and **47** exchanges more rapidly with t-BuOD/DMSO-d_6 than the *endo* proton, by a factor of 6. Also, the proton at C2 exchanges at about the same rate as the *endo* proton at C4 [51]. However, a complete interpretation of the kinetic results is not possible with the available data without making assumptions about rates and isotope effects in deprotonation and reprotonation at the various sites. It is interesting that protonation of **46** by quenching a solution in tetrahydrofuran with CD_3SOCD_3 occurs predominantly *exo*, while quenching with CD_3OD occurs predominantly in *endo* fashion [52]. This suggests that hydrogen bonding as well as stereoselectronic effects must be operative in these systems (cf. refs. 37–39).

Analogous with the homoaromatic analog of the cyclopentadienide anion is the work reported on the homo analog of the cyclooctatetraene dianion [53, 54]:

126

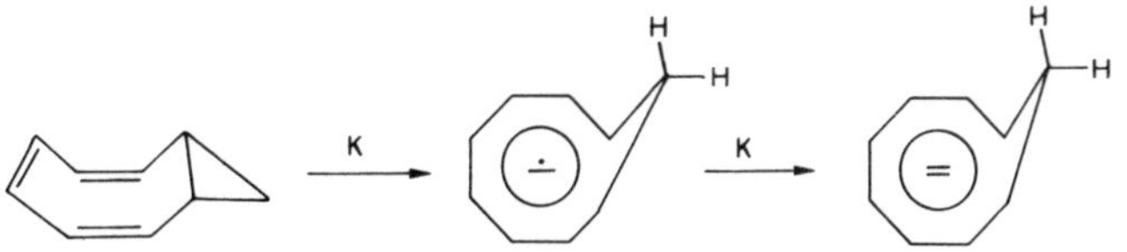

49   50   51

Treatment of the cyclopropane derivative **49** with potassium in dimethoxyethane at $-80°$ yields consecutively the anion radical **50** and the dianion **51**. The former is identified by ESR and the latter by NMR spectroscopy. In each case the data are completely in accord with the delocalized structures shown, designated as the monohomocycloocta-tetraene anion radical and dianion.

Having discussed the *positive* evidence for the occurrence of the phenomenon of homoaromaticity in several carbanionic systems, we now mention some studies of systems in which homoaromaticity was sought but *not* experimentally detected.

Proton exchange in **52** was examined (in DMSO-d_6 containing *t*-BuOK) [55] and found to result in *no* rate enhancement compared with cyclohexene, whereas it will be recalled that **45** showed an enhancement of ca. 10^4. This shows that the cyclopropyl σ-bond does not interact with the transannular allylic carbanion, and provides a contrast with the corresponding *carbonium* ion system, in which positive interaction was found [56]. Thus, the trishomocyclopentadienyl-type delocalized anionic system (**53**) has not yet been detected.

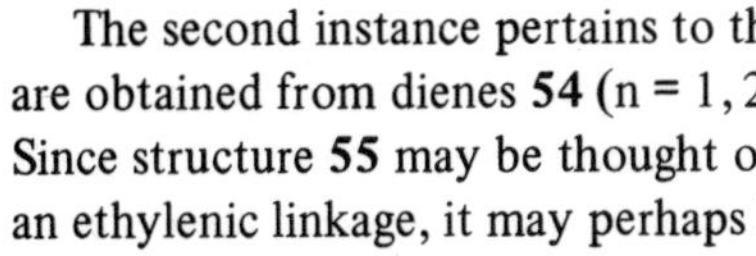

52   53

The second instance pertains to the cyclohexadienyl anions **55**, which are obtained from dienes **54** (n = 1, 2, 3) by the action of base [57]. Since structure **55** may be thought of as an allylic anion interacting with an ethylenic linkage, it may perhaps have been expected to show homo-aromatic character as the homocyclopentadienyl anion **56**. The conclu-sion from an NMR study of the anions **55**, however, was that these are characterized by approximately zero ring current. (More accurately, *if*

127

there is a ring current, it is apparently equal for the cases n = 1, 2, 3. However, since the 1,5-distances and orientations must be unequal in the three cases, this is unlikely.)

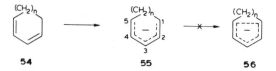

54   55   56

We conclude this section with discussion of a further type of non-classical carbanion interaction. Anion **58** is obtained when the 4-methoxynonatriene **57** (R = OMe) is treated with sodium-potassium alloy in dimethoxyethane [58, 59], and NMR data suggest that it is a delocalized aromatic system (the vinyl protons on C6, C9 undergo a downfield shift of 1.3 ppm in **58** relative to **57**, but of this shift 0.8 ppm has been attributed to a field effect). That delocalization here takes place over the whole ring system is indicated by the observation that hydrogen-deuterium exchange of triene **57** (R = Me) occurs 750 times faster than that of the corresponding 2,6-diene, in lithium cyclohexyl-amide-cyclohexylamine as base. Anion **58** may be designated as *bicyclo-aromatic* according to a novel proposal of π bicycloaromaticity, based on orbital symmetry considerations [60]. Although the relative importance of bicycloaromaticity and of homoaromaticity in **58** appears somewhat uncertain [58b], this novel concept should prove to be challenging in future work.

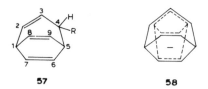

57   58

A most interesting result was obtained when the precursor of anion **58** was the 4-deuterio-4-methoxytriene **57** (R = OMe). Quenching of the derived anion with methanol after 2 days yielded the hydrocarbon **57** (R = H), together with the corresponding dimer, with the deuterium distributed statistically throughout the molecule. This result is in accord with occurrence of a series of rearrangements of the type shown with the

128

aid of the valence bond structures **59a, b** (although alternative schemes involving symmetry allowed cyclopropane openings and closures may also be written [58b]):

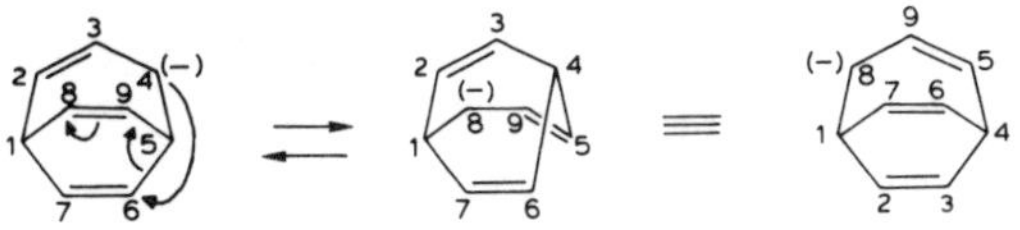

59a   **59b**

Since the process of rearrangement merely reproduces the structurally identical anion one may call this a degenerate process. Other degenerate isomerisations are discussed in Chapter 6.

6. CARBANIONS AND ANTIAROMATICITY

The aromatic character of cyclically delocalized systems containing $4n + 2$ π-electrons has been one of the foundation stones of chemical theory since the 1930's. The concept of homoaromaticity is a more recent one, dating to the late 1950's. A still more recent concept for which evidence started accumulating in the 1960's is that of antiaromaticity [61]. Most of the evidence on which this concept is based has been gathered with cyclic carbanionic systems containing $4n$ π-electrons. The study of hydrogen-deuterium exchange has provided particularly valuable evidence in this regard.

In essence, the term antiaromaticity applies to the situation that *cyclic delocalization results in destabilization*, which is in direct contrast to the generally accepted idea of delocalization leading to stabilization, as so well exemplified by the aromatic $4n + 2$ π-electron systems. Now the delocalization energy of a molecule is considered to be the increased stability, resulting from an additional π-bonding energy, when originally non-interacting p_z-electrons (generally in isolated double bonds) are allowed to delocalize over a number of atomic centers. In the case of a cyclic system one obtains the delocalization energy from comparison of the derived π-electron energy with an acyclic analog. Thus, benzene, compared with hexatriene, is found to have a delocalization energy of 2β (from Hückel theory the π-electron energy of benzene is 8β while that of hexatriene is 6β, corresponding to three isolated double bonds) and is considered to be aromatic by the criterion of a very large delocal-

129

ization energy. Aromaticity may also be defined by other criteria to be discussed later in this section (see also refs. 62–69), but the theoretically derivable delocalization energy provides, at least conceptually, a straightforward approach to the question. The antiaromaticity counterpart will then be the situation in which cyclic delocalization leads to an *increase* in energy of the system relative to an open-chain analog. Thus *antiaromatic species* will be associated with a negative delocalization energy.

The cyclopropenyl carbanion provides the simplest, as well as the most convincing case for the concept of antiaromaticity in a cyclic system containing 4n π-electrons in conjugation [70]. For application of the delocalization energy criterion one compares the energy of the cyclopropenyl carbanion with that of the allyl carbanion, which is the open-chain analog. Now Hückel molecular orbital (HMO) calculations for the allyl carbanion yield a π-bonding energy of 2.82β (DE = 0.82β), while for the cyclopropenyl carbanion the HMO π-energy is 2β (DE = 0β, since no interaction is predicted between the double bond and the negative charge in the HMO method). On this basis the cyclopropenyl anion is destabilized relative to the allyl anion and is thus antiaromatic.

DE = 0.82 β   DE = 0β

Calculations by the more sophisticated Pariser-Parr-Pople (PPP) method also indicate that the cyclopropenyl anion is destabilized relative to the allyl anion, to the extent of 46 kcal/mole for the singlet state [71]. Similarly the PPP method predicts that the cyclopentadienyl cation is destabilized relative to the acyclic pentadienyl cation, by 21 kcal/mole in the singlet state. Resonance destabilization in cyclic systems containing 4n π-electrons is predicted to decrease as the integer n increases, so that the cyclopropenyl anion should be the most favourable system for observation of the antiaromaticity phenomenon. It is interesting that the PPP calculations predict that the cyclopropenyl anion in the triplet state (two electrons unpaired) is of lower energy than in the singlet state, by 19 kcal/mole. A similar result is obtained for the cyclopentadienyl cation, though the energy difference between the two electronic states

130

is predicted to be smaller in this case (7 kcal/mole). Experimental observations on the pentachlorocyclopentadienyl cation confirm the triplet ground state in this system [72].

An alternative measure of antiaromaticity has been suggested as being the increase in energy for the process **60a**→**60b**, corresponding to the delocalization of electrons in the doubly-occupied orbital of π symmetry at C1 into the C2–C3 double bond system [73].

60a   60b

Using non-empirical (LCAO-MO-SCF) calculations with Gaussian type functions, the antiaromaticity of the planar cyclopropenyl anion is obtained as 143 kcal/mole [73]. It is noteworthy, however, that these calculations indicate that the planar cyclopropenyl anion is of considerably higher energy than the non-planar structure. The cyclopropenyl anion is hence predicted by this method to be non-planar, with the C1–H bond at 68° to the plane of the ring, and a barrier to inversion of 52 kcal/mole (cf. Chapter 2, Section 1). The proposal that antiaromaticity can be relieved by deviation from coplanarity is interesting, and its consequences worthy of further study.

What is the experimental evidence for antiaromaticity in the cyclopropenyl carbanion? Undoubtedly, the most convincing evidence would be based on observed molecular ground state properties of the species. Traditionally the thermochemical method (based on heats of combustion or hydrogenation), yielding the empirical resonance energy, was accepted as the normal criterion of aromaticity. More recently, however, it has been realized that the calculated heat of formation value is subject to considerable uncertainty, depending on the reference state chosen in calculation of bond constants and the correction terms for environmental effects (bond compression, hybridization effects, ring strain, etc.) applied [64]. Thus, estimates of the resonance energy of benzene vary between 13 kcal and 112 kcal! At present the thermochemical method, whatever its merit or otherwise, is inapplicable since there is as yet no report of the isolation and physical examination of the cyclopropenyl anion.

131

Another kind of ground state property, which would have provided a more suitable criterion in this case, is the magnetic anisotropy criterion, i.e. the ability of aromatic systems to sustain a ring current. It has been shown from theoretical considerations that in ring structures containing 4n π-electrons the ring current induced by an external magnetic field should flow in the opposite direction to that in systems with 4n + 2 π-electrons. The former should sustain a paramagnetic ring current, the latter a diamagnetic ring current, and will be, respectively, *paratropic* and *diatropic* substances [74–77]. This criterion has been substantiated for the case of the 4n and 4n + 2 annulenes [76, 77], and should be a most appropriate one to test the concept of antiaromaticity in due course.

Most of the experimental evidence on which antiaromaticity is currently based is derived from reactivity studies. The chemical reactivity criterion, reflecting the transition state energy relative to the ground state, is also subject to uncertainty in interpretation. Nevertheless, by the use of some plausible assumptions and with considerable care on the choice of system, a very strong case for antiaromaticity has been made on the basis of kinetic measurements of reactions involving the cyclo-propenyl anionic species as a transient intermediate. The relevant studies will now be considered.

Base-catalyzed hydrogen-deuterium exchange is a reaction known to proceed via carbanionic intermediates. Since the kinetics of isotopic exchange reactions can readily be measured it is not surprising that much of the evidence concerning antiaromaticity of the cyclopropenide system is derived from such exchange studies. In practice the isotopic exchange experiments have been performed on suitably substituted cyclopropenyl derivatives, in part because exchange in unactivated cyclopropene would be extremely slow, and in part because the vinylic hydrogens of cyclopropene are known to be more acidic than the methylene hydrogens, both from experimental observations [78] and from theoretical calculations [73]. Typically, then, the vinylic hydrogens are substituted by phenyl groups and one of the methylene bridge hydrogens is substituted by groups such as –COR, CN etc.

The first intimation of destabilization of the cyclopropenyl anion was obtained from the hydrogen-deuterium exchange process [79]:

132

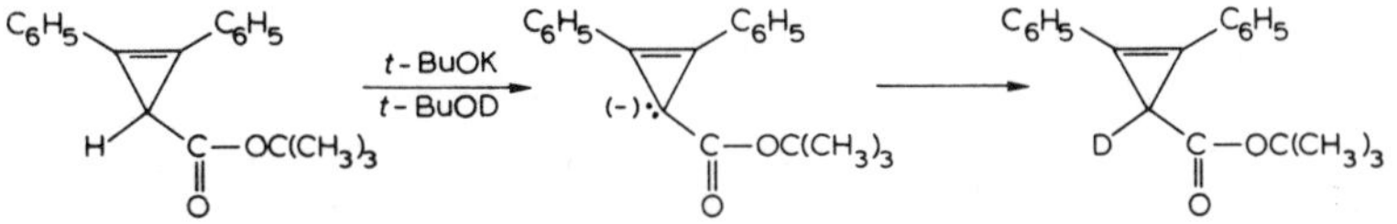

The rate of exchange of the diphenyl cyclopropenyl ester was found to be 300 times slower than that of cyclopropyl *t*-butyl carboxylate. As the electron withdrawing effect of the phenyls should lead to enhancement of rate in the carbanion mechanism, it is clear that proton abstraction from the cyclopropenyl derivative occurs with much greater difficulty than from cyclopropyl. Since that time a number of systems have been examined in this manner and even more remarkable rate differences have been observed. The relative rates of these hydrogen exchange reactions under base catalysis are summarised below [71, 80]. It is also noteworthy that the rate differences are reflected almost entirely in the enthalpies of activation and not in the entropies.

Table 1

RELATIVE RATES OF BASE-INDUCED HYDROGEN-DEUTERIUM EXCHANGE OF CYCLOPROPENYL AND CYCLOPROPYL DERIVATIVES

61	1	300	
62	1	1000	
63	1	6000	2000
64	1	10,000	

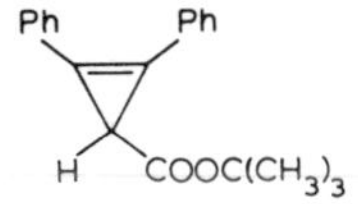

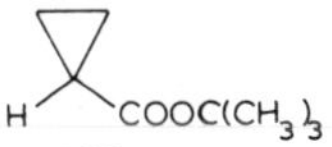

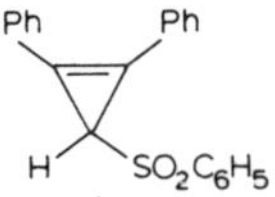

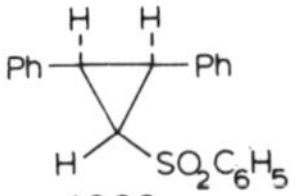

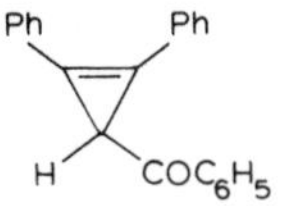

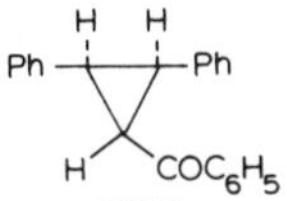

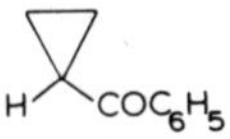

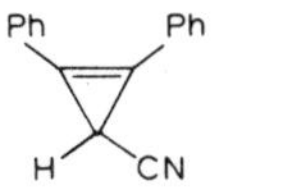

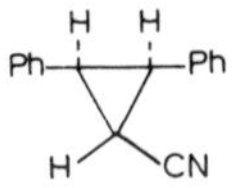

The results given in Table 1 are in complete accord with the anti-aromatic destabilization of the cyclopropenyl anion. However, other possible explanations must be considered as alternatives to the concept of antiaromaticity. Probably the most plausible alternative explanation would be based on steric considerations, as it is well substantiated that there is an unfavourable steric effect associated with formation of an exocyclic double bond in the cyclopropyl system (I-strain). Since contribution from an enolate resonance structure is expected to be appreciable, for example with the benzoyl derivative, the resulting I-strain effect should discourage carbanion formation. This steric factor manifests itself in a reduction in rate of H/D exchange by a factor of 170 in benzoylcyclopropane compared to 2-benzoylpropane [81] (see also Chapter 3, Section e, for other examples). The I-strain factor is expected to affect the cyclopropene system more adversely than the cyclopropane case, since the internal bond angle at C3 is reduced to $50°$ in cyclopropene from a value of $60°$ in cyclopropane [82]. Furthermore, whereas formation of the cyclopropyl carbanion should result in relief of eclipsing interactions, the opposite will hold for the cyclopropenyl case:

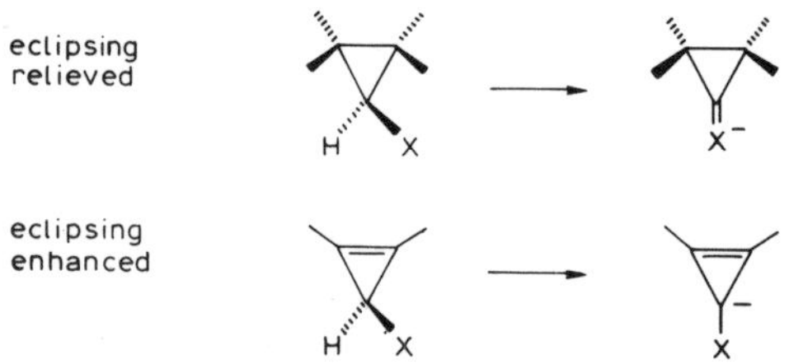

The cyanocyclopropene derivative (64), which provides the largest rate factor for hydrogen exchange, would be least expected to fit into the steric explanation. On the other hand this system fits well the electronic explanation, since the PPP calculations indicate in this case that cyano withdraws negative charge less effectively than benzoyl, so that the antiaromaticity effect becomes further enhanced. Neither does the benzenesulfonyl system (62) fit the I-strain explanation, since carbanion stabilization by the sulfonyl group occurs largely by electrostatic and stereoelectronic effects (see Chapter 2, Section 4), so that very little double bond character should be present in this case. It is

134

relevant that **62** undergoes *more* rapid hydrogen exchange (by a factor of 12 in EtOK/EtOD at 20°) than 2-phenylsulfonylpropane, the open-chain derivative [70]. Thus, it is unlikely that angle strain will be a major factor in destabilization of the cyclopropenyl anionic systems.

At the beginning of this chapter it was argued that just as the logical construction of an acidity scale for hydrocarbons is fundamental to the understanding of carbanion stability-structure relationships, so the finding of abnormality in the pK_a-structure relationship becomes one of the criteria of nonclassical behaviour. It is therefore particularly striking that an estimate of the pK_a of triphenylcyclopropene (based on electro-chemical measurements with the triphenylcyclopropenyl cation and triphenylmethane and using a thermodynamic "pseudocycle") is obtained as 51 [83]. Although a comparison with triphenylmethane (pK_a 31.5) is not strictly valid on steric grounds, a substantial portion of the pK difference is most reasonably assignable to antiaromaticity.

Cyclobutadiene, as the next member of the series of cyclic hydro-carbons containing 4n π-electrons in conjugation, had for many years been believed to be simply non-aromatic, and its instability was thought to be the result of internal strain [84]. Recently, however, evidence has been presented in support of the view that it is actually antiaromatic. Thus quantitative measurements of the electrochemical oxidation of quinol **65** to quinone **66** (R = Me or Ph), and comparison with the corresponding cyclobutane analog of **65** as well as with naphthahydro-quinone, show that the fused cyclobutadiene ring in **66** has an adverse effect on $\Delta G°$ for this oxidation to the extent 12—16 kcal/mole [85]. This free energy difference must be due almost entirely to the anti-aromaticity associated with cyclobutadiene, since steric changes in the process are minimal. Furthermore, since the fused four-membered ring in the initial state (**65**) is expected to have partial dienoid character, the derived value of $\Delta G°$ of 12—16 kcal/mole must represent a minimum value of the destabilization energy.

65   66

135

The antiaromaticity of cyclobutadiene is also indicated from measurements on the ionization of **67** in tetrahydrofuran containing lithium cyclohexylamide and tetramethylethylenediamine [86]:

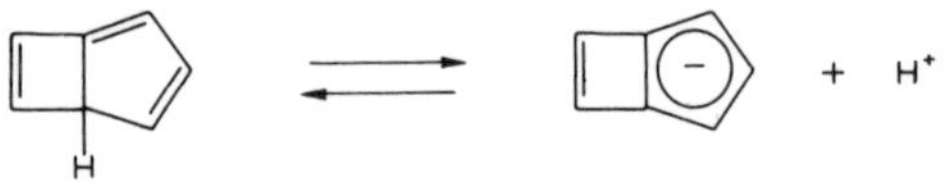

The pK_a for **67** was obtained by an extrapolation procedure (see Chapter 1, Section 2) as ca. 29, and could be related to cyclopentadiene for which the pK_a was 18 in the reaction medium. The difference of 11 pK units (15 kcal/mole) again reflects in large measure destabilization associated with the cyclobutadiene structure of **68**, though some contribution from a steric effect will also be present.

It should be noted that the pK_a studies considered above, as well as the electrochemical oxidation, constitute thermodynamic measurements referring to ground state properties. It is therefore particularly important that the thermodynamic measurements relating to the cyclopropenyl anion and to cyclobutadiene substantiate strikingly the available kinetic data on the effect of antiaromaticity on reactivity. Quantum-mechanical predictions on the destabilizing effect of electron delocalization in these cyclic 4n π-electron systems are thus upheld. (Use of the PPP method for cyclobutadiene leads to a resonance destabilization of 18 kcal/mole [87] while the Hückel method yields 33 kcal/mole [88], exclusive of ring strain.)

Assuming that we accept the evidence for antiaromaticity in the aforegoing cases, it is still pertinent to enquire whether antiaromatic properties will prevail in 4n π-electron systems when the integer n assumes yet higher values. Of obvious interest is the cycloheptatrienyl (tropenide) anion, $C_7H_7^-$, which HMO theory places in sharp contrast to the tropylium cation [64]. Whereas $C_7H_7^+$ is predicted to possess a complete electron shell configuration, the additional two electrons in $C_7H_7^-$ are placed in antibonding degenerate molecular orbitals so that a triplet ground state is predicted for the latter species.

Hydrogen exchange in cycloheptatriene (with Et_3CO^-/Et_3COD at 115°, or in DMSO rich media at 25°) implicates the intermediacy of the tropenide anion [89]. The anion is obtained, as a blue solution in tetra-

136

hydrofuran, on treatment of 7-methoxy-1,3,5-cycloheptatriene with sodium-potassium alloy [90]:

69

That the tropenide anion possesses comparative instability is indicated from the pK_a value of cycloheptatriene, which is estimated to be 36 [90, 91]. However, magnetic resonance studies on heptaphenylcyclo-heptatrienyl anion point to a singlet ground state [92], contrary to HMO theory. On the other hand MO theory predicts greater stability for the cycloheptatrienyl radical [64, 93] than for the anion and it is noteworthy that the heptaphenylcycloheptatrienyl anion and cation interact to produce the heptaphenylcycloheptatrienyl radical (**70**) [92].

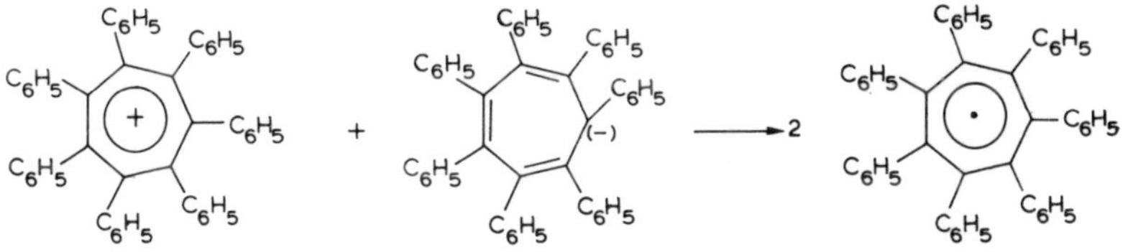

70

Thus the currently available experimental evidence does not indicate that the tropenide anion possesses antiaromatic properties. However, the benzocycloheptatrienyl anion has recently been characterized by NMR as a paratropic, antiaromatic, species [97]. Another approach which should be of promise in this regard is based on the measurement of reduction potentials of carbonium ions by cyclic voltammetry [20], via the reaction scheme:

$$R^+ \xrightarrow{e} R \cdot \xrightarrow{e} R^- \xrightarrow{H^+} RH$$

The results suggest a most interesting relationship, namely that there is a proportionality between the stabilization of an aromatic species and the destabilization of the corresponding antiaromatic species [20].

In closing this chapter we may point to one other direction in which work in this area is heading, namely towards establishing the concept of

137

homoantiaromaticity [94]. In fact there is already some evidence that in the 7-norbornenyl anion, **71**, the interaction between the negative charge and the double bond is destabilizing [95]; hence **71** may be called *bishomoantiaromatic* (cf. ref. 96). It will be recalled that the very large rate enhancement in formation of the 7-norbornenyl cation, **10**, may be ascribed to its stabilization as the bishomocyclopropenyl species [7].

10

71

REFERENCES

1a E. Hückel, Z. Physik., 70 (1931) 204; 76 (1932) 628.
1b L. Pauling and G.W. Wheland, J. Chem. Phys., 1 (1933) 362.
1c J.D. Roberts, A. Streitwieser, Jr. and C.M. Regan, J. Amer. Chem. Soc., 74 (1952) 4579.
2   C.W. Shoppee, J. Chem. Soc., 1147 (1946).
3   S. Winstein and R. Adams, J. Amer. Chem. Soc., 70 (1948) 838.
4   B. Capon, Quart. Rev., 18 (1964) 45.
5   A. Ehret and S. Winstein, J. Amer. Chem. Soc., 88 (1966) 2048.
6   M. Simonetta and S. Winstein, J. Amer. Chem. Soc., 76 (1954) 18;
    R.J. Piccolini and S. Winstein, Tetrahedron, 19, Suppl. 2, (1963) 423.
7   S. Winstein, in "Aromaticity", Chem. Soc. Spec. Publ., No. 21 (1967);
    S. Winstein, Quart. Rev., 23 (1969) 141.
8   H.C. Brown, Chemistry in Britain, 2 (1966) 199.
9   A.N. Bourns and E. Buncel, Ann. Rev. Phys. Chem., 12 (1961) 1.
10 P.D. Bartlett, "Nonclassical Ions. Reprints and Commentary", Benjamin, New York, 1965.
11 D. Bethell and V. Gold, "Carbonium Ions. An Introduction", Academic Press, London, 1967.
12 G.A. Olah and P.v.R. Schleyer (Editors), "Carbonium Ions. Vol. 3. Classical and Nonclassical Ions", Interscience, New York, 1970.
13 S. Winstein, J. Amer. Chem. Soc., 81 (1959) 6524; S. Winstein and J. Sonnenberg, ibid., 83 (1961) 3235, 3244.
14 T.J. Katz and E.H. Gold, J. Amer. Chem. Soc., 86 (1964) 1600.
15 G.A. Olah, P.R. Clifford, Y. Halpern and R.G. Johanson, J. Amer. Chem. Soc., 93 (1971) 4219.
16 S. Winstein and C. Ordronneau, J. Amer. Chem. Soc., 82 (1960) 2084;
    S. Yoneda, Z. Yoshida and S. Winstein, Tetrahedron, 28 (1972) 2395.

17   C.E. Keller and R. Pettit, J. Amer. Chem. Soc., 88 (1966) 606.
18   S. Winstein, C.G. Kreiter and J.I. Brauman, J. Amer. Chem. Soc., 88 (1966) 2047.
19   L.A. Paquette, J.R. Malpass and T.J. Barton, J. Amer. Chem. Soc., 91 (1969) 4714.
20   M. Feldman and W.C. Flythe, J. Amer. Chem. Soc., 93 (1971) 1547.
21   G. Schröder, U. Prange, B. Putze, J. Thio and J.F.M. Oth, Chem. Ber., 104 (1971) 3406.
22   H.A. Corver and R.F. Childs, J. Amer. Chem. Soc., 94 (1972) 6201.
23a P. Ahlberg, D.L. Harris, M. Roberts, P. Warner, P. Seidl, M. Sakai, D. Cook, A. Diaz, J.P. Dirlam, H. Hamberger and S. Winstein, J. Amer. Chem. Soc., 94 (1972) 7063.
23b L.A. Paquette, M.J. Broadhurst, P. Warner, G.A. Olah and G. Liang, J. Amer. Chem. Soc., 95 (1973) 3386.
24   R.A. Clark and R.A. Fiato, J. Amer. Chem. Soc., 92 (1970) 4736.
25a R. Gleiter, E. Heilbronner and A. de Meijere, Helv. Chim. Acta, 54 (1971) 1029.
25b P. Bischof, R. Gleiter and E. Heilbronner, Helv. Chim. Acta, 53 (1970) 1425.
26a M.J. Goldstein and R. Hoffmann, J. Amer. Chem. Soc., 93 (1971) 6193.
26b J.F. Chiang and C.F. Wilcox, Jr., J. Amer. Chem. Soc., 95 (1973) 2885.
27   P. Rys, P. Skrabal and H. Zollinger, Tetrahedron Lett., 1797 (1971); P. Rys and R. Vogelsanger, Helv. Chim. Acta, 55 (1972) 2844.
28   H. Prinzbach, Pure Appl. Chem., 28 (1971) 281.
29   P. Radlick and S. Winstein, J. Amer. Chem. Soc., 85 (1963) 344; K.G. Untch, J. Amer. Chem. Soc., 85 (1963) 345.
30   P. Skrabal and H. Zollinger, Helv. Chim. Acta, 54 (1971) 1069.
31   L.A. Paquette, M.R. Short and J.F. Kelly, J. Amer. Chem. Soc., 93 (1971) 7179; L.A. Paquette, R.E. Wingard, Jr. and R.K. Russell, J. Amer. Chem. Soc., 94 (1972) 4739.
32   A. Nickon and J.L. Lambert, J. Amer. Chem. Soc., 84 (1962) 4604; ibid., 88 (1966) 1905.
33   D.H. Hunter, A.L. Johnson, J.B. Stothers, A. Nickon, J.L. Lambert and D.F. Covey, J. Amer. Chem. Soc., 94 (1972) 8583.
34   S.K. Hsü, C.K. Ingold and C.L. Wilson, J. Chem. Soc., 78 (1938).
35   A. Nickon, J.L. Lambert and J.E. Oliver, J. Amer. Chem. Soc., 88 (1966) 2787.
36a N.H. Werstiuk and R. Taillefer, Can. J. Chem., 48 (1970) 3966.
36b R. Taillefer, Ph.D. Thesis, McMaster University, 1973.
37   T.T. Tidwell, J. Amer. Chem. Soc., 92 (1970) 1448; S.P. Jindal, S.S. Sohoni and T.T. Tidwell, Tetrahedron Lett., 779 (1971).
38   S.P. Jindal and T.T. Tidwell, Tetrahedron Lett., 783 (1971).
39   J.K. Stille, W.A. Feld and M.E. Freeburger, J. Amer. Chem. Soc., 94 (1972) 8485.
40   G.C. Joshi and E.W. Warnhoff, J. Org. Chem., 37 (1972) 2383.
41   K.W. Turnbull, S.J. Gould and D. Arigoni, Chem. Commun., 597 (1972).

42   B. Miller, J. Amer. Chem. Soc., 91 (1969) 751.
43   D.P.G. Hamon and R.W. Sinclair, Chem. Commun., 890 (1968).
44   J.P. Freeman and J.H. Plonka, J. Amer. Chem. Soc., 88 (1966) 3662.
45   P. Yates, G.D. Abrams, M.J. Betts and S. Goldstein, Can. J. Chem., 49 (1971) 2850.
46   J.K. Stille and K.N. Sannes, J. Amer. Chem. Soc., 94 (1972) 8494.
47   J.M. Brown and J.L. Occolowitz, Chem. Commun., 376 (1965).
48   J.M. Brown and J.L. Occolowitz, J. Chem. Soc. (B), 411 (1968).
49   S. Winstein, M. Ogliaruso, M. Sakai and J.M. Nickolson, J. Amer. Chem. Soc., 89 (1967) 3656.
50   J.M. Brown, Chem. Commun., 638 (1967).
51   J.M. Brown, E.N. Cain and M.C. McIvor, J. Chem. Soc. (B), 730 (1971).
52   J.M. Brown and E.N. Cain, J. Amer. Chem. Soc., 92 (1970) 3821.
53   T.J. Katz and C. Talcott, J. Amer. Chem. Soc., 88 (1966) 4732.
54   M. Ogliaruso, R. Rieke and S. Winstein, J. Amer. Chem. Soc., 88 (1966) 4731; M. Ogliaruso and S. Winstein, ibid., 89 (1967) 5290.
55   P.K. Freeman and T.A. Hardy, Tetrahedron Lett., 3939 (1971).
56   A.F. Diaz, D.L. Harris, M. Sakai and S. Winstein, Tetrahedron Lett., 303 (1971).
57   H. Kloosterziel and J.A.A. Van Drunen, Rec. Trav. Chim. Pays-Bas, 89 (1970) 368.
58a   J.B. Grutzner and S. Winstein, J. Amer. Chem. Soc., 90 (1968) 6562.
58b   J.B. Grutzner and S. Winstein, J. Amer. Chem. Soc., 94 (1972) 2200.
59   S.W. Staley and D.W. Reichard, J. Amer. Chem. Soc., 91 (1969) 3998.
60   M.J. Goldstein, J. Amer. Chem. Soc., 89 (1967) 6357.
61   R. Breslow, Chem. Eng. News, No. 26, (1965) 90.
62   J.-F. Labarre and F. Crasnier, Fortschr. Chem. Forsch., 24 (1971) 33.
63   D. Lloyd and D.R. Marshall, Angew. Chem. Inter. Ed., 11 (1972) 404.
64   A. Streitwieser, Jr., "Molecular Orbital Theory for Organic Chemists", Wiley, New York, 1961, Chapters 9, 10.
65   L. Salem, "The Molecular Orbital Theory of Conjugated Systems", Benjamin, New York, 1965, Chapter 3.
66   C.K. Ingold, "Structure and Mechanism in Organic Chemistry", 2nd Ed., Cornell University Press, Ithaca, N.Y., 1969, Chapter 4.
67   J.P. Snyder, Ed., "Non-Benzenoid Aromatics", Vol. 1. Academic Press, New York, 1969; Vol. 2, 1971.
68   P.J. Garratt, "Aromaticity", McGraw-Hill, Maidenhead, 1971.
69   E.D. Bergmann and B. Pullman, "Aromaticity, Pseudoaromaticity and Anti-aromaticity", Academic Press, New York, 1971.
70   R. Breslow, J. Brown and J.J. Gajewski, J. Amer. Chem. Soc., 89 (1967) 4383.
71   R. Breslow, Angew. Chem. Intern. Ed., 7 (1968) 565; Chemistry in Britain, 4 (1968) 100; Accounts Chem. Res., 6 (1973) 393.
72   R. Breslow, H.W. Chang, R. Hill and E. Wasserman, J. Amer. Chem. Soc., 89 (1967) 1112.
73   D.T. Clark, Chem. Commun., 637 (1969); D.T. Clark and D.R. Armstrong, ibid., 850 (1969).

74 J.A. Pople and K.G. Untch, J. Amer. Chem. Soc., 88 (1966) 4811.
75 H.C. Longuet-Higgins, in "Aromaticity", Chem. Soc. Spec. Publ., No. 21 (1967), p. 109.
76 F. Sondheimer, in "Aromaticity", Chem. Soc. Spec. Publ., No. 21 (1967), p. 75; Accounts Chem. Res., 5 (1972) 81.
77 H.J. Dauben, Jr., J.D. Wilson and J.L. Laity, in "Nonbenzenoid Aromatics", Vol. 2, edited by J.P. Snyder, Academic Press, New York, 1971.
78a K.B. Wiberg, R.K. Barnes and J. Albin, J. Amer. Chem. Soc., 79 (1957) 4994.
78b E.A. Dorko and R.W. Mitchell, Tetrahedron Lett., 341 (1968).
78c A.J. Schipperijn, Rec. Trav. Chim. Pays-Bas, 90 (1971) 1110.
79 R. Breslow and M. Battiste, Chem. Ind., 1143 (1958).
80 R. Breslow and M. Douek, J. Amer. Chem. Soc., 90 (1968) 2698.
81 W.Th. Van Wijnen, H. Steinberg and Th.J. De Boer, Rec. Trav. Chim. Pays-Bas, 87 (1968) 844.
82 P. Kasai, R. Myers, D. Eggers, Jr. and K.B. Wiberg, J. Chem. Phys., 30 (1959) 512.
83 R. Breslow and K. Balasubramanian, J. Amer. Chem. Soc., 91 (1969) 5182.
84 M.P. Cava and M.J. Mitchell, "Cyclobutadiene and Related Compounds", Academic Press, New York, 1967.
85 R. Breslow, R. Grubbs and S.-I. Murahashi, J. Amer. Chem. Soc., 92 (1970) 4139.
86 R. Breslow and W. Washburn, J. Amer. Chem. Soc., 92 (1970) 427.
87 M.J.S. Dewar, M.C. Kohn and N. Trinajstíc, J. Amer. Chem. Soc., 93 (1971) 3437.
88 B.A. Hess, Jr. and L.J. Schaad, Tetrahedron Lett., 5113 (1972).
89 W. von E. Doering and P.E. Gaspar, J. Amer. Chem. Soc., 85 (1963) 3043.
90 H.J. Dauben, Jr. and M.R. Rifi, J. Amer. Chem. Soc., 85 (1963) 3041.
91 R. Breslow and W. Chu, J. Amer. Chem. Soc., 92 (1970) 2165; ibid., 95 (1973) 411.
92 R. Breslow and H.W. Chang, J. Amer. Chem. Soc., 87 (1965) 2200.
93 G. Vincow, H.J. Dauben, Jr., F.R. Hunter and W.V. Holland, J. Amer. Chem. Soc., 91 (1969) 2823.
94 R. Breslow, Pure Appl. Chem., 28 (1971) 111.
95 J.K. Stille and K.N. Sannes, J. Amer. Chem. Soc., 94 (1972) 8489.
96 D.D. Davis and W.B. Bigelow, Tetrahedron Lett., 149 (1973).
97 S.W. Staley and A.W. Overdal, J. Amer. Chem. Soc., 95 (1973) 3382.

Chapter 5

ENOLATE AND HOMOENOLATE REARRANGEMENTS

1. SCOPE OF THE PRESENTATION

The present chapter, and the one following, are concerned with rearrangement reactions of carbanionic species. However, the coverage will be restricted in scope to certain selected topics which will be considered in some depth. It is especially our purpose to unify various concepts, and to present in different light certain themes which have been discussed elsewhere in this work. A wider range of carbanion rearrangements has been reviewed by other authors [1].

Enolate ions have received attention in this volume in Chapter 1 and in Chapter 3. In the former chapter the ability of the carbonyl function to delocalize negative charge on an adjacent carbon was considered quantitatively within the context of the strengths of carbon acids. In Chapter 3, on the other hand, we considered the keto-enol system within the context of the general phenomenon of tautomerism, or prototropy. Evidence for the intermediacy of carbanionic species in these 1,3-proton transfer processes was discussed at that time, and various catalytic pathways were evaluated.

In the present chapter we examine the occurrence of skeletal rearrangements in systems for which enolate ions have been proposed as reaction intermediates. The case of the Favorskii rearrangement is considered in detail, for several reasons: (a) here we have a system for which there has been a careful evaluation of evidence for the intermediacy of carbanions, as opposed to a concerted pathway; (b) contrary to expectations, there is strong evidence for the intervention of a second reaction intermediate; and (c) despite extensive investigations of this reaction many questions concerning its mechanism remain unanswered. Some related enolate rearrangements are considered next, though more briefly. The remainder of the chapter is assigned to rearrangement processes of homo-enolate ions, a concept which was first introduced in Chapter 4 in relation to nonclassical interactions in anionic systems.

2. THE FAVORSKII REARRANGEMENT

The reaction of α-halogenoketones with bases, resulting in the formation of carboxylic acids or esters, involves a structural reorganization of the carbon skeleton, known as the Favorskii rearrangement [2]. For example, treatment of α-chlorocyclohexanone with sodium ethoxide in ethanol yields ethyl cyclopentanecarboxylate (eq. (1)), while with aqueous sodium hydroxide the corresponding free acid is obtained. The synthetic utility of such a ring contraction process has been well recognized [3].

(1)

A number of possible mechanisms have been proposed for the Favorskii rearrangement [3–5], and of these the main contenders are the four represented in eq. (2)–(5).

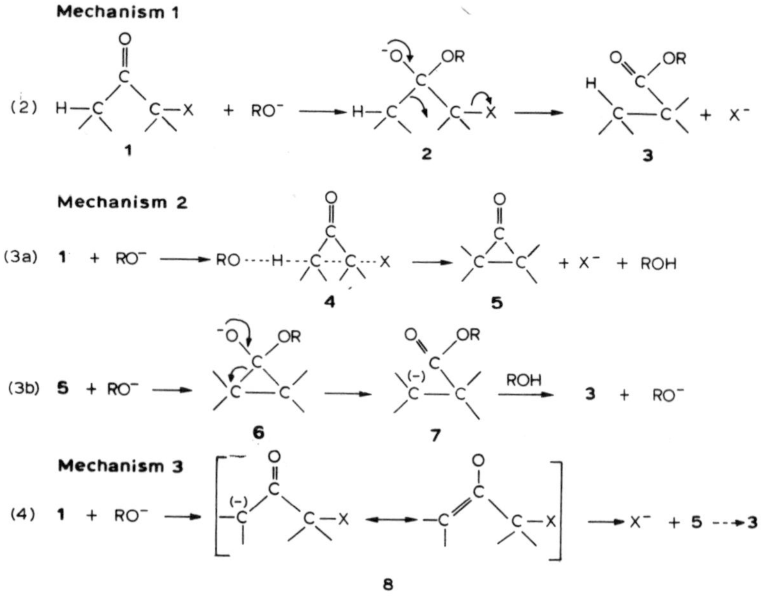

Mechanism 1

(2)

Mechanism 2

(3a)

(3b)

Mechanism 3

(4)

144

Mechanism 4

$$(5) \quad 1 + RO^- \longrightarrow 8 \longrightarrow X^- + \begin{smallmatrix} O^- \\ \parallel \\ C \end{smallmatrix} \quad \longrightarrow 5 \dashrightarrow 3$$

9

The first mechanism (eq. (2)) is characterized by initial addition of alkoxide to carbonyl to form the tetrahedral intermediate 2, followed by a 1,3-alkyl shift with simultaneous expulsion of halide ion, to yield ester 3. This is known as the semi-benzylic acid pathway since it is somewhat similar to the mechanism of the benzylic acid rearrangement [6]; in the latter reaction, C=O takes the place of C–X and becomes transformed into C–O$^-$ in the process. Mechanism 1 is clearly differentiated from the other three, in that the *alpha* C–H remains intact during the rearrangement, so that neither a carbanion nor a cyclopropanone-like intermediate is involved. Since there is considerable evidence for intermediates of these types in the reactions of a large number of α-halo-ketones, it appears that the first mechanism cannot be a general one. However, in systems in which the *alpha* C–H is relatively nonacidic, or in which some unusual steric factors operate, it has been found [7] that certain substrates can in fact react by this mechanism in the Favorskii transformation; a notable example is given by ring contraction in the cyclobutanone system [7d].

In the second mechanism there is concerted, base-induced, 1,3-elimination of HX (via transition state 4) to give a cyclopropanone derivative 5 as the key reaction (eq. (3a)). This is followed by an addition of RO$^-$ to C=O, and then ring opening of 6; protonation of the resulting transient carbanion 7 yields the ester product (eq. (3b)). The ring opening of cyclopropanones through the action of base is known from independent studies with these recently characterized compounds [4, 8].

The third and fourth mechanisms feature initial formation of a carbanion-enolate ion intermediate, 8. In Mechanism 3 (eq. (4)), loss of halide ion from 8 yields the cyclopropanone 5, which suffers ring opening to give the product as before (eq. (3b)). In Mechanism 4 (eq. (5)) the loss of halide from 8 gives initially a dipolar ionic intermediate 9, which then collapses to cyclopropanone; reaction is completed as before. These aspects will be discussed in detail later in this section but first we consider the evidence for the intermediacy of a cyclopropanone structure in Favorskii rearrangements.

The potentiality of an intermediate cyclopropanone in imparting symmetry suggests that an isotopic tracer experiment would provide valuable evidence in support of mechanisms involving such species. That such is indeed the case is seen from the following results obtained in the reaction of EtO⁻ with 2-chlorocyclohexanone labelled with ^{14}C on C1 and C2 [9].

(6)

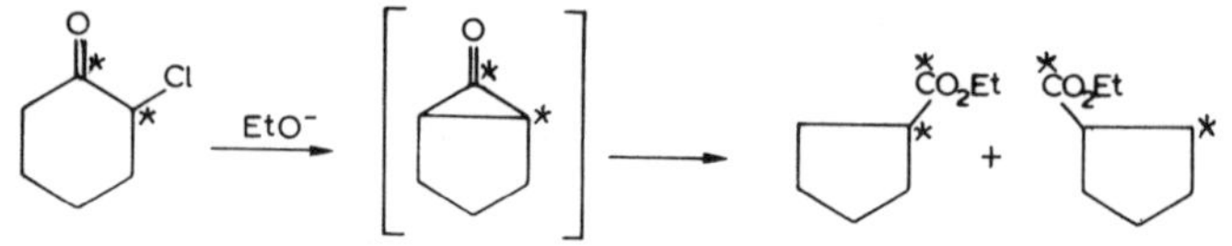

The cyclopentane ethyl ester product contained radioactivity on the carboxyl carbon (50%), on the *alpha* (25%) *and* on the *beta* carbon (25%), indicating that the two *alpha* carbons of the cyclohexanone had been equalised. The starting material, recovered from partial reaction, had its label undisturbed, showing that no prior halogen migration had occurred.

Evidence of a structural nature concerning intermediates in the Favorskii rearrangement is given by the observation that the bromoketones **10** and **11** both yield trimethyl acetate (**16**) in the reaction with NaOMe, as shown in Scheme 1 [10a]. This is indicative of the formation of a common intermediate, plausibly **12**, in the two reactions.

Scheme 1

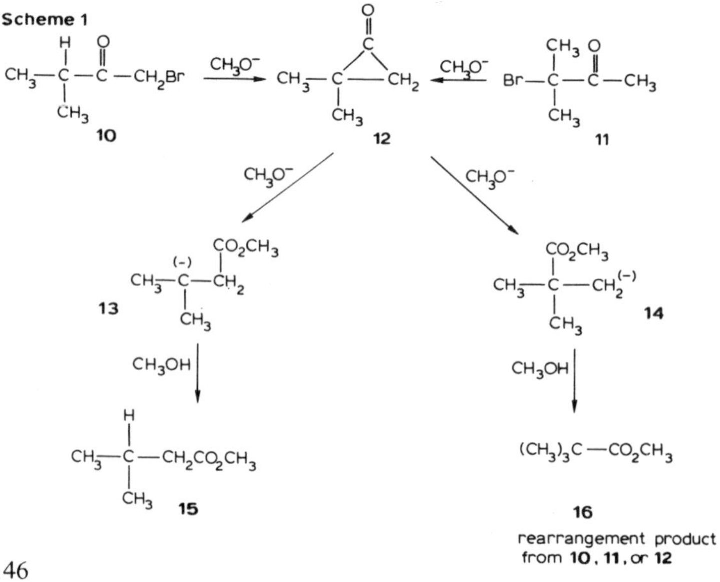

rearrangement product
from **10**, **11**, or **12**

146

Even firmer evidence for the intermediate is available in this particular case, since the 2,2-dimethylcyclopropanone 12 could be prepared independently, and on treatment with methoxide was found to give the same product 16 [8]. (The hemiketal of 12 behaves in similar fashion.) That 16 should be obtained exclusively (rather than 15) is in accord with formation of the more stable primary carbanion 14 in the base cleavage of 12 (or more accurately, that in the ring-opening of 12, the transition state leading to 14 is of lower energy than that leading to 13). However, in the case of more highly substituted α-haloketones, such as $(CH_3)_2CXCOC_3H_7$, both the Favorskii esters are formed in the reaction with alkoxide, and the ratio of these is different when methoxide or when t-butoxide is used as base [10b]. These results are explicable in part on the basis of a smaller difference in stabilities of secondary and tertiary carbanions (compared with primary and tertiary in Scheme 1), but also point to the operation of a steric factor in such systems. (For an account of the stereochemistry of opening of cyclopropanone hemi-ketals, see ref. 10c; the ring opening of cyclopropanimines is also a regiospecific process [10d].)

As one more piece of evidence in support of formation of cyclo-propanone intermediates in such systems, one can cite the known 1,3-elimination reaction of α,α'-dibromoketones. For example, α,α'-di-bromodibenzyl ketone on treatment with triethylamine yields the stable diphenylcyclopropenone, evidently formed via a 1,3-elimination [11]:

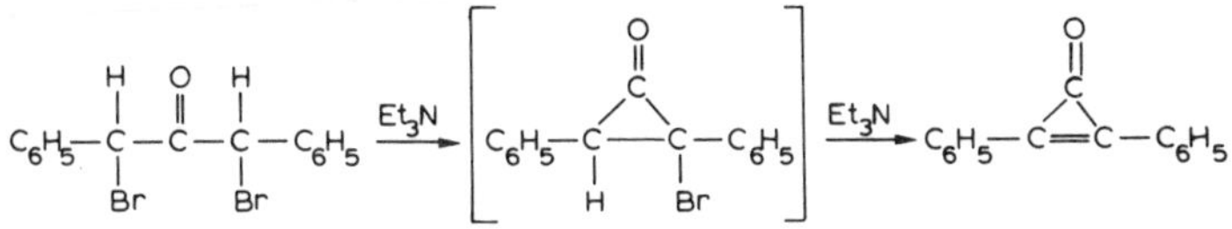

1,3-Elimination also occurs in the base-induced Ramburg-Backlund reaction of α-halosulfones, but here the intermediate three-membered episulfone decomposes by loss of SO_2 to alkene. The reaction mechanism is believed to involve reversible formation of a carbanion, followed by loss of halide with ring closure [5]:

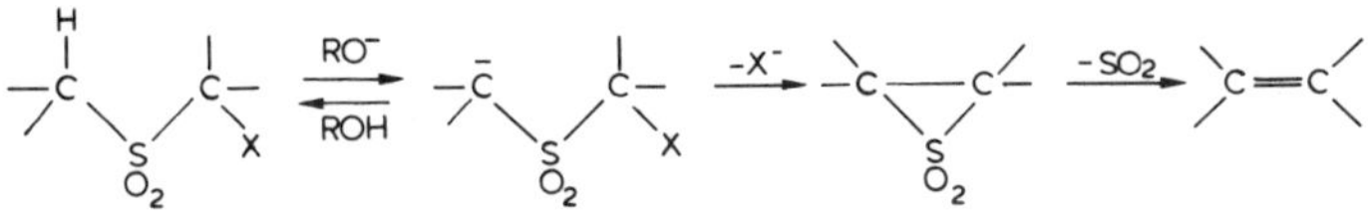

If, as a result of the forementioned evidence, one can consider the formation of a cyclopropanone intermediate in the Favorskii rearrangement as established, it is by no means clear yet whether such an intermediate is formed in a one-stage, two-stage, or three-stage process, as represented in Mechanisms 2, 3, and 4, respectively. We now examine the question of carbanionic species as reaction intermediates in these systems.

The formation of carbanionic species from α-haloketones under Favorskii rearrangement conditions has been established from deuterium exchange experiments. For example, the 2-halocyclohexanones undergo hydrogen isotope exchange when the reaction is carried out in CH_3ONa/CH_3OD, and such exchange also occurs in the case of the benzylic haloketones $ArCH_2COCH_2X$, $ArCH_2COCHXCH_3$, and $ArCHXCOCH_3$ [12], as well as with 2-α-halocholestan-3-one steroidal substrates [13]. It still remains to be shown, however, that such carbanions are actually present on the Favorskii reaction pathway, as distinct from being formed independently in a reversible proton abstraction process [14], in which case their presence would merely be co-incidental as far the Favorskii rearrangement process is concerned.

Evidence indicating that carbanions *are* actually on the reaction path is derived from the observed relationship between the extent of deuterium exchange in the recovered ketone and the k_{Br}/k_{Cl} rate constant ratio for loss of halide ion, as seen in the following data [12a]:

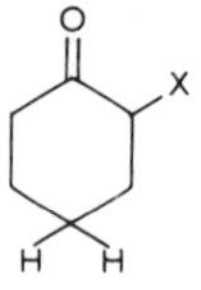

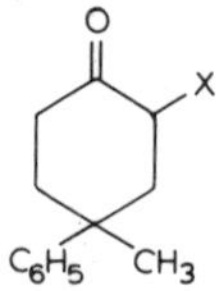

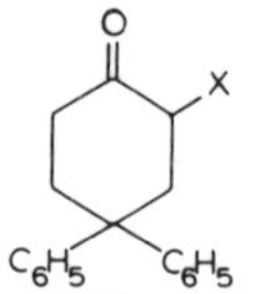

	17	18	19
% H/D exchange(X=Cl)	19	50	100
k_{Br}/k_{Cl}	ca. 30	52	116

The correlation between the extent of exchange and the effect of changing the leaving group, given by k_{Br}/k_{Cl}, follows from consideration of eq. (9), it being assumed that the subsequent product-yielding steps are relatively fast.

A large k_{Br}/k_{Cl} rate ratio indicates that C—X bond rupture must be largely rate-determining, and in a two-stage carbanion mechanism

148

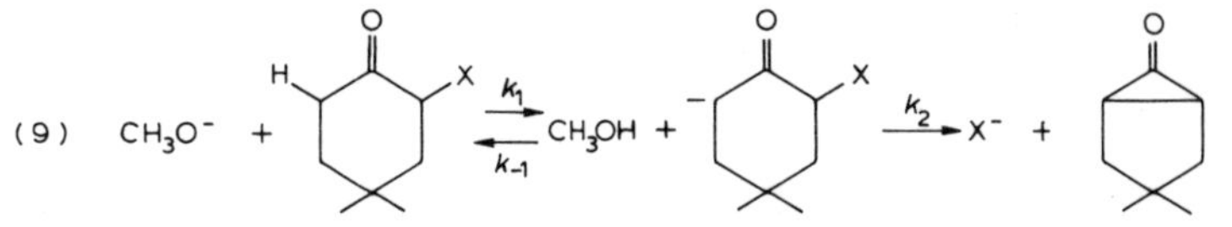

$$(9) \quad CH_3O^- + \text{[structure]} \xrightarrow[k_{-1}]{k_1} CH_3OH + \text{[structure]} \xrightarrow{k_2} X^- + \text{[structure]}$$

(eq. (9)) this will require that $k_{-1} > k_2$. It then necessarily follows from the condition $k_{-1} > k_2$ that there will be extensive deuterium exchange when the reaction is performed in CH_3OD (provided that "internal return" can be neglected — see Chapter 1, Section 5 and Chapter 2, Section 3). The important point here is the observation of a *variable* extent of deuterium exchange and a parallelism with the leaving group tendency. (If the study had been limited to the finding of a large k_{Br}/k_{Cl} ratio, one could have explained that equally on the basis of the concerted Mechanism 2.) It would be difficult to visualise any connection between the % exchange and the k_{Br}/k_{Cl} ratio if carbanions did not occur as an integral part of the reaction pathway.

The effect of substituents at C4 in the series of 2-halocyclohexanones **17−19** on deuterium exchange and the k_{Br}/k_{Cl} ratio can be understood on the basis of stereo-electronic factors [12]. The ensuing discussion also has a bearing on differentiation between Mechanisms 3 and 4.

On the one hand we can have an essentially internal displacement of the C_α-halogen by the $C_{\alpha'}$-electron pair, a process requiring the rotation of $C_{\alpha'}$ out of conjugation with C=O and rehybridization from sp^2 to sp^3:

internal C-X displacement in carbanion

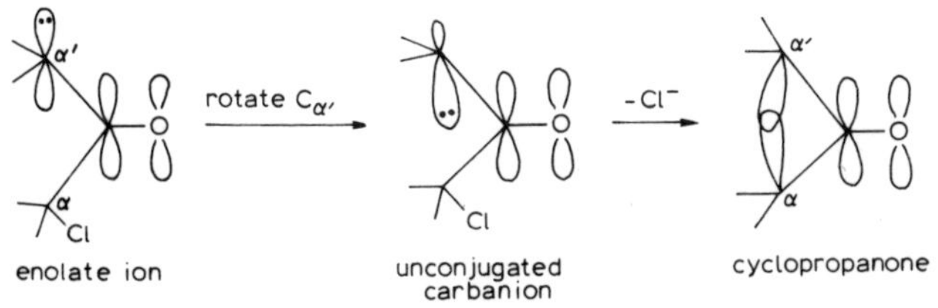

enolate ion   unconjugated   cyclopropanone
                     carbanion

On the other hand, halide ion release can occur directly from the enolate ion, but presumably the C–Cl bond will now be oriented parallel to the p-orbital at $C_{\alpha'}$, so that in the transition state **20** for C–Cl dissociation the developing p-orbital at C_α would overlap with $C_{\alpha'}$ as

149

well as with the carbonyl group; the overall process is seen to be one of π-participation:

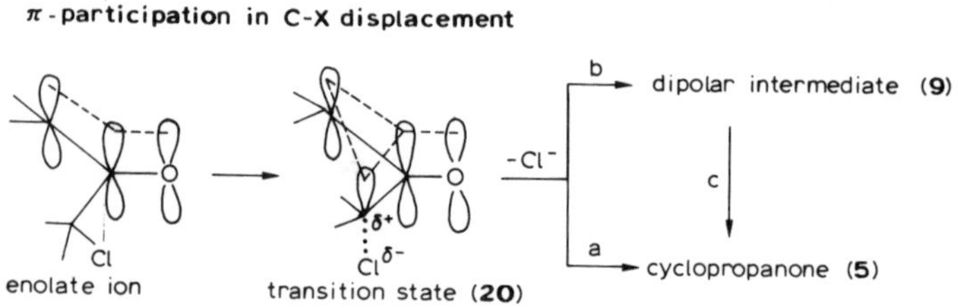

π - participation in C-X displacement

enolate ion   transition state (20)   dipolar intermediate (9)   cyclopropanone (5)

Since formation of 20 implies an axially oriented halogen, it is apparent that when another substituent in the molecule is forced to be axial (or pseudoaxial), as in 18 or 19, the formation of 20 would be inhibited. This would result in a decreased rate of halide release, so that in terms of the symbolism of eq. (9) the k_{-1}/k_2 ratio would increase. Thus the π-participation mechanism satisfactorily accounts for the observed trend in the percent of H/D exchange in the series 17—19.

Now whereas the internal displacement mechanism has been considered as applicable only to Mechanism 3, the π-participation process may in principle apply to either of Mechanisms 3 or 4. In the former case product formation from transition state 20 would occur via route (a), and in the latter via (b) and (c). In (b) dissociation of the chloride ion becomes complete without C–C rotation and one obtains the delocalized dipolar intermediate 9. In the following step, (c), cyclopropanone is formed by a disrotatory process about the C–C$_\alpha$ and C–C$_{\alpha'}$ bonds. In route (a), 5 is formed directly from 20, so that halide dissociation and bond rotation become coalesced into a single concerted process.

The delocalized intermediate 9 may be represented by the valence bond structures 9a and 9b:

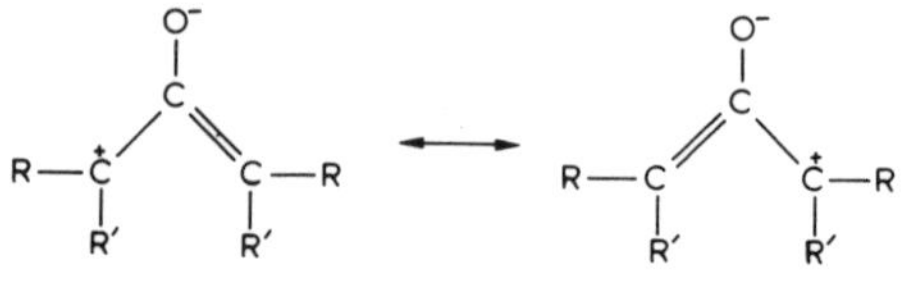

9a   9b

150

One can readily visualise collapse within the dipolar intermediate **9** to cyclopropanone **5**, or, alternatively to an allene oxide **21**; the latter could subsequently rearrange to **5**:

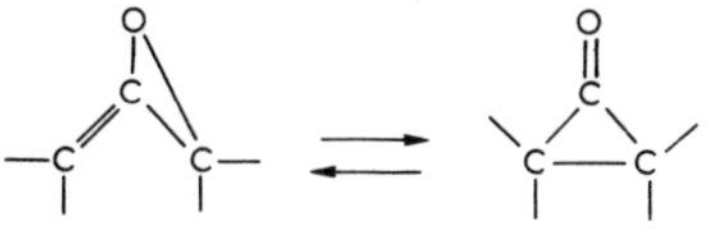

21   5

The electronic structure of the delocalized intermediate (**9**) and its relative stability compared to cyclopropanone (**5**) and allene oxide (**21**) has been discussed by a number of workers [4, 5, 15–24] and only a few comments need be made here pertaining to some of the recent work.

The microwave spectrum of cyclopropanone itself is fully in accord with the ring-closed structure **5** rather than the valence tautomer **21** or an acyclic dipolar structure [20]. The highly hindered 1,3-di-*t*-butyl-allene oxide has actually been isolated [21, see also 22], and observed to undergo isomerisation to *trans*-2,3-di-*t*-butylcyclopropanone at 100°, as shown in eq. (10) (R = *t*-butyl):

(10)

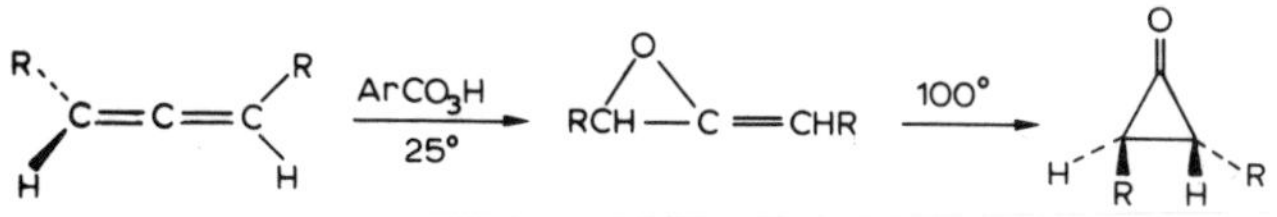

5a

Interestingly, when resolved, the cyclopropanone product **5a** can be observed to undergo racemization on heating (80°), and the rate of racemization increases with solvent polarity [23]. These facts may be accommodated in terms of the formation of a dipolar species as a reaction intermediate:

(11)   (+) - **5a** $\xrightarrow{80°}$ $\longrightarrow$ (±) - (**5a**)

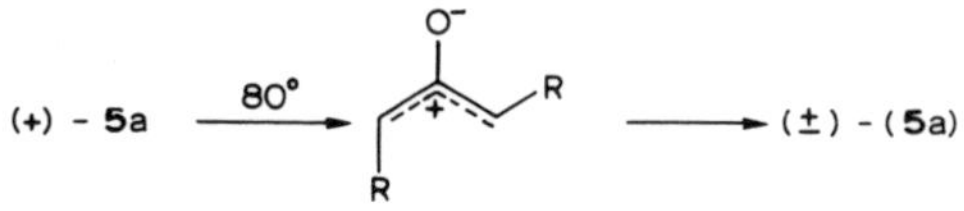

The kinetics of cycloaddition reactions of cyclopropanones have likewise been interpreted in terms of a large degree of zwitterion character in the transition state [24].

151

The stereochemical pathway in the Favorskii rearrangement is still the subject of considerable controversy [25–28], a situation no doubt due largely to the complexity of many of the systems investigated. (The steroids, for example, have provided fertile territory for Favorskii reactions.) Hence one should regard the detailed geometry of the transition state configuration **20** as only tentative. Consistent with an appreciable degree of C–Cl bond rupture in the rate-limiting transition state is the observation of a large negative ρ value (−5.0) in the reactions of $ArCH_2COCH_2Cl$ compounds, as well as the effects of medium changes and of added salts on the rate of reaction [12c]. These observations can also be interpreted, however, in terms of resonance stabilization in the transition state **22**, without the requirement of π-participation.

22

An unusual case of a Favorskii reaction has been described recently [29]. The diphenyl α-halopropanones **23** and **24** react with CH_3ONa/CH_3OH to yield the rearranged ester **27** and indanone **29** as main products, as well as some alkoxyketone, $Ph_2C(OMe)COCH_3$. The proposed mechanism of formation of **27** and **29**, as outlined in Scheme 2, features initial formation of enolate ions **23a** and **24a**, followed by loss of halide ion to give the common dipolar ion intermediate **25**. Collapse within the dipolar ion gives cyclopropanone **26**, which on reaction with methoxide yields the Favorskii ester **27**. On the other hand, bonding between the terminal carbon of **25** and an *ortho* carbon of one of the phenyls yields the intermediate **28**, which on tautomerisation gives the indanone product **29**. The methoxyketone $Ph_2C(OMe)COCH_3$, is considered to be derived by methanolysis of the enol allylic chloride, $Ph_2CCl-C(OH)=CH_2$, rather than from intermediate **25**[*].

---

[*]While methoxyketones are the common side products in Favorskii reactions, yet other by-products may also be obtained, depending on the reacting system. Detailed consideration of this aspect has been given by several authors [3, 10, 12].

152

Scheme 2

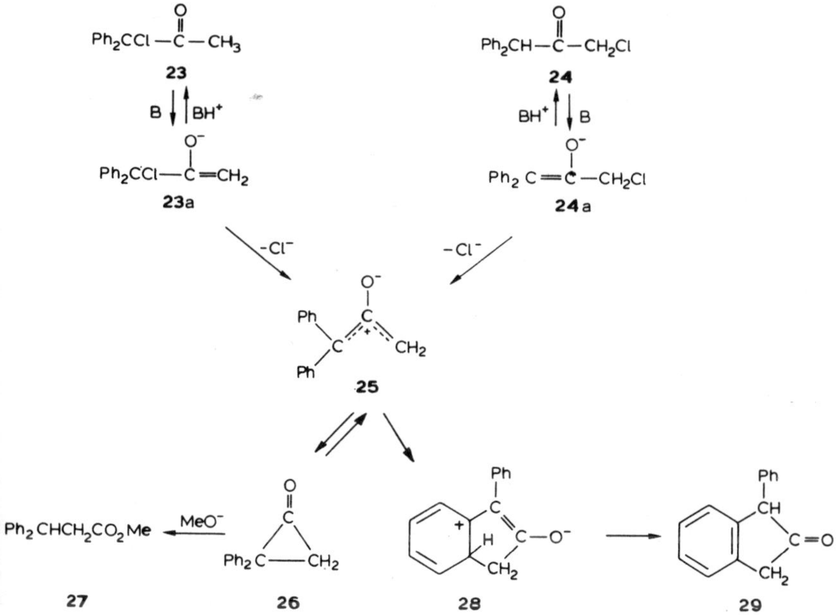

The essential point is that the dipolar ion **25** is plausibly established as a common intermediate for formation of **27** *and* of **29**, and that there appears to be no need to postulate a π-participation mechanism in the pathways leading to formation of **25**.

The value of an alternative approach to a problem is aptly illustrated by recent work on the electroreduction of α,α'-dibromoketones [30]. The objective of the experiments was to test the possibility of the synthesis of cyclopropanones through 1,3-electroreductive ring closure:

(12)

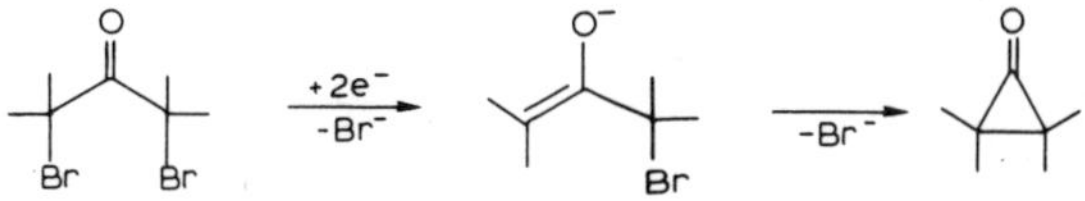

An intensive investigation has led to the formulation of Scheme 3 [30a]. Thus electroreduction of 2,4-dibromo-2,4-dimethyl-3-pentanone (**30**) proceeds readily in dimethylformamide or acetonitrile, by a two electron process. In the presence of acetic acid, ethanol, or water as

153

proton donor, the major product is 2-acetoxy-2,4-dimethyl-3-pentanone (37, R = Ac) or the corresponding ethoxy and hydroxy derivatives; these products are probably produced from the initially formed enolate ion 31 by solvolysis. Also formed is a small amount of the enone 38, which can be accounted for by elimination, as indicated.

Scheme 3

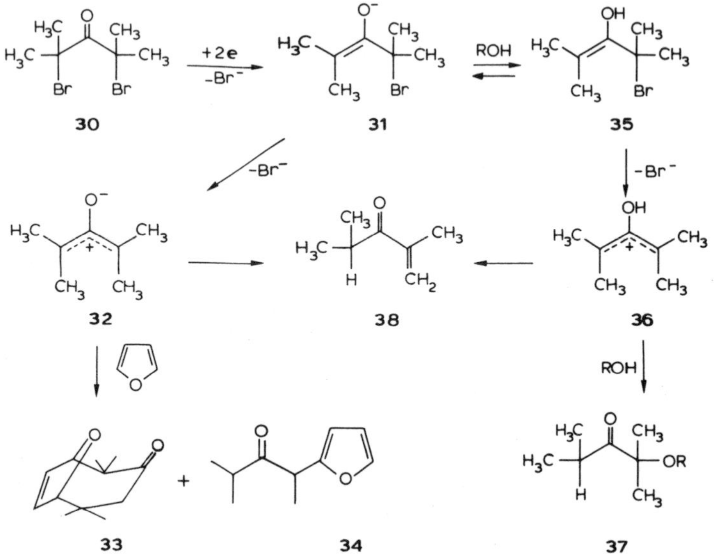

Significantly, electroreduction of 30 in the presence of furan results in formation of 33 and 34 as the major products; it is proposed [30a] that these adducts arise from trapping of the dipolar, or zwitterionic, intermediate 32 (cf. ref. 24, in which adduct 33 is reported to be formed from tetramethylcyclopropanone and furan). An ionic mechanism for formation of 33 and 34, via 32, may readily be formulated.

In other work [30b], the electrochemical reduction of 2,4-dibromo-2-methyl-3-pentanone (in dimethylformamide containing tetraethyl-ammonium bromide) gave evidence for formation of 2,2-dimethylcyclo-propanone (12) as a transient intermediate (a transient infrared absorption at 1825 cm^{-1} was seen). Electrolysis in methanol gave rise to the corresponding hemiketal. Though further work on solvent and electrolyte effects should help to delineate the equilibrium between

154

cyclopropanone and the zwitterion, the electrochemical approach has already given new insight onto this challenging problem.

3. SOME FURTHER ENOLATE REARRANGEMENTS

(a) Base-induced rearrangements of α-epoxy ketones

α-Epoxy ketones, which are readily obtainable from the corresponding α,β-unsaturated ketones by epoxidation, are known to undergo base-induced rearrangement with formation of α-hydroxycarboxylic acids. The reaction sequence, shown in eq. (13), is of value in synthesis as well as being of mechanistic interest [31].

(13)

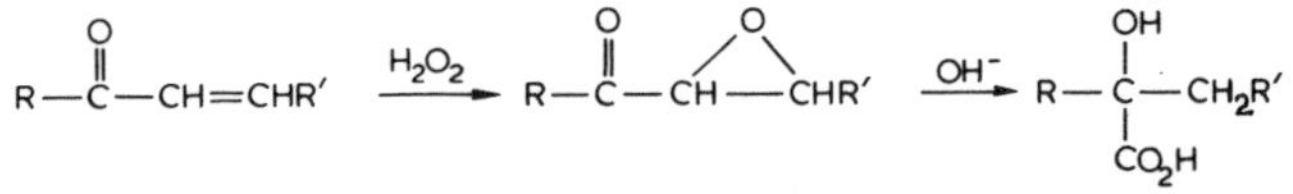

The probable reaction pathway for the rearrangement of **40** is as follows:

(14)

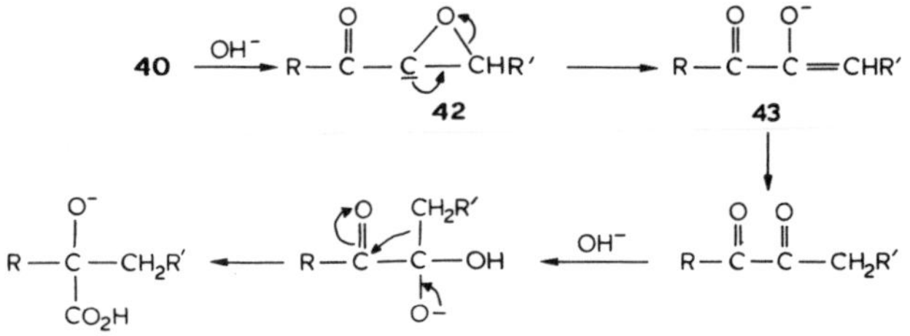

Reaction is initiated by removal of the proton *alpha* to carbonyl, which is followed by opening of the epoxy ring and formation of a second enolate ion (**43**). The ensuing sequence of reactions involving conversion of **44** into the product **46** corresponds to the well known benzilic acid rearrangement of α-diketones. The benzylic acid rearrangement has been studied extensively, typically with aryl *alpha* diketones [6].

Contrasting observations were made in study of epoxy ketones in the series 1-aryl-3-cycloalkyl-2,3-epoxypropanones, i.e. **40** with R' = phenyl and R = cycloalkyl [31]. Whereas the *cyclopropyl* system yielded the normal rearranged product (**41**, R' = phenyl and R = cyclopropyl), the

155

cyclobutyl, cyclopentyl and *cyclohexyl* derivatives followed a different reaction pathway. The novel reaction, illustrated by the cyclohexyl case in eq. (15), yields a rearranged unsaturated carboxylic acid **48** and lactone **49**.

(15)

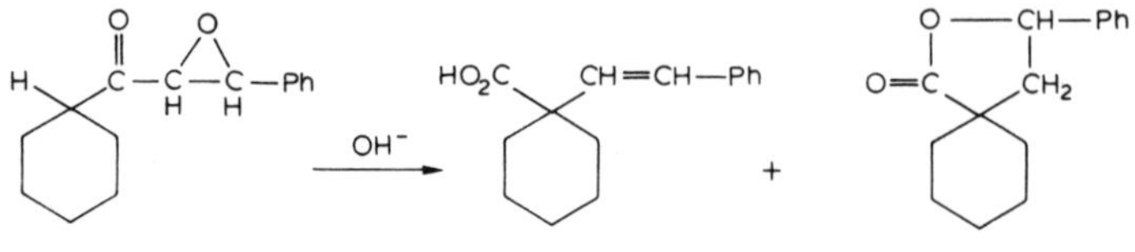

47   48   +   49

The proposed mechanism for the rearrangement **47→48** is shown in eq. (16).

(16)

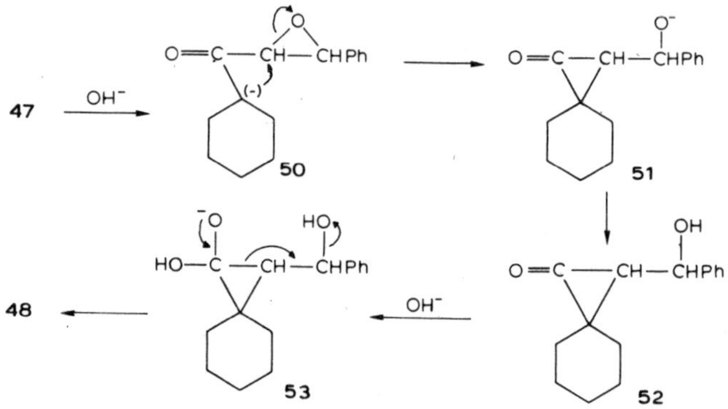

The initial step now is abstraction of the tertiary cyclohexyl proton by base, followed by formation of cyclopropanone with concomitant epoxide ring opening. The cyclopropanone ring in **52** is opened by way of the hemiketal function (cf. [10]), leading to formation of **48** through elimination of hydroxide, as indicated (**53**). Alternatively, the carbanion formed on ring opening of hemiketal **53** becomes protonated, yielding a γ-hydroxy carboxylic acid, which would readily lactonize to produce **49**.

This reaction sequence for rearrangement of epoxy derivative **47** is basically similar to that in the Favorskii rearrangement, since the formation of an intermediate cyclopropanone and its subsequent cleavage is common to both systems. The cyclopentyl and cyclobutyl analogs probably also follow this mechanism.

156

Why then does the cyclopropyl derivative follow a different reaction (eq. (14)) pathway? It should be recalled (see Chapter 3, Section 1e) that abstraction of a hydrogen atom situated *alpha* to carbonyl is less favourable for a cyclopropyl than for an acyclic system as a result of steric inhibition of delocalization in the transition state for enolization. The effect of such ring strain should be much smaller for enolization in the cyclobutyl case, and virtually absent in the cyclopentyl and cyclohexyl systems. As a result, the cyclopropyl system avails itself of the alternative pathway (eq. (14)), through the base abstraction of the other (C2) *alpha* hydrogen, whose acidity will be enhanced by the electron withdrawing effect of the epoxy-oxygen. Clearly there is a driving force for both rearrangement pathways, corresponding to eqs. (14) and (16), so that the balance of factors may shift from one pathway to the other depending on structural and electronic factors in a particular system. Several other α-epoxy ketones have been shown to follow the Favorskii type rearrangement pathway corresponding to eq. (16) [32–34].

(b) The rearrangements of halogeno amides

Reaction of an α-halogeno amide with base might well be expected to lead to formation of a three-membered lactam, by analogy with the reaction of α-halogeno ketones. The first successful report of this type of reaction is shown below.

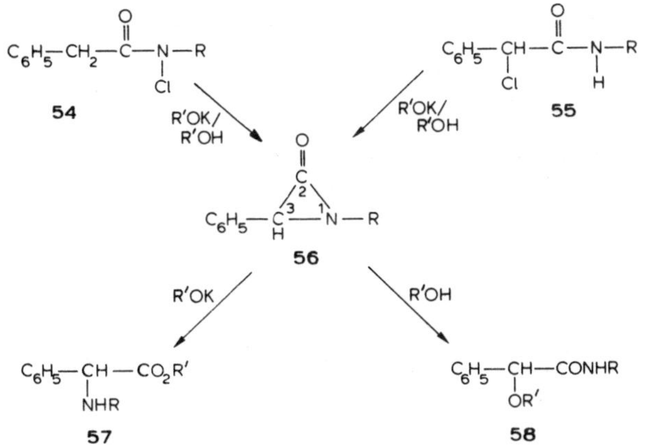

The halogeno amides **54** and **55** (R = *t*-butyl) on treatment with potassium *t*-butoxide in *t*-butyl alcohol yield an intermediate α-lactam, or aziridone, **56**, which undergoes ring opening processes to give the products **57** and **58** [35]. The overall reactions show interesting analogy to the Favorskii rearrangement.

Although α-lactams had been postulated in the past as reaction intermediates, their isolation had previously been unsuccessful because of their high reactivity. Isolation of **56** was accomplished in the first instance [35b] by using N-*t*-butylphenylacetamide as the precursor of **54** and causing it to react with *t*-butyl hypochlorite and potassium *t*-butoxide in toluene; in the absence of *t*-butyl alcohol or of excess of *t*-butoxide, **56** could be isolated in the pure state and characterized. Appearing almost simultaneously with this pioneering work, were reports of similar preparation of yet other α-lactams (generally by way of the α-chloroamides) [36]. Formation of α-lactam presumably proceeds via the (enolate) anions produced by proton abstraction from sites *alpha* to carbonyl, although a concerted 1,3-elimination process has not been ruled out. (An SNi type of cyclization mechanism has been suggested for the formation of steroidal α-lactams (1-*t*-butyl- and 1-adamantyl-3-bisnorcholanyl aziridone) [37], on the basis of inversion of configuration at the *alpha* carbon, though whether or not this stereochemistry is specific to this system remains to be seen.)

The ring opening reactions of α-lactams are remarkably sensitive to the nature of substituents. For example, 1-*t*-butyl-3,3-dimethylaziridone reacts with ionic nucleophiles solely by attack at the carbonyl centre (acyl-nitrogen bond scission), while uncharged nucleophiles attack at the *alpha* carbon (alkyl-nitrogen bond scission) [36]. This is analogous to the reactions **56→57** and **56→58**, respectively. On the other hand, 1-*n*-propyl-3,3-dimethylaziridone on reaction with *t*-butoxide yields products of both ring-opening modes simultaneously. Moreover, 1-*t*-butyl-3-methylaziridone reacts with *t*-butoxide almost exclusively by alkyl-nitrogen bond scission.

As in the case of the Favorskii rearrangement, there is some doubt about the nature and structure of the intermediate(s) formed in these processes. In addition to the α-lactam structure, other possibilities which have been considered include the uncharged species **59** and **60**, as well as the dipolar ionic species **61** and **62**.

158

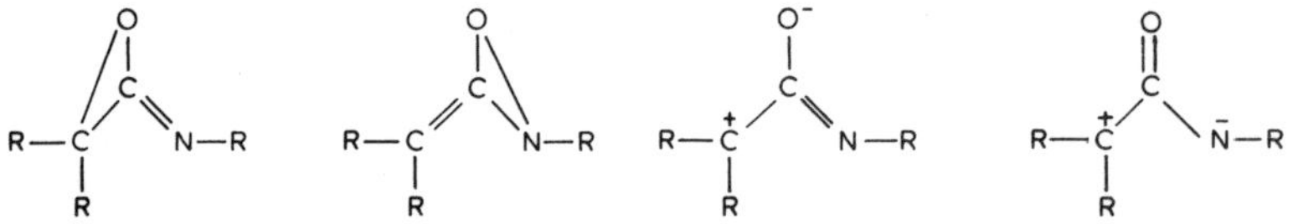

$$
\begin{array}{cccc}
\underset{59}{R-\underset{R}{\overset{O}{\underset{|}{C}}}-C<^{O}_{N-R}} &
\underset{60}{R-\underset{R}{\overset{|}{C}}\overset{O}{\underset{}{C}}N-R} &
\underset{61}{R-\overset{+}{\underset{R}{\underset{|}{C}}}-C<^{O^-}_{N-R}} &
\underset{62}{R-\overset{+}{\underset{R}{\underset{|}{C}}}-\overset{O}{\overset{||}{C}}\overset{}{N}-R}
\end{array}
$$

Although the spectral characteristics of the product obtained from the haloamides **54** and **55** are in accord with the α-lactam structure **56**, evidence exists for another type of intermediate in these systems [36]. The structure-reactivity dependence of aziridones is relevant to this question. (See, however, ref. 38 for a criticism of the following argument.) Thus, 1,3-di-*t*-butylaziridone (**63**) is the least chemically reactive as well as the most thermally stable α-lactam yet isolated [36b]. On the other hand, 1,3,3-triphenylaziridone (**65**) is so unstable that its observation has not proved feasible [36d], while 1-*t*-butyl-3,3-diphenylaziridone (**64**) is of intermediate stability [39].

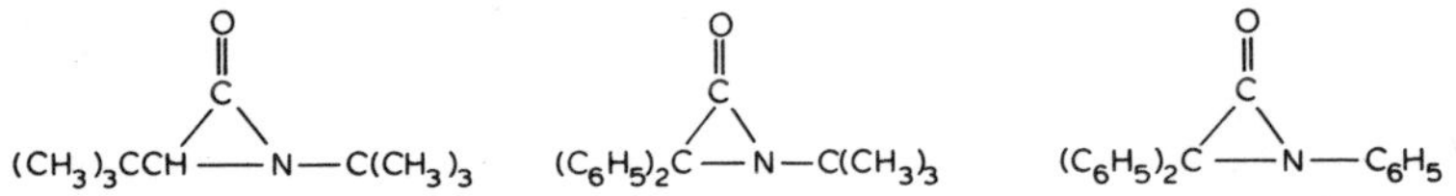

$$
\underset{63}{(CH_3)_3CCH-N-C(CH_3)_3} \qquad
\underset{64}{(C_6H_5)_2C-N-C(CH_3)_3} \qquad
\underset{65}{(C_6H_5)_2C-N-C_6H_5}
$$

Thus, while the low reactivity of **63** points to steric hindrance towards nucleophilic attack, the high reactivities of **64** and **65** imply the operation of an electronic effect, which destabilizes the ground state aziridone in these systems. It is proposed that the acyclic charge-delocalized species **66** intervenes in the reactions of α-lactams, so that any factor which stabilizes such species will enhance reactivity.

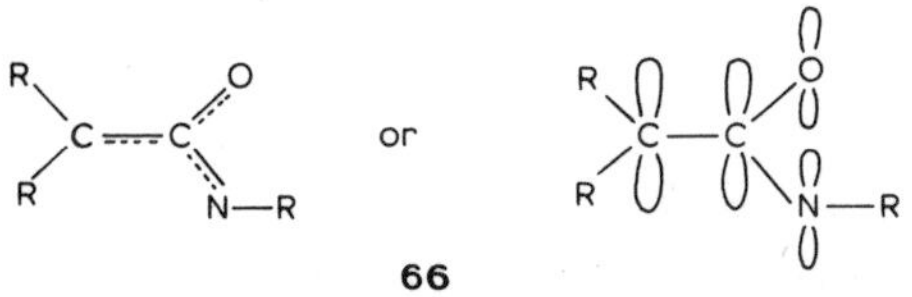

On this basis the *t*-butyl substituent at C3 will cause a decrease in reactivity not only through a steric effect operating in the initial state, but also through the inability of *t*-butyl to stabilize by a hyperconjugative

159

effect the developing positive charge on the *alpha* carbon in the transition state of the reaction.

The course of the thermal decomposition of α-lactams is in agreement with the proposal of the delocalized intermediate **66**. For example, 1-*t*-butyl-3,3-dimethylaziridone (**67**) on heating yields N-*t*-butylmethacrylamide, acetone, and *t*-butyl isocyanide [36b]:

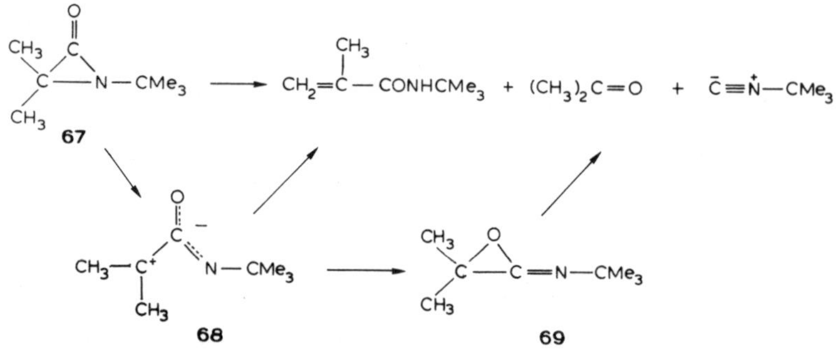

The thermolysis products may be explained through initial formation of the dipolar-ionic intermediate **68**, which on the one hand isomerises to the iminolactone **69** (cf. **59**), this being followed by fragmentation to give ketone plus isonitrile, and on the other hand undergoes rearrangement to give the unsaturated amide. Formation of **68** in this system is favoured by the electron releasing methyl groups at C3. Analogous results are obtained in the thermal decomposition of spiro-α-lactams, e.g. [36c]:

(17)

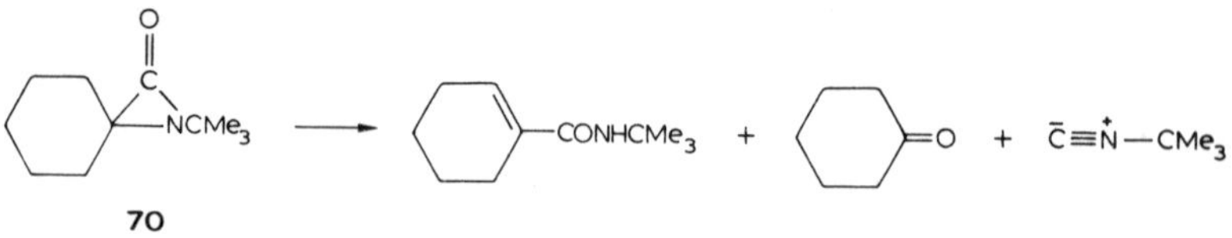

70

In this connection it is noteworthy that the spiro-α-lactam **71** (R = *t*-butyl or 1-adamantyl) is unusually stable towards thermal decomposition [40a]. It may be seen that elimination of a β-hydrogen in **71** is precluded by the adamantane bridgehead structure, so that a reaction pathway analogous to that taken by **70** becomes blocked.

160

(18)

71

However, other α-lactams containing the adamantyl moiety as substituents [40b] are also especially stable, so that presumably a steric effect is also operative. An X-ray crystal-structure analysis [40c] of 1,3-diadamantylaziridone has led to the remarkable finding of a pyramidal configuration at nitrogen, pointing to a large energy barrier to inversion (cf. Chapter 2, Section 1). It will be interesting to see whether this structural feature can be related to reactivity parameters for these α-lactams.

In conclusion of this section it is apparent that there are similarities but also differences in the base-induced reactions of *alpha* halogeno ketones and *alpha* halogeno amides. In both cases there is evidence for the formation of a delocalized dipolar-ionic species, but the significance of such species in the overall reaction sequence is different in the two systems.

4. REARRANGEMENTS OF HOMOENOLATE IONS

In Chapter 4 we introduced the phenomenon of homoenolization, and presented evidence for interaction between a carbonyl function and a non-contiguous carbanionic centre. We now consider some systems for which the homoenolization concept and the occurrence of skeletal rearrangements are intertwined.

Since recognition of homoenolization as a possible mechanism of carbanion stabilization first came about in the work of Nickon and his coworkers, it seems appropriate that our first example be taken from research in their laboratory. Thus it has been reported [41] that the tricyclic ketone (brexan-2-one) **72** on treatment with potassium *t*-butoxide in *t*-butyl alcohol at 185° is transformed into an isomeric ketone, whose structure was established as **74** by independent synthetic evidence. The proposed mechanism of rearrangement is shown in eq. (19):

161

(19)

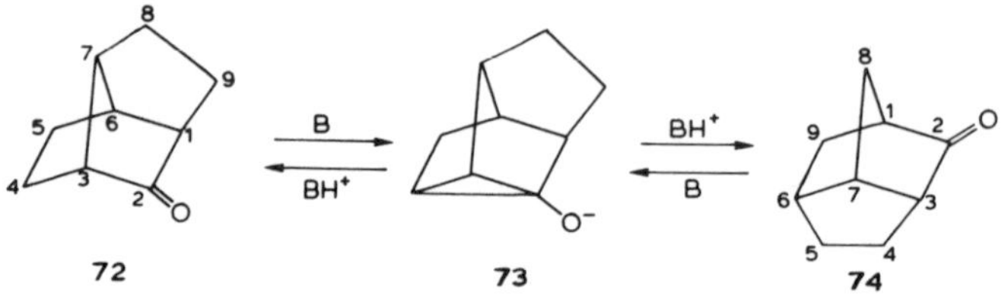

72   73   74

Since the *alpha* hydrogens in 72 are situated on bridgehead carbons (C1 and C3), the formation of a normal enolate ion in the base abstraction process is improbable (cf. Chapter 3, Section 1e). On the other hand the relative spatial configurations in the cyclic system are especially suited for abstraction of the *beta* hydrogen from C4 or C9 to form the homoenolate ion 73. (For brevity a single valence bond structure is used here to describe the homoenolate ion, it being understood that the charge is delocalized over several centers — see also Chapter 4, Section 4.) Opening of the three-membered ring in 73 and protonation leads to formation of 74. Presumably part of the driving force for the rearrangement lies in the relief of steric strain in formation of 74 from its precursor.

The "bird-cage" system provides a fascinating structural environment for the study of nonclassical interactions. It is found that the "half-cage" ketone 75 undergoes base-catalyzed conversion to the less strained half-cage ketone 78 [42]. Kinetic and other evidence indicates that formation of the homoenolate transition state 76 is the rate-limiting process. The highly strained bird-cage alcohol 77 is transformed rapidly under the reaction conditions into the half-cage ketone 78.

75   76

(20)

78   77

162

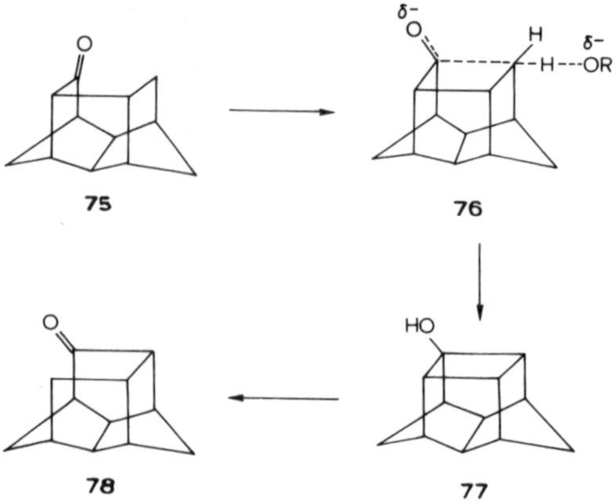

It should be emphasized (see also Chapter 4) that the carbonyl group here *aids* in the proton abstraction step, by delocalization of the developing negative charge (participation by carbonyl in the rate-determining step follows from the unusually high observed reaction rates). Also noteworthy is that abstraction of a γ-hydrogen is involved in this transannular keto-enol isomerisation, in contrast to β-hydrogen abstraction in the other cases of homoenolization discussed.

Homoketonization in a homocubane system has recently been described [43]. The homocubane bridgehead alcohol **79** on treatment with sodium methoxide yields the half-cage ketone **80**, in excellent yield. An alternative possible mode of ring opening, leading to ketone **81**, is not observed, probably because **81** is more strained than **80**.

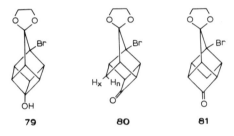

It is interesting that the homoketonization occurs with a high degree of stereospecificity, as shown by the fact that the deuterated ketone obtained on treatment of **79**-acetate with CH_3ONa/CH_3OD has the *endo* proton (H_n) replaced by deuterium. In contrast, the homoketonization of 1-hydroxynortricyclene results in virtually stereospecific *exo* protonation [44a]. The two processes are indicated in structures **82** and **83**, respectively. The different behaviour in the two cases is not yet fully understood.

163

Homoketonization is also observed when the acetate **84** is treated at room temperature with potassium *t*-butoxide in *t*-butyl alcohol [44b]. The product **86** is obtained quantitatively, evidently via homoenolate ion **85**.

(21)

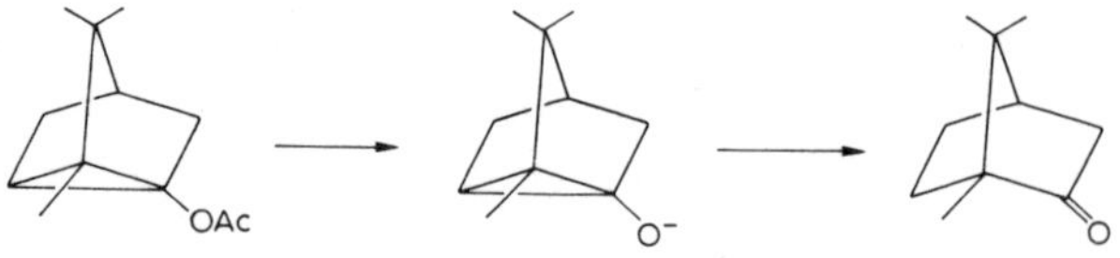
OAc
O⁻
O

84 **85** **86**

An acyclic system for which good evidence for homoenolization is available is the rearrangement of γ-diketone **87** into the isomeric diketone **88**, which is accomplished by the action of sodium methoxide in ether [45]:

(22)

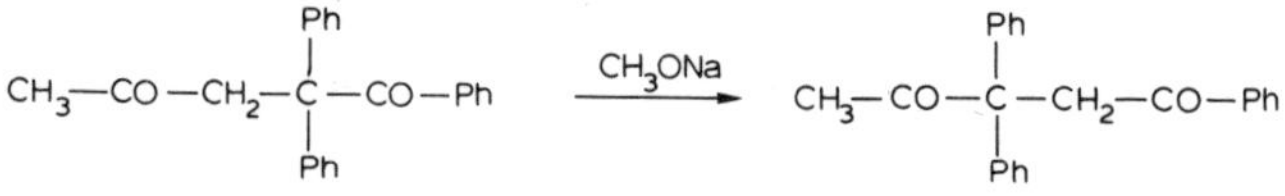
CH3—CO—CH2—C(Ph)(Ph)—CO—Ph
CH3ONa
CH3—CO—C(Ph)(Ph)—CH2—CO—Ph

87 **88**

One possible mechanism for the rearrangement, in which homoenolization is not invoked, is based on two consecutive 1,2-phenyl migrations within enolate ions, (**89** and **91**), as shown in eq. (23):

(23)

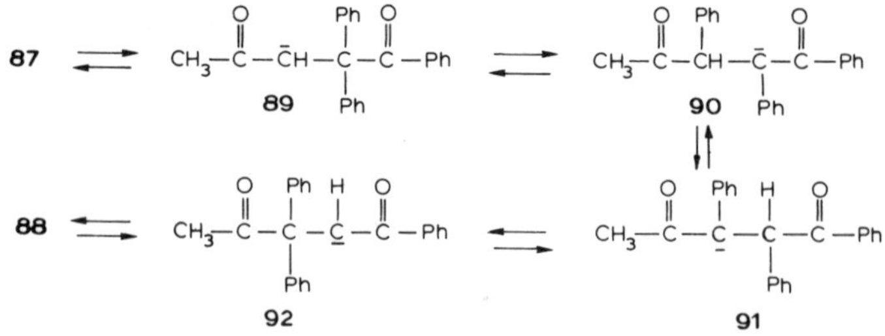
87
89
90
91
92
88
Ph
CH3
Ph

There are, of course, precedents for 1,2-phenyl migrations to carbanionic centres [46]. However, the case for rearrangement of **90** to **91** is less convincing, since a sigmatropic 1,2-hydrogen shift in the anionic system would be symmetry forbidden [47] (see Chapter 6 for further consideration of orbital symmetry in carbanionic rearrangements). Alternatively the transformation $90 \rightarrow 91$ could proceed by a non-concerted intramolecular rearrangement involving a radical ion mechanism, somewhat similar to the Wittig and Stevens rearrangements [48–50]. A third possibility is an intermolecular process, involving consecutive protonation and deprotonation.

A fundamentally different mechanism for the rearrangement $87 \rightarrow 88$ invokes formation of homoenolate species, as shown in eq. (24):

$$
\begin{array}{c}
87 \rightleftharpoons \underset{\substack{\\ \text{Ph} \ \ \text{Ph} \\ \mathbf{89}}}{CH_3-C(=O)-\overset{-}{C}H \ \ C(=O)-Ph} \rightleftharpoons \underset{\substack{\\ \text{Ph} \ \ \text{Ph} \\ \mathbf{93}}}{CH_3-C(=O)-CH-\overset{O^-}{C}-Ph}
\end{array}
$$

(24)

$$
\underset{\substack{\\ \text{Ph} \ \ \text{Ph} \\ \mathbf{94}}}{CH_3-\overset{O^-}{C}-CH-C(=O)-Ph}
$$

$$
\begin{array}{c}
88 \rightleftharpoons \underset{\substack{\\ \text{Ph} \ \ \text{Ph} \\ \mathbf{92}}}{CH_3-C(=O) \ \ \overset{-}{C}H-C(=O)-Ph} \rightleftharpoons \underset{\substack{\\ \text{Ph} \ \ \text{Ph} \\ \mathbf{95}}}{CH_3-C(\overset{-O}{})-CH-C(=O)-Ph}
\end{array}
$$

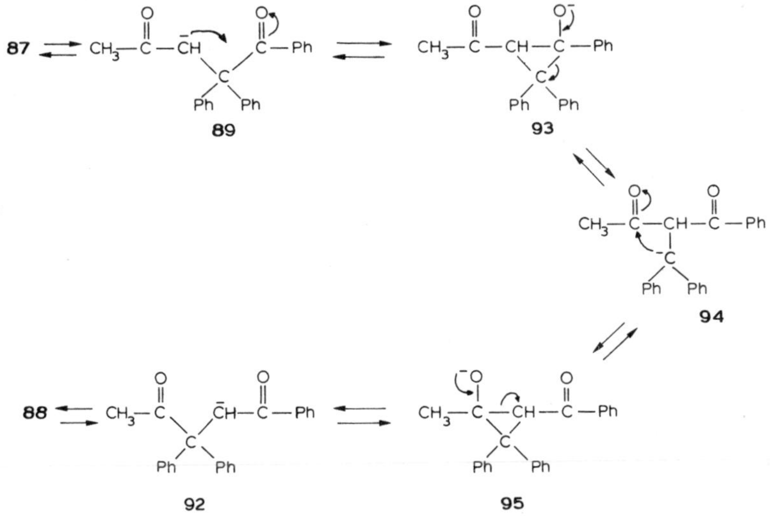

An elegant isotopic tracer experiment was devised to differentiate between the alternative pathways of eqs. (23) and (24). It is apparent that whereas in the mechanism of eq. (23) the integrity of the backbone carbon skeleton remains undisturbed, that is not the case in eq. (24). Thus if one could synthesize the substrate 87 isotopically labelled at C2 with carbon-13, then according to eq. (24) the reaction product 88 should contain the label at C3, since the geminal phenyl groups remain attached to the same carbon. In practice, the isotopic synthesis was achieved for the structurally related γ-diketone 96, and its course of reaction was shown by ^{13}C-NMR spectroscopy to be as follows [51]:

(25) 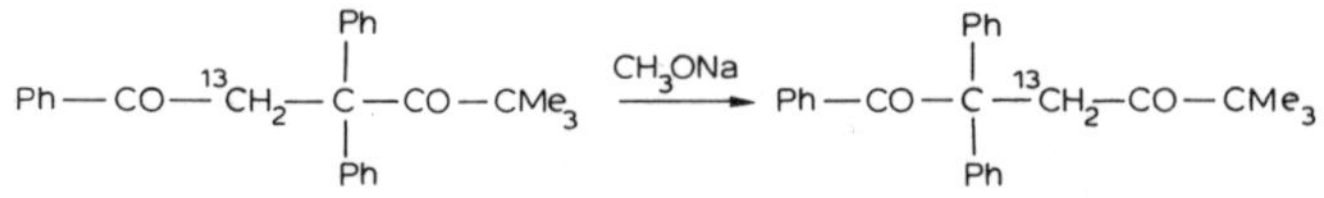 $Ph-CO-^{13}CH_2-\underset{\underset{Ph}{|}}{\overset{\overset{Ph}{|}}{C}}-CO-CMe_3 \xrightarrow{CH_3ONa} Ph-CO-\underset{\underset{Ph}{|}}{\overset{\overset{Ph}{|}}{C}}-^{13}CH_2-CO-CMe_3$

96   **97**

The result of this isotopic experiment is in complete accord with the homoenolate ion mechanism in the γ-diketone rearrangement.

It seems appropriate to conclude this section with the recent report of a homo-Favorskii rearrangement, applicable to the transformation of β-haloketones:

(26) 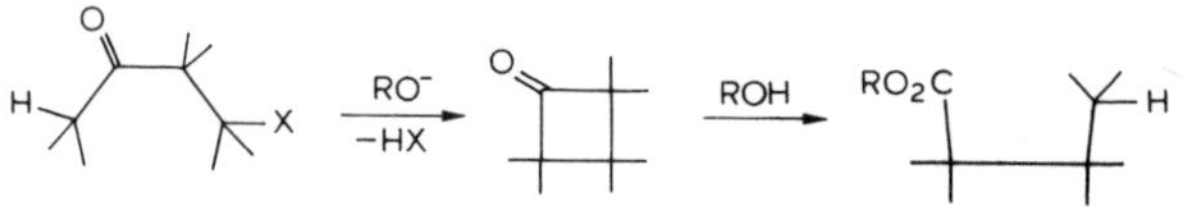   [structure] $\xrightarrow[-HX]{RO^-}$ [structure] $\xrightarrow{ROH}$ [structure]

A number of β-haloketones have been shown to follow the reaction pathway corresponding to eq. (26), an example being the hydroxide ion rearrangement of 3,6-dimethyl-6-dichloromethyl-2-cyclohexenone **(98)** [52]:

(27) 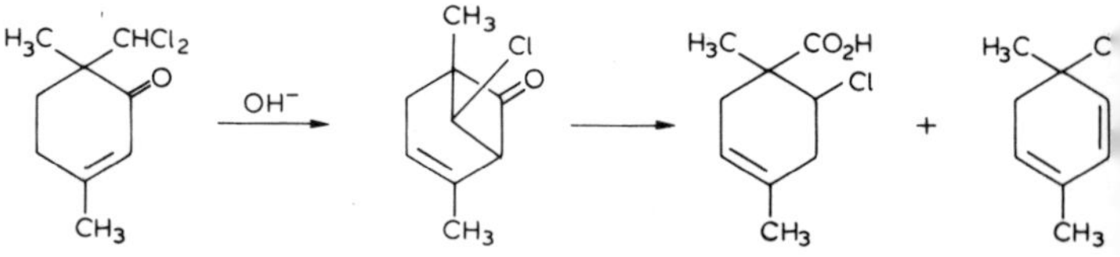   [structure **98**] $\xrightarrow{OH^-}$ [structure **99**] $\longrightarrow$ [structure **100**] $+$ [structure **101**]

98   **99**   **100**   **101**

The reaction products, identified as acids **100** and **101**, plausibly arise via the intermediate cyclobutanone **99**. This postulate receives support from the finding that the related saturated β-haloketone **102** reacts with alkoxide to give the stable cyclobutanone derivative **103** [53]:

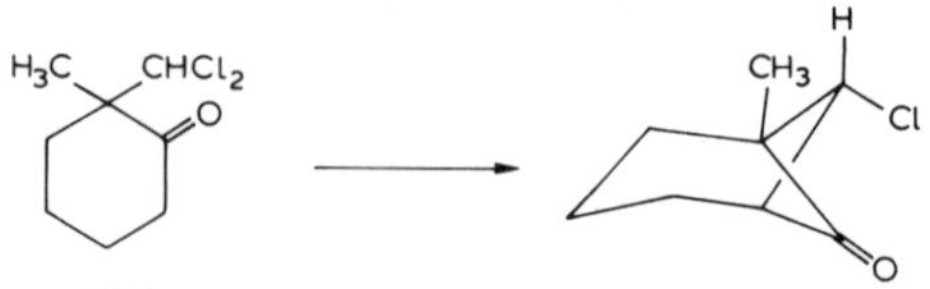

[structure **102**] $\longrightarrow$ [structure **103**]

102   **103**

It can be hoped that future work on the homo-Favorskii process will be as rewarding as have been the studies of the ordinary Favorskii-reaction.

166

REFERENCES

1a D.J. Cram, "Fundamentals of Carbanion Chemistry", Academic Press, New York, 1965.

1b H.E. Zimmerman, in "Molecular Rearrangements", Vol. 1, edited by P. de Mayo, Interscience, New York, 1963.

2 A. Favorskii, J. Russ. Phys. Chem. Soc., 26 (1894) 559; J. Pract. Chem., 51(2) (1895) 533.

3a A.S. Kende, Org. Reactions, 11 (1960) 261.

3b A.A. Akhrem, T.K. Ustynyuk and Yu.A. Titov, Uspek. Khim., 39 (1970) 1560.

3c D. Redmore and C.D. Gutsche, Adv. Alicyclic Chem., 3 (1971) 1.

4 N.J. Turro, Accounts Chem. Res., 2 (1969) 25.

5 F.G. Bordwell, Accounts Chem. Res., 3 (1970) 281.

6 S. Selman and J.F. Eastham, Quart. Rev., 14 (1961) 221.

7a B. Tchoubar and O. Sackur, Compt. Rend., 208 (1939) 1020.

7b E.W. Warnhoff, C.M. Wong and W.T. Tai, J. Amer. Chem. Soc., 90 (1968) 514.

7c W.C. Fong, R. Thomas and K.V. Scherer, Jr., Tetrahedron Lett., 3789 (1971).

7d J.M. Conia and J.R. Salaun, Accounts Chem. Res., 5 (1972) 33.

8 N.J. Turro and W.B. Hammond, J. Amer. Chem. Soc., 87 (1965) 3528; W.B. Hammond and N.J. Turro, ibid., 88 (1966) 2880.

9 R.B. Lotfield, J. Amer. Chem. Soc., 73 (1951) 4707.

10a N.J. Turro, R.B. Gagosian, C. Rappe and L. Knutsson, Chem. Commun., 270 (1969); C. Rappe and L. Knutsson, Acta Chem. Scand., 21 (1967) 2205.

10b C. Rappe, L. Knutsson, N.J. Turro and R.B. Gagosian, J. Amer. Chem. Soc., 92 (1970) 2032.

10c P.S. Wharton and A.R. Fritzberg, J. Org. Chem., 37 (1972) 1899.

10d H. Quast, E. Schmitt and R. Frank, Angew. Chem. Intern. Ed., 10 (1971) 651.

11 R. Breslow, T. Eicher, A. Krebs, R.A. Peterson and J. Posner, J. Amer. Chem. Soc., 87 (1965) 1320.

12a F.G. Bordwell, R.R. Frame, R.G. Scamehorn, J.G. Strong and S. Meyerson, J. Amer. Chem. Soc., 89 (1967) 6704.

12b F.G. Bordwell and R.G. Scamehorn, J. Amer. Chem. Soc., 90 (1968) 6751.

12c F.G. Bordwell, R.G. Scamehorn and W.R. Springer, J. Amer. Chem. Soc., 91 (1969) 2087.

12d F.G. Bordwell and M.W. Carlson, J. Amer. Chem. Soc., 92 (1970) 3370.

12e F.G. Bordwell and J. Almy, J. Org. Chem., 38 (1973) 575.

13 H.R. Nace and B.A. Olsen, J. Org. Chem., 32 (1967) 3438.

14a E. Buncel and A.N. Bourns, Can. J. Chem., 38 (1960) 2457.

14b R. Breslow, Tetrahedron Lett., 399 (1964).

15 J.G. Burr, Jr. and M.J.S. Dewar, J. Chem. Soc., 1201 (1954); N. Bodor, M.J.S. Dewar, A. Harget and E. Haselbach, J. Amer. Chem. Soc., 92 (1970) 3854.

16 H.O. House and W.F. Gilmore, J. Amer. Chem. Soc., 83 (1961) 3980.

17 A.W. Fort, J. Amer. Chem. Soc., 84 (1962) 2620, 2625.

167

18 R.C. Cookson and M.J. Nye, J. Chem. Soc., 2009 (1965); R.C. Cookson, M.J. Nye and G. Subrahmanyam, J. Chem. Soc. (C), 473 (1967).

19a R. Hoffmann, J. Amer. Chem. Soc., 90 (1968) 1475.

19b A. Liberles, S. Kang and A. Greenberg, J. Org. Chem., 38 (1973) 1922.

20 J.M. Pochan, J.E. Baldwin and W.H. Flygare, J. Amer. Chem. Soc., 90 (1968) 1072; ibid., 91 (1969) 1896.

21 R.L. Camp and F.D. Greene, J. Amer. Chem. Soc., 90 (1968) 7349.

22 J.K. Crandall and W.H. Machleder, J. Amer. Chem. Soc., 90 (1968) 7347; J.K. Crandall, W.H. Machleder and S.A. Sojka, J. Org. Chem., 38 (1973) 1149.

23 D.B. Sclove, J.F. Pazos, R.L. Camp and F.D. Greene, J. Amer. Chem. Soc., 92 (1970) 7488.

24 S.S. Edelson and N.J. Turro, J. Amer. Chem. Soc., 92 (1970) 2770.

25 G. Stork and I.J. Borowitz, J. Amer. Chem. Soc., 82 (1960) 4307.

26 H.O. House and G.A. Frank, J. Org. Chem., 30 (1965) 2948; H.O. House and F.A. Richey, Jr., ibid., 32 (1967) 2151.

27 A. Skrobek and B. Tchoubar, Compt. Rend., C 263 (1966) 80.

28 E.W. Garbisch, Jr. and J. Wohllebe, Chem. Commun., 306 (1968).

29 F.G. Bordwell and R.G. Scamehorn, J. Amer. Chem. Soc., 93 (1971) 3410.

30a J.P. Dirlam, L. Eberson and J. Casanova, J. Amer. Chem. Soc., 94 (1972) 240.

30b A.J. Fry and R. Scoggins, Tetrahedron Lett., 4079 (1972).

31 G.R. Treves, H. Stange and R.A. Olofson, J. Amer. Chem. Soc., 89 (1967) 6257.

32 H.O. House and W.F. Gilmore, J. Amer. Chem. Soc., 83 (1961) 3972.

33 W. Baker and R. Robinson, J. Chem. Soc., 1798 (1932).

34 S.A. Achmad and G.W.K. Cavill, Australian J. Chem., 16 (1963) 858.

35a H.E. Baumgarten, R.L. Zey and U. Krolls, J. Amer. Chem. Soc., 83 (1961) 4469.

35b H.E. Baumgarten, J. Amer. Chem. Soc., 84 (1962) 4975.

35c H.E. Baumgarten, J.F. Fuerholzer, R.D. Clark and R.D. Thompson, J. Amer. Chem. Soc., 85 (1963) 3303.

36a J.C. Sheehan and I. Lengyel, J. Amer. Chem. Soc., 86 (1964) 1356.

36b J.C. Sheehan and J.H. Beeson, J. Amer. Chem. Soc., 89 (1967) 362.

36c J.C. Sheehan and J.H. Beeson, J. Amer. Chem. Soc., 89 (1967) 366.

36d I. Lengyel and J.C. Sheehan, Angew. Chem. Intern. Ed., 7 (1967) 25.

37 S. Sarel, B.A. Weissman and Y. Stein, Tetrahedron Lett., 373 (1971).

38 E.R. Talaty and C.M. Utermoelen, Tetrahedron Lett., 3321 (1970).

39 H.E. Baumgarten, R.D. Clark, L.S. Endres, L.D. Hagemeier and V.J. Elia, Tetrahedron Lett., 5033 (1967).

40a E.R. Talaty and A.E. Dupuy, Jr., Chem. Commun., 790 (1968).

40b E.R. Talaty, J.P. Madden and L.H. Stekoll, Angew. Chem. Intern. Ed., 10 (1971) 753.

40c A.H.-J. Wang, I.C. Paul, E.R. Talaty and A.E. Dupuy, Jr., Chem. Commun., 43 (1972).

41 A. Nickon, H. Kwasnik, T. Swartz, R.O. Williams and J.B. DiGiorgio, J. Amer. Chem. Soc., 87 (1965) 1615.

42a R. Howe and S. Winstein, J. Amer. Chem. Soc., 87 (1965) 915.

42b T. Fukunaga, J. Amer. Chem. Soc., 87 (1965) 916.

43   A.J.H. Klunder and B. Zwanenburg, Tetrahedron, 29 (1973) 1683.

44a A. Nickon, J.L. Lambert, R.O. Williams and N.H. Werstiuk, J. Amer. Chem. Soc., 88 (1966) 3354.

44b D.H. Hunter, A.L. Johnson, J.B. Stothers, A. Nickon, J.L. Lambert and D.F. Covey, J. Amer. Chem. Soc., 94 (1972) 8582.

45   P. Yates, G.D. Abrams and S. Goldstein, J. Amer. Chem. Soc., 91 (1969) 6868; P. Yates, G.D. Abrams, M.J. Betts and S. Goldstein, Can. J. Chem., 49 (1971) 2850.

46a E. Grovenstein, Jr. and L.C. Rogers, J. Amer. Chem. Soc., 86 (1964) 854; E. Grovenstein, Jr. and Y.-M. Cheng, J. Amer. Chem. Soc., 94 (1972) 4971.

46b H.E. Zimmerman and A. Zweig, J. Amer. Chem. Soc., 83 (1961) 1196.

47   R.B. Woodward and R. Hoffmann, "The Conservation of Orbital Symmetry", Academic Press, New York, 1970.

48   U. Schöllkopf, U. Ludwig, G. Ostermann and M. Patsch, Tetrahedron Lett., 3415 (1969).

49a J.E. Baldwin, J. DeBernardis and J.E. Patrick, Tetrahedron Lett., 353 (1970).

49b P.T. Lansbury, V.A. Pattison, J.D. Sidler and J.B. Bieber, J. Amer. Chem. Soc., 88 (1966) 78.

50   A.R. Lepley and A.G. Giumanini in "Mechanisms of Molecular Migrations", Vol. 3, edited by B.S. Thyagarajan, Wiley-Interscience, New York, 1970.

51   M.J. Betts and P. Yates, J. Amer. Chem. Soc., 92 (1970) 6982; P. Yates and M.J. Betts, J. Amer. Chem. Soc., 94 (1972) 1965.

52   E. Wenkert, P. Bakuzis, R.J. Baumgarten, C.L. Leicht and H.P. Schenk, J. Amer. Chem. Soc., 93 (1971) 3208.

53   E. Wenkert, P. Bakuzis, R.J. Baumgarten, D. Doddrell, P.W. Jeffs, C.L. Leicht, R.A. Mueller and A. Yoshikoshi, J. Amer. Chem. Soc., 92 (1970) 1617.

Chapter 6

ORBITAL SYMMETRY CONTROL IN CARBANION REARRANGEMENTS

1. INTRODUCTION

The recognition of orbital symmetry control as the major factor governing reactivity and stereochemistry in pericyclic reactions came about in 1965 as a result of a series of communications by Woodward and Hoffmann [1], and, almost simultaneously, Longuet-Higgins and Abrahamson [2]. There followed in rapid succession a number of publications by other workers [3–10], in which the characteristic stereoselectivity observed in such multi-centred reactions was treated by theoretical methods which are complementary to this concept.

The theory of the conservation of orbital symmetry, like other scientific theories, explained and correlated a number of known facts and predicted the discovery of others. Reactions which could thus be classified included cycloadditions, fragmentations, and numerous rearrangements of hydrocarbons and of uncharged functional organic compounds, as well as carbonium ion processes. In the area of carbanion chemistry, only a limited number of observations had been recorded at that time which could be classified as symmetry-controlled reactions. However, in the intervening years an increasing number of carbanion reactions have been discovered which fit into this category.

The purpose of this chapter is to bring together and discuss those carbanion rearrangements which fall under the common theme of orbital symmetry-controlled reactions. The concepts used will generally be illustrated by means of examples from non-carbanion processes, including hydrocarbon and carbonium ion rearrangements. Theoretical principles will be presented within a non-mathematical framework. Since the subject is developing at a rapid rate, coverage is not intended to be comprehensive.

Carbanion rearrangements in general have been collectively discussed by Zimmerman [11] and by Cram [12],and for a general background the reader is referred to those accounts.

2. ELECTROCYCLIC REARRANGEMENTS

(a) Rearrangement of a dienyl anion

The rearrangement which forms the basis of our discussion in this section is shown in eq. (1). It was reported in 1969 by Bates and McCombs [13] and almost simultaneously by Kloosterziel and Van Drunen [14, see also 15]. The overall reaction is a transformation of 1,3-cyclooctadiene, 1, to *cis*-bicyclo[3.3.0]oct-2-ene, 4, in the presence of strong base, *viz.* butyllithium in tetrahydrofuran/hexane [13] or potassium amide in liquid ammonia [14]. 1,4-Cyclooctadiene reacts in a similar manner.

(1)

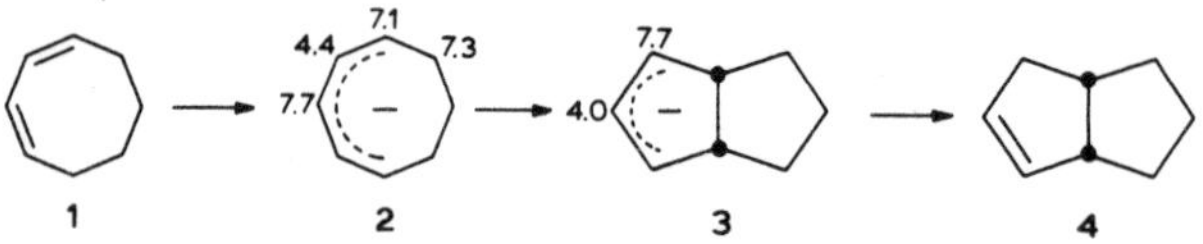

In these basic systems either reactant gives rise initially to the delocalized cyclooctadienyl anion 2, which was characterized by means of its NMR spectrum. The NMR parameters for this species at $-78°$ (chemical shifts given in ppm) are as shown[*]. Between $-78°$ and $-20°$ reversible changes in the NMR spectrum are observed, which can be ascribed to conformational changes; models show that not even the five sp^2 carbons of 2 can be coplanar. Irreversible changes in the NMR spectrum occur when a solution of 2 is allowed to warm to room temperature. In the THF/hexane medium signals characteristic of the formation of 3 appear, with a first order rate constant of 8.7×10^{-3} min^{-1} ($T_{1/2}$ 80 min) for this process at $35°$. (The reaction was conveniently followed by observing the rate of disappearance of the quartet at $\tau 4.4$ in 2 and the emergence of the triplet at $\tau 4.0$ due to 3.) However, subsequent protonation of 3 takes place in this medium, yielding 4, which was also characterized by NMR as well as by other techniques. In the liquid ammonia medium signals due to 3 cannot be observed at all, since 3 is now immediately protonated. This indicates that the bicyclooctene product 4 is a weaker acid than ammonia, whereas the parent cyclooctadienes are stronger acids.

---

[*]For a discussion of the charge distribution in delocalized anions see ref. (16).

172

The pentadienylic anion **2** can be represented as a resonance hybrid of the contributing valence bond structures **2a**, **2b**, and **2c**:

The allylic ion **3** may similarly be represented by the canonical forms **3a** and **3b**:

It should be noted, before proceeding with the discussion, that in many of the cases dealt with in these sections the carbanions are stable with respect to the environment and can hence be present in high concentration. This contrasts with, say, the Favorskii rearrangement where carbanionic species are formed as unstable reaction intermediates, present in small concentration (Chapter 5).

The approach which will be taken in our discussion of the fundamentals of the skeletal rearrangement, $2 \rightarrow 3$, is as follows. We shall consider, in the first place, rearrangement in an electronically equivalent uncharged system and cover some of the theoretical principles which govern that reaction; then we shall return to the process represented by $2 \rightarrow 3$, and also examine some related anionic processes. Application to cationic rearrangements will also be illustrated. The determining factor in the present approach is, of course, that anionic rearrangements constitute one domain in the application of principles which historically were proposed for systems in the other major areas of organic chemistry. However, as will be seen, the principles to be considered are independent of the charge type.

(b) Orbital symmetry and the rearrangement of 1,3,5-trienes

In principle the cyclooctadienyl–bicyclo-octenyl carbanion isomerisation, $2 \rightarrow 3$, bears analogy to isomerisation of the electrically neutral hydrocarbon 1,3,5-cyclooctatriene to the bicyclo[4.2.0]diene, **5→6**, and the reverse process:

(2)

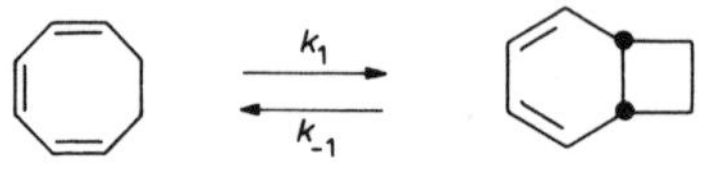

5   **6**

The isomerisation in eq. (2) occurs on heating alone at ca. $100°$ ($\Delta H_1^{\ddagger} = 25.7$ kcal/mole, $\Delta S_1^{\ddagger} = -3$ e.u.; $\Delta H_{-1}^{\ddagger} = 24.7$, $\Delta S_{-1}^{\ddagger} = -2$), with the monocyclic olefin predominating at equilibrium [17]. It will be seen, moreover, that the isomerisation **5⇌6** is a specific example of thermally induced interconversions of the type **7⇌8**, which are well documented for a variety of structural variations in the X moiety:

(3)

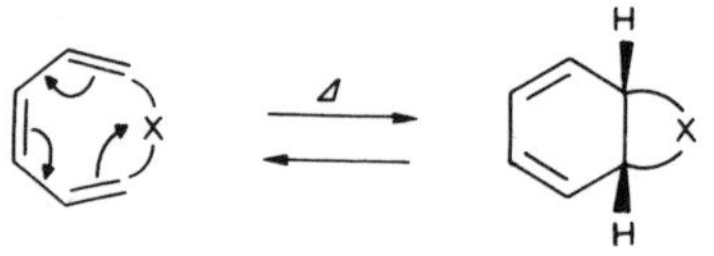

7   **8**

Systems for which this isomerisation has been studied include those with $X = CH_2$ (or CR_2), $(CH_2)_2$, $(CH_2)_3$, and CH=CH for carbocyclic series, and $X = O$, NR, SO_2 for heterocyclic series [18–24]. The various structural modifications do, of course, affect the reactivities and positions of equilibria within the series of cyclic systems. The parent case of the transformation **7⇌8** is that in which X is absent, so that one is dealing simply with the interconversion of an acyclic triene and a monocyclic diene:

(4)

9   **10**

This reaction is known to occur for a variety of substituents, R, including hydrogen. For example, when $R = CH_3$, reaction occurs readily at $130°$, and the equilibrium lies almost completely over to the cyclic form, *cis*-1,2-dimethylcyclohexadiene [25].

The transformations represented by equations (2)–(4) may be viewed as intra-annular valence bond isomerisations, or alternatively as intra-

174

molecular cycloadditions. Thus, a molecule with n π-electrons is converted into a ring system with $n-2$ π-electrons and one new σ-bond. The reactions are uncatalyzed thermal processes in which the various bond changes occur in a concerted manner. They can be classified, together with a large number of other pericyclic reactions, as being governed by the principle of orbital symmetry, enunciated by Woodward and Hoffmann [1], and by others [2–10] in somewhat alternative forms. In brief, the principle states that *for a reaction to occur in one concerted step the participating orbitals of the reactant must transform by way of the transition state into the product orbitals with conservation of orbital symmetry*. Many different reaction types are encompassed by this principle (see refs. 26–30 for applications).

The reaction type which we are considering may be termed an *electrocyclic transformation*. Thus the isomerisations $7 \rightarrow 8$ or $9 \rightarrow 10$ are

Table 1

ELECTROCYCLIZATION IN SOME CHARGED AND NEUTRAL SYSTEMS

Reaction		Δ	$h\nu$
(cation, butadienyl)	$\rightleftharpoons$ (cyclopropenyl$^+$)	dis	con
(anion, butadienyl)	$\rightleftharpoons$ (cyclopropenyl$^-$)	con	dis
(cation, pentadienyl)	$\rightleftharpoons$ (cyclopentenyl$^+$)	con	dis
(butadiene)	$\rightleftharpoons$ (cyclobutene)	con	dis
(hexatriene)	$\rightleftharpoons$ (cyclohexadiene)	dis	con
(anion, pentadienyl)	$\rightleftharpoons$ (cyclopentenyl$^-$)	dis	con
(anion, heptatrienyl)	$\rightleftharpoons$ (cycloheptadienyl$^-$)	con	dis

175

essentially electrocyclic processes in which a single bond is formed between the termini of a linear conjugated system containing 6 π-electrons. For such systems, which fall under the $4n+2$ π-electron grouping ($n = 0, 1, 2 \ldots$), the thermally effected process is orbital symmetry-allowed and predicted to occur in disrotatory fashion by a low energy pathway. However, in the case of $4n$ π-electron systems, the thermal electrocyclic isomerisation will occur in conrotatory manner. The relationships are reversed for photochemically induced reactions. These predictions are summarised for several $4n$ and $4n+2$ π-electron systems in Table 1 for both the thermal and the photochemical electrocyclization processes.

It is important to note that, as a result of orbital symmetry control, the electrocyclizations proceed stereospecifically, and the thermal and photochemical reactions lead to different stereochemical results. We may illustrate for the 1,6-dimethylcyclohexadienes [25]:

(5) m_2 [structure: diene with CH$_3$ groups] $\xrightarrow[\Delta]{\text{disrotatory}}$ [structure: product with CH$_3$, H groups]

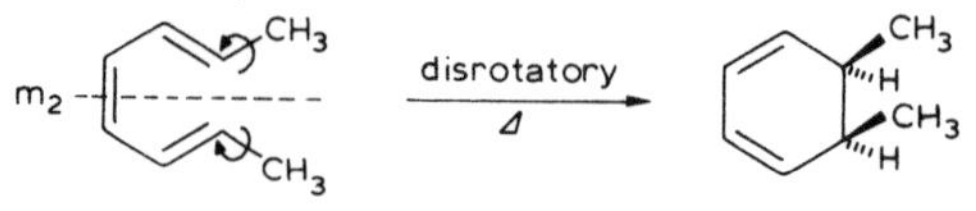

(6) C_2 [structure: diene with CH$_3$ groups] $\xrightarrow[h\nu]{\text{conrotatory}}$ [structure: product with CH$_3$, H, CH$_3$ groups]

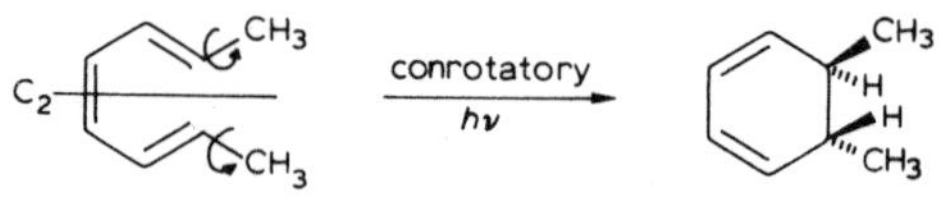

In eq. (5) the groups attached to the terminal carbons rotate in opposite directions, or counter-phase, while in eq. (6) these groups rotate in the same direction, or in-phase. The symmetry element which is maintained in the course of reaction (5) is a mirror plane (m_2) while in reaction (6) it is a two-fold axis of symmetry (C_2). In practice it is found that the thermal reaction (eq. (5)) gives rise to >99.9 of the *cis* isomer, even though the *trans* isomer is thermodynamically the more stable. Conversely, the photochemical reaction (eq. (6)) gives rise only to the *trans* isomer.

A simplified orbital representation for reaction (2) is given in Figure 1. Clearly for σ-bond formation to occur between the two termini (C1 and C6), the π-lobes of the terminal atoms must interact. Overlap of these π-lobes becomes feasible through a concerted rotation about the C1—C2 and C5—C6 bonds. It should be noted, however, that a bonding interaction will result only if there is no sign inversion (node) in the region of overlap. In the present case involving 6 π-electrons between the termini, this rotation must occur by a disrotatory mode, so that the product is formed in a stereospecific manner as the *cis-cis* isomer.

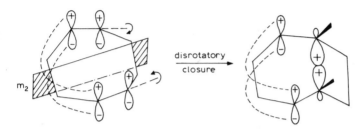

disrotatory
closure

Figure 1. Orbital illustration of σ-bond formation in the disrotatory ring-closure of 1,3,5-cyclooctatriene. Also shown is the symmetry plane which characterizes the transition state for cyclization.

It will be seen that the reacting orbital indicated schematically in Figure 1 is the highest occupied molecular orbital (HOMO)[*]. The lobes on the terminal carbons of 4n + 2 π-electron systems will be in phase and σ-bond formation will occur via the disrotatory mode of ring closure. In contrast, a polyene with a 4n π-electron system will have the terminal lobes out of phase in the HOMO, so that a bonding interaction of these orbitals will require conrotatory closure. The above considerations are applicable to ground state transformations. In photochemical (first excited state) reactions an electron is promoted to the lowest unoccupied molecular orbital (LUMO), which has opposite symmetry to the ground state HOMO. In this situation a bonding interaction between the termini will result from conrotatory closure in 4n + 2 π-electron systems, and disrotatory closure in 4n π-electron systems (Table 1).

A more complete molecular orbital picture of the electrocyclic transformation results from consideration of the *correlation diagram* for the

---

[*]It is recognized [2] that the representation of Figure 1, which emphasizes formation of the σ-bond of the diene as originating from the HOMO of the triene system, does not readily provide a fully consistent pathway for the reverse process. This deficiency is remedied by the correlation diagram approach (*vide infra*) which utilizes the various occupied and unoccupied MO's of the reactant *and* of the product system.

process [2]. This is given in Figure 2 for the structurally simpler case of the thermal isomerisation **9⇌10** with R = H. (It should be noted, however, that the two-carbon bridge in **5**, joining the ends of the conjugated system, provides no constraint on disrotatory closure.) On the left hand

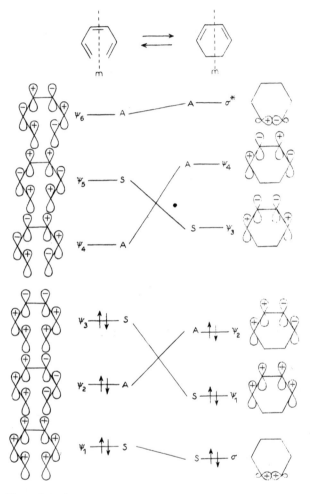

Figure 2. Molecular orbital correlation diagram for disrotatory ring closure of 1,3,5-hexatriene to 1,3-cyclohexadiene and the reverse process (energy levels not drawn to scale).

178

side are shown the molecular orbitals for triene **9** in order of increasing energy, the occupied levels in the ground state being ψ_1, ψ_2, and ψ_3. The orbitals are designated as symmetric (S) or antisymmetric (A) with respect to the symmetry plane which characterizes the transition state in disrotatory closure. On the right hand side are shown the corresponding molecular orbitals for the cyclic structure **10**. It is seen that disrotatory ring closure results in a smooth correlation of reacting bonding levels with product bonding levels, while maintaining the symmetry relationships of corresponding states. A low energy transition state for this process will result, and so the reaction is said to be symmetry-allowed.

We may indicate, very briefly, what the consequences of conrotatory ring closure would be. The symmetry element corresponding to this mode would be the C_2 axis of rotation. Now the symmetries of the molecular orbitals (focussing on the three lowest energy levels which are occupied in the ground state) are: for the acyclic triene (ψ_1, ψ_2, ψ_3), A, S, A; and for the cyclic diene (σ, ψ_1, ψ_2), S, A, S. In this case there is no longer a smooth correlation between the ground state occupied orbitals of the reactant and those of the product, so that a high energy transition state is predicted for the process. However, if we consider the photochemical reaction, as represented by eq. (6), then an electron in the triene is promoted from ψ_3 to ψ_4, and it follows that the conrotatory process becomes orbital symmetry-allowed, since now a smooth correlation between occupied orbitals is possible.

This completes our brief treatment of the theoretical principles relevant to electrocyclic transformations. However, it should be mentioned that several other theoretical models which have been proposed also explain successfully the experimental findings in pericyclic reactions, for example the *frontier orbital approach*, and the *aromatic transition state approach*. These concepts have been treated in several excellent reviews [3–5].

Finally, it should be stated that the very concepts "orbital-symmetry-allowed" and "orbital-symmetry-forbidden", and the relationship to energetics of concerted and non-concerted processes, are currently receiving critical evaluation, by both theoretical and experimental methods [31, 32].

(c) Carbanion electrocyclizations

Now we return to the carbanion cyclization reaction of equation (1). For illustrative purposes we represent the reaction as occurring between the valence bond structures **2a** and **3a**:

(1a)

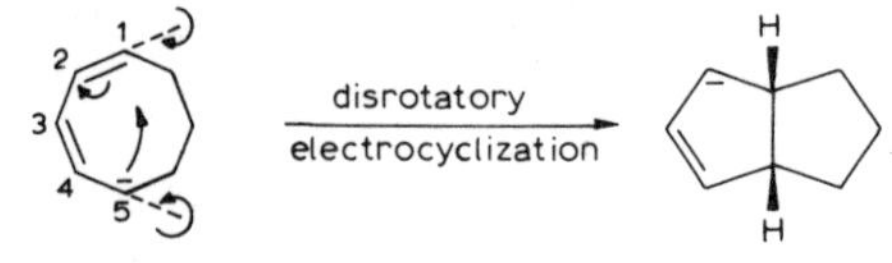

disrotatory electrocyclization

2a   **3a**

Since it is obvious that we have here 6 π-electrons between the termini, our previous discussion requires that the electrocyclization occur by the disrotatory mode. Thus, the product is formed in a stereospecific manner, as the *cis, cis*-isomer.

The correlation diagram applicable to this process may readily be constructed. The illustrations in Figure 3 are limited to the pentadienylic-pentenylic system, for simplicity, it being understood that the three-carbon bridge is attached to the termini. The five molecular orbitals of the pentadienyl system are shown on the left, the lower three levels doubly occupied while the upper two (antibonding) levels are vacant in the ground state. On the right hand side are shown the molecular orbitals of the product, consisting essentially of a σ-bond and the allyl π-system superposed. For disrotatory ring closure a plane of symmetry must be maintained throughout the reaction; accordingly the orbitals are labelled as symmetric or antisymmetric. It is seen that there is a smooth correlation between the bonding orbitals of the reactant and those of the product, so that the thermal process is expected to occur with a low activation energy. One can readily show that these relationships would not hold for the conrotatory process, which is thus orbital symmetry-disallowed.

Let us now generalize the carbanion electrocyclization, within the 6 π-electron system. Just as the corresponding reaction of the uncharged species depicted in eq. (2) may be generalized as in eq. (3), so the general pentadienylic-pentenylic carbanion electrocyclization may be represented by eq. (7):

(7)

11   **12**

180

The relative sizes of the ring systems become the dominant factor in consideration of relative stabilities of the two isomers. With $X = CH_2$, or CR_2, the bicyclic structure is expected to be relatively unstable as a result of ring strain. Accordingly, it has been observed that 6,6-diphenyl-bicyclo[3.1.0]hex-2-ene (13) undergoes isomerisation to the 1,3- and 1,4-hexadienes (eq. (8)) when heated with potassium t-pentoxide in t-pentyl

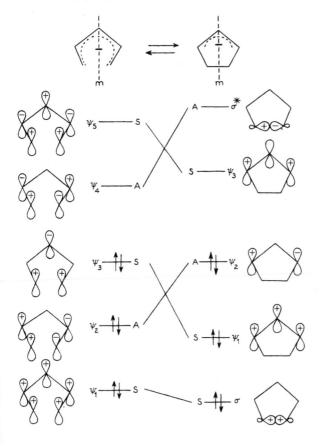

Figure 3. Molecular orbital correlation diagram for disrotatory ring closure of a pentadienyl anion to a cyclopentenyl anion and the reverse process (energy levels not drawn to scale).

alcohol [33]. Evidently the initially formed bicyclohexenyl anion 14 undergoes facile valence isomerisation to the cyclohexadienyl anion 15.

(8)

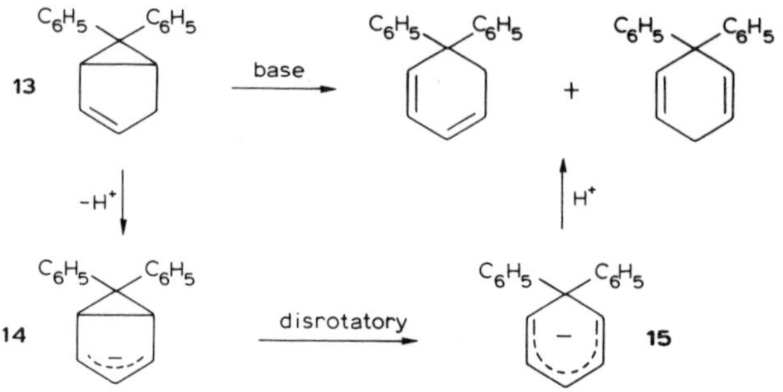

It may be noted that the present case involves only a very small concentration of carbanion unlike the situation in reaction 1. It appears also that 14, once formed, is immediately transformed into 15; this conclusion follows from the observation that recovery of 13 from deuterated (ROD) medium under the reaction conditions revealed no incorporation of deuterium into the substrate.

The cycloheptadienyl anion, structure 11 with $X = (CH_2)_2$, has been prepared in liquid ammonia by proton abstraction from cyclohepta-1,3-diene [14]. Isomerisation to the bicycloheptenyl anion (12, $X = (CH_2)_2$) was not detected, indicating that even that structure is relatively strained with respect to the monocyclic isomer. The situation in the case of $X = (CH_2)_3$ has already been considered in detail (eq. (1)).

Electrocyclization of the parent unsubstituted pentadienyl anion has not been observed [34]. However, an acyclic diaza-derivative, on examination under both thermal and photochemical conditions, has yielded results which are highly instructive, and will be considered in detail [35].

Hydrobenzamide (16) reacts with phenyllithium in THF solution at $-70°$ to yield a blue solution (λ_{max} 555, 592 nm). On warming to room temperature this is transformed to a pale yellow solution, which on treatment with acetic acid yields the *cis*-amarine, 20, only. This result is in accord with initial formation of the delocalized carbanion 17 (blue), which undergoes disrotatory ring closure to yield 18 (pale yellow), with a *cis* geometry imposed in the electrocyclization. On the other hand, irradiation of **17** in the 577–579 nm region followed by the acetic acid treatment yields the *trans* isomarine **21**, in increasing proportions the lower the temperature. This indicates that on irradiation carbanion **17** undergoes conrotatory cyclization to **19**, with a *trans* geometry. The relative thermodynamic stabilities of **20** and **21** were established by treatment with base (*t*-BuOK/*t*-BuOH at 100°); the equilibrium mixture contained 4% of **20** and 96% of **21**, irrespective of whether the starting material was **20** or **21**. Thus the thermal and photochemical ring closures of the anion **17** are shown to proceed by different stereochemical pathways, the former resulting in the thermodynamically less stable product.

SCHEME 1

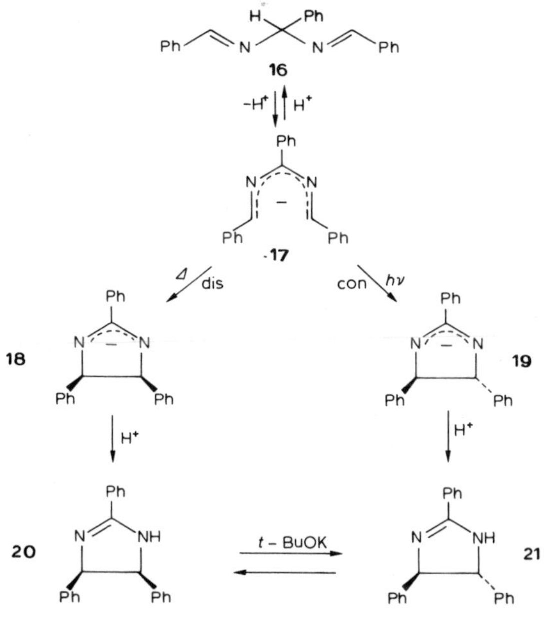

Cyclization in the hydrobenzamide system is even more noteworthy in view of the observation [36] that the anion **23**, derived from 2,5-dihydrofuran **22**, undergoes a facile ring opening to the butadienolate anion **24**, a six π-electron system. Analogous processes are observed for the sulfur-containing heterocyclic derivative corresponding to **22**. Thus there is apparently no tendency for anions such as **24** to revert to cyclic forms. This may indicate that the charge in **24** is concentrated largely on the electronegative oxygen and that the open anion is thereby stabilized in this case.

(9)

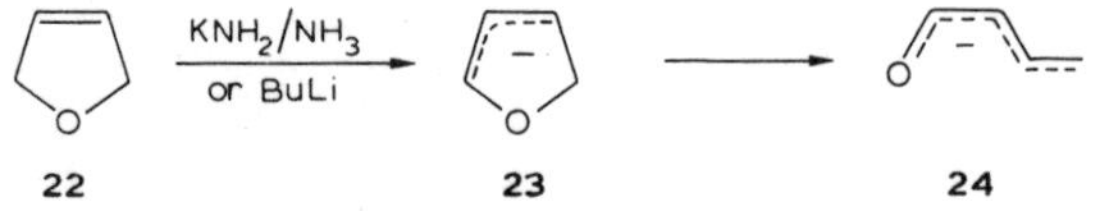

| **22** | **23** | **24** |

Hitherto we have discussed electrocyclizations of the 6 π-electron pentadienylic-pentenylic system. Now we consider cyclization in a 8 π-electron case, and finally deal with the 4 π-electron system.

The 8 π-electron heptatrienyl anion **27** is generated by proton abstraction from 1,3,5-heptatriene (**25**) or 1,3,6-heptatriene (**26**) by the action of strong base (KNH$_2$ in NH$_3$ or BuLi in THF-hexane) at low temperature [37]:

(10)

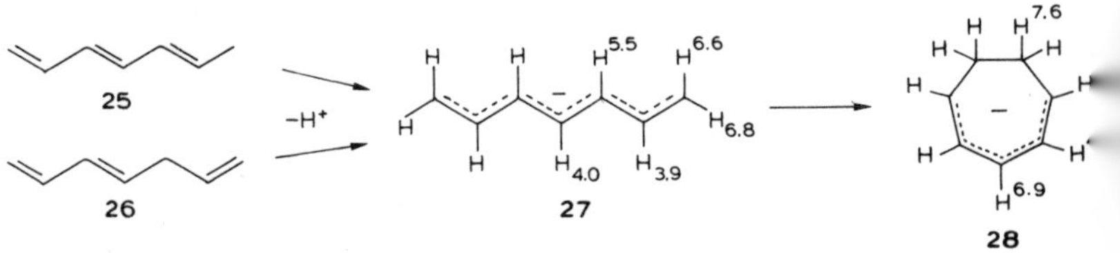

When a solution of anion **27** is allowed to rise in temperature, characteristic changes occur in the NMR spectrum resulting from isomerisation to **28**. In the liquid ammonia medium the half-life for conversion to **28** is 60 min at 0°C. The transformation **27**→**28** clearly falls within the scope of electrocyclization of 4n π-electron systems, for which the thermal process occurs in a conrotatory manner. Similar results are obtained with a number of substituted hexatrienes. It is interesting that cyclization of **27** requires initial conversion from the experimentally more stable all-*trans* (extended-W) conformation to the helical all-*cis* (U-shaped) conformation **27a** [38]:

(11) **27** ⟶

| **27a** | **28a** |

184

In accord with the above discussion, the anion **28** is also obtained from cyclohepta-1,3-diene or cyclohepta-1,4-diene by the action of potassium amide in liquid ammonia.

Now we consider electrocyclization of the 4 π-electron system. Actually, of the various electrocyclic transformations given in Table 1, which are "symmetry-allowed" on Woodward-Hoffmann theory, the last to have been verified experimentally is the interconversion of the cyclopropyl and allyl anions. A substrate which has allowed the feasibility of this process to be demonstrated is 1-methyl-3,3-diphenylcyclopropane-1,2-*trans*-dicarboxylate, i.e. structure **29** in Scheme 2 [39]. It is apparent that not only is **29** a suitable substrate in that it has available an acidic cyclopropyl proton, but also that ring opening should be facilitated by the presence of electron-withdrawing groups on the remaining carbons, which become the terminal centres of the allylic anion.

SCHEME 2

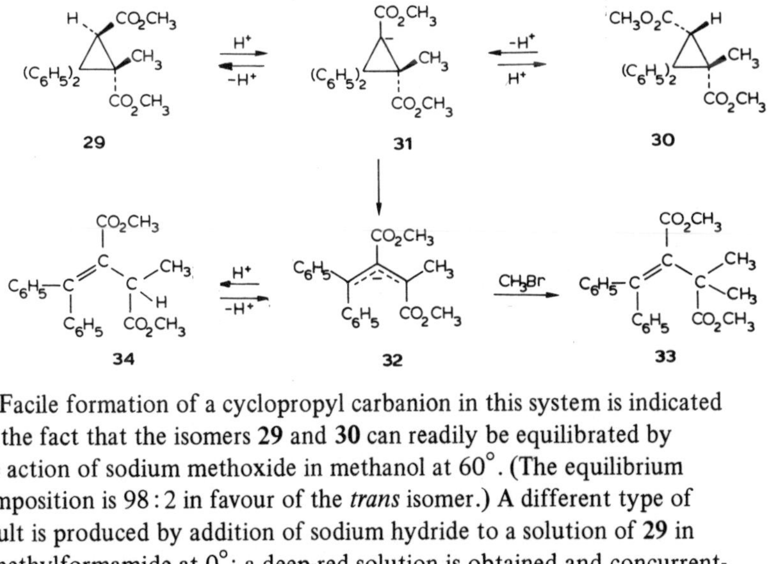

Facile formation of a cyclopropyl carbanion in this system is indicated by the fact that the isomers **29** and **30** can readily be equilibrated by the action of sodium methoxide in methanol at 60°. (The equilibrium composition is 98 : 2 in favour of the *trans* isomer.) A different type of result is produced by addition of sodium hydride to a solution of **29** in dimethylformamide at 0°: a deep-red solution is obtained and concurrently 0.9 equivalents of hydrogen are evolved. Strong evidence for formation of the acyclic delocalized anion **32** in this system is given by the isolation of the alkylation product **33** in high yield after the red solution had been treated with methyl bromide, and of the olefinic diester **34** after

185

acidification with methanolic hydrogen chloride, again in high yield. The allyl anion **32** was also obtained by treatment of **34** with sodium hydride in DMF [39].

It may be pointed out that the transformation **31→32** occurs irreversibly, since the cyclic form is thermodynamically less stable (ring strain). Also noteworthy is the fact that, whereas the 1-cyano-2,2-diphenylcycloprop-1-yl anion does not undergo ring opening [40], the 1-cyano-2,3-diphenylcycloprop-1-yl anion isomerises to the allylic anion (red violet) at a conveniently measurable rate at 25° (half life ~30 min) [41]. Similarly, the 1,2,3-triphenylcyclopropyl anion undergoes ring opening to the 1,2,3-triphenylallyl anion [42]. Apparently an essential condition for isomerisation to occur is the presence of an electron withdrawing substituent on each of the terminal carbons of the ring-opened allylic anion.

The steric course of the cyclopropyl anion ring opening is not discernible from the results given in Scheme 2. Relevant information on this question has been obtained from experiments performed with the nitrogen-containing ring system **35**, viz. N-lithio-2,3(*cis*)-diphenylaziridine [43]. It is found that the initially colorless solution of **35** in THF turns red on warming to 40–60° under nitrogen. The colored solution plausibly contains the 1,3(*cis, trans*)-diphenyl-2-azaallyl anion **36**, formed from **35** by conrotatory ring-opening. Thus the trapping reaction with *trans*-stilbene yielded as major products (73%) the tetraphenylpyrrolidine **38**, in which the aza-allyl portion was shown to have

SCHEME 3

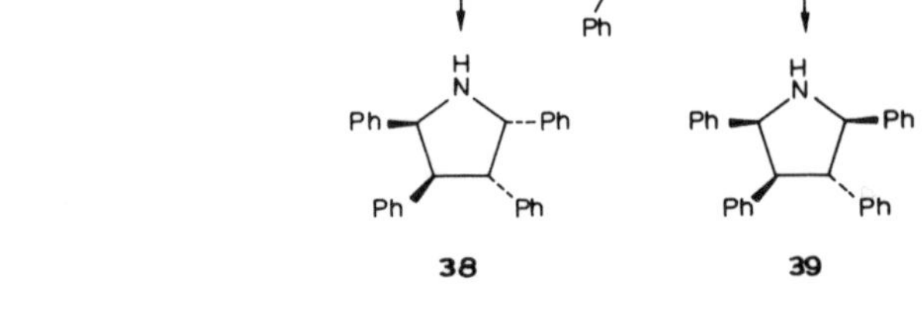

186

a *trans* orientation by NMR. Interestingly, when the product of ring-opening was cooled to $0°$ before treatment with *trans*-stilbene, the pyrrolidine **39** was obtained, indicating that a prior rearrangement of **36** had occurred to the *trans, trans* anion **37**.

Thus conrotatory ring opening in this aziridine system (**35**) is established, and it is reasonable to assume that this result may be generalized to include the ring-opening of cyclopropyl anions*.

(d) Carbonium ion electrocyclizations

For balance of presentation and general interest, we conclude this section with brief mention of electrocyclic rearrangements of carbonium ions. Orbital symmetry considerations have successfully elucidated many carbonium ion rearrangements, as described in several accounts [26–29, 45–47]. A pertinent example [48] concerns the thermal transformation of the dienylic cation **41**, obtained from alcohol **40** in an acid medium, to the cyclopentenyl cation **43**, a process occurring by conrotatory cyclization via the U-conformation **42**.

SCHEME 4

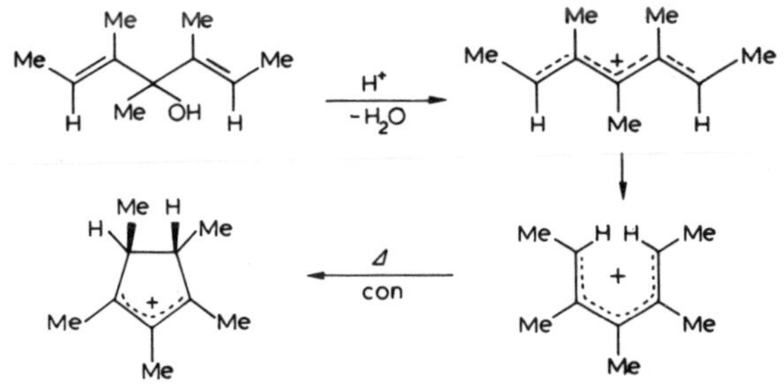

A carbonium ion reaction which provides a counterpart to the carbanion isomerisation of eq. (1) is the transformation of tropylium ion **44** to the bicycloheptadienyl cation **45** [49, 50]. Ring closure must occur in a disrotatory manner, which is allowed in the photochemical

---

*In certain fused ring systems, in which a symmetry-allowed conrotatory opening of the cyclopropyl anion would lead to highly strained products, the observed reaction is one of overall disrotatory opening [44]. This illustrates the principle that when a symmetry-allowed process becomes prohibitive energetically an alternative pathway will be followed by the reacting system (see also Section 3b).

process for the case of 4 π-electrons participating (cf. Scheme 4 where a constraint on conrotatory closure is not present).

(12)

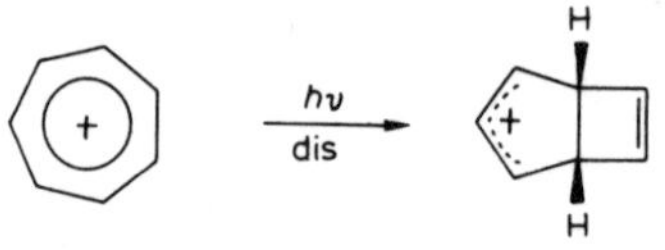

44 **45**

3. SIGMATROPIC REARRANGEMENTS

(a) [1,j] Hydrogen migrations

A sigmatropic change involves migration of a σ-bond, which is flanked by one or more π-electron systems, to a new position within a molecule. An example of such a process is the [1,5]-hydrogen shift in cis-1,3-pentadiene, in an uncatalyzed thermally induced intramolecular process:

(13)

46

Though this process is degenerate, that is one of self-interconversion, deuterium labelling experiments prove that it does in fact occur [51]. In the transition state of this reaction the hydrogen is joined simultaneously to both termini of the π-system, and it is expected that in the case of **46** a highly symmetrical transition state configuration should occur. The observation of a large kinetic isotope effect ($k(47)/k(48)$) has a value of 5 at $200°$ or 12 at $25°$) is consistent with this expectation [51].

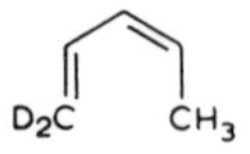

47 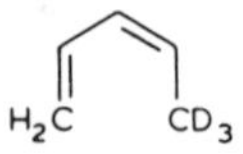 **48**

Sigmatropic rearrangements related to that of eq. (13), which may be generalized as [1,j]-migrations, are found in abundance in the reactions of hydrocarbons [30a, 52], and also of carbonium ions [45–47], though there are relatively few examples reported to date relating to carbanions

188

[53–55]. Selection rules pertaining to thermally and photochemically induced sigmatropic [1,j]-shifts, and the stereochemical consequences of migration, have been proposed [29]. A summary for the thermally induced processes is given in Table 2 for polyenyl systems of various charge types. For photochemical reactions reverse rules will apply, as a result of the different symmetry properties of the first excited state orbitals.

Table 2

THERMALLY ALLOWED SIGMATROPIC HYDROGEN MIGRATIONS: SELECTION RULES

Participating electrons	Neutral polyene	Anionic polyene	Cationic polyene	Stereochemical course
2	–	–	[1,2]	suprafacial
4	[1,3]	[1,2]	[1,4]	antarafacial
6	[1,5]	[1,4]	[1,6]	suprafacial
8	[1,7]	[1,6]	[1,8]	antarafacial
4n+2	[1,(4n+1)]	[1,4n]	[1,(4n+2)]	suprafacial
4n	[1,(4n–1)]	[1,(4n–2)]	[1,4n]	antarafacial

The [1,5]-hydrogen shift given in eq. (13) is predicted to have *suprafacial* character, that is the migrating atom is associated at all times with the same face of the π-system. This was elegantly shown to be the case by the demonstration [51] that **49** is converted stereospecifically into **50** and **51**. An *antarafacial* process would have the hydrogen atom passing from the top face of one carbon terminus to the bottom face of the other, which is not in accord with the experimental result.

(14)

49 → **50**

(15)

49 → **51**

A simplified explanation of the selection rules for [1,j]-sigmatropic hydrogen shifts may be given using the concept of the aromatic transition state [4, 5, 56]*. Thus Figure 4a illustrates the transition state in a suprafacial shift, while Figure 4b is applicable to the antarafacial shift. In the former there is a continuous overlap between the 1s orbital and the top face of the π-system, without a phase dislocation; but in the latter type of overlap a sign inversion occurs in the p-orbital array at the migration centre. Figure 4a denotes a *Hückel system* while Figure 4b denotes an *anti-Hückel* or *Möbius system*. (In the general case, conjugated systems associated with an even number of sign inversions are classified as of the Hückel type, and those with an odd number of sign inversions as of the Möbius type.)

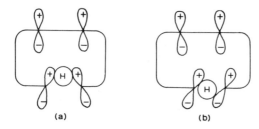

(a)　　　　　　　　(b)

Figure 4. Transition states in [1,j] hydrogen migrations: (a) suprafacial shift (Hückel system), (b) antarafacial shift (Möbius system).

The Hückel transition state will have aromatic character for 4n + 2 π-electron systems and antiaromatic character for 4n π-electron systems. The opposite relationship holds for Möbius transition states. Hence thermal pericyclic reactions with 4n π-electrons will be favourable when the transition state is of the Möbius type but unfavourable for a Hückel transition state. On the other hand, a process in which there are 4n + 2 participating π-electrons will be favourable in a Hückel system and unfavourable in a Möbius system. These relationships will be reversed in

---

*The aromatic transition state approach is emphasized in this section, for balance of presentation, in view of the fact that the previous section emphasized the orbital symmetry-correlation diagram approach. Alternative treatments of sigmatropic reactions have been given, and include discussions based on orbital symmetry arguments [1] and the frontier orbital method [3]. The reader is referred to the relevant references for these viewpoints.

photochemical processes. The selection rules summarised in Table 2 for thermal migrations follow directly from the above considerations.

As indicated previously, only few examples of [1,j]-hydrogen shifts in carbanionic systems have been reported hitherto. One such case involves the thermal interconversion of the dienylic anions 52 and 53 (shown in eq. (16) as the charge-localized species for simplicity) [54].

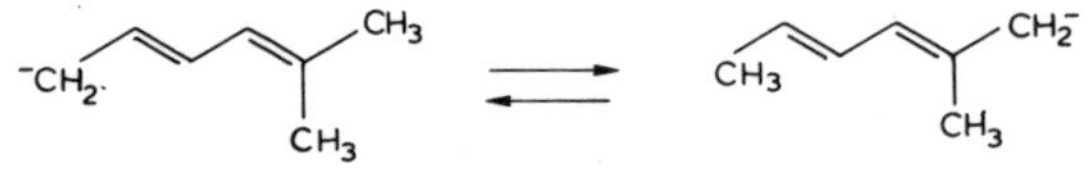

Deuterium labelling experiments show that this [1,6]-migration occurs intramolecularly.

According to the selection rules given in Table 2, the [1,6]-hydrogen shift in an anionic system should occur in antarafacial manner in the thermally induced process; conversely the photochemical process should occur in suprafacial manner. This prediction was ingeniously demonstrated [54] for the case of the cycloheptadienyl anion 54. On irradiation of a solution of the anion in THF/hexane at $0°$ a smooth conversion to 55 was achieved. On the other hand, heating of the solution at $150°$ could not effect the transformation of 54 to 56, or the reverse process. It is seen that the ring system precludes an antarafacial migration mode but allows occurrence of the suprafacial shift.

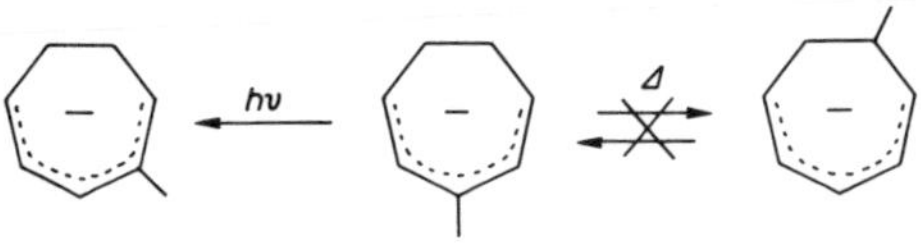

(b) [1,j] Alkyl and aryl migrations

Migration of alkyl groups follow principles which are analogous to those considered for hydrogen shifts, provided the bonding type in the transition state is similar, i.e. when a σ-type orbital is used for overlap with the π-system. This is illustrated in Figure 5a for [1,3] alkyl migration in an allyl system $(R-C-C=C \rightarrow C=C-C-R)$. Since we have here a Hückel type transition state with four participating electrons the thermal process is disallowed. However, the alternative type of overlap shown in

Figure 5b leads to a Möbius system, so that this type of process is thermally allowed. In the case of Figure 5b we have a suprafacial shift with *inversion* of configuration at the migrating alkyl centre (cf. Figure 5a for which retention of configuration would be expected).

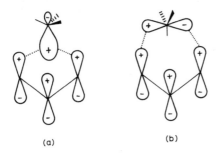

(a)

(b)

Figure 5. Transition states for 1,3-alkyl migration in an allyl system; (a) suprafacial shift with retention, (b) suprafacial shift with inversion.

1,2-Rearrangements in carbonium ion systems are very common, as is well known. Such a process may be represented by the orbital diagrams of Figure 6a, b, c, representing the reactant, transition state, and product of rearrangement, respectively.

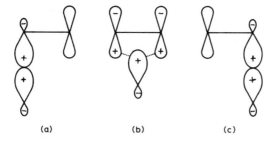

(a)

(b)

(c)

Figure 6. Orbital representations for 1,2-rearrangement of a carbonium ion: (a) reactant, (b) transition state, (c) product.

Since the transition state with 2 π-electrons is of the Hückel type, the process is favoured in a thermal reaction. The overall process is suprafacial with retention of configuration. Migrations involving other electron deficient systems (carbenes, nitrenes, and nitrenium ions) can be treated in analogous manner.

Let us now proceed to consider the 1,2 rearrangement to an anionic centre. It will be seen that the transition state corresponding to Figure 6b would contain 4 π-electrons, and hence should be thermally disallowed. On the other hand, a transition state analogous to that of Figure 5b, in which opposite faces of the p-orbital are used for overlap (Möbius system), would be symmetry allowed. However, the suprafacial shift with inversion of configuration is structurally not feasible for the case of 1,2-alkyl migration. Likewise one can rule out the possibility of an antarafacial shift with retention, which would be symmetry-allowed but structurally improbable. Thus thermally induced 1,2-migrations in anionic systems are not expected to be favourable. This prediction is borne out by detailed calculations of the energetics for concerted 1,2-rearrangements in a number of carbonium ion and carbanion processes [57].

In view of the above considerations, it is somewhat paradoxical that 1,2-rearrangements do in fact occur in certain carbanion reactions known as the Zimmerman-Grovenstein rearrangement [58, 59]. Two examples of the reaction will be discussed here:

(18)  $\quad$ Ph$_2$C—CH$_2^-$ $\longrightarrow$ Ph$_2\bar{\text{C}}$—CH$_2$—Ph
$\qquad\quad$ |
$\qquad\quad$ Ph

(19)  $\quad$ Ph$_2$C—CH$_2^-$ $\longrightarrow$ Ph$_2\bar{\text{C}}$—CH$_2$—CH$_2$Ph
$\qquad\quad$ |
$\qquad\quad$ CH$_2$Ph

Considering in the first place the phenyl migration shown in eq. (18), there is another pathway by which reaction can occur, involving two separate stages:

(20)

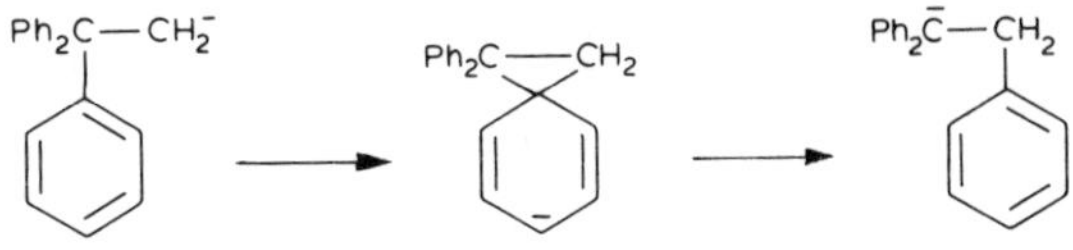

Ph$_2$C—CH$_2^-$ $\qquad$ Ph$_2$C——CH$_2$ $\qquad$ Ph$_2\bar{\text{C}}$—CH$_2$

$\qquad$ 57 $\qquad\qquad\qquad$ 58 $\qquad\qquad\qquad$ 59

Thus, instead of a sigmatropic shift (disallowed), we have here a nucleophilic attack by the carbanion on the electrophilic center of the

193

benzenoid ring, viz. addition, followed by elimination. Evidence has recently been presented that cyclic structures analogous to **58** may be present as reaction intermediates in 1,4-aryl migrations [60, 61].

Proceeding to the case of benzyl migration shown in eq. (19), since we are now concerned with an alkyl shift it is clear that an argument such as used above for the phenyl case can no longer hold. It is therefore noteworthy that further investigation of this reaction (e.g. by isotope tracer studies) has revealed that the reaction is not intramolecular [62]. Two possible intermolecular mechanisms are:

(21)
$$\underset{\substack{|\\ CH_2\\ |\\ Ph}}{Ph_2C-CH_2Li} \longrightarrow \underset{\substack{|\\ CH_2^-\ Li^+\\ |\\ Ph}}{Ph_2C=CH_2} \longrightarrow \underset{\substack{Li^+\ |\\ CH_2\\ |\\ Ph}}{Ph_2C-CH_2}$$

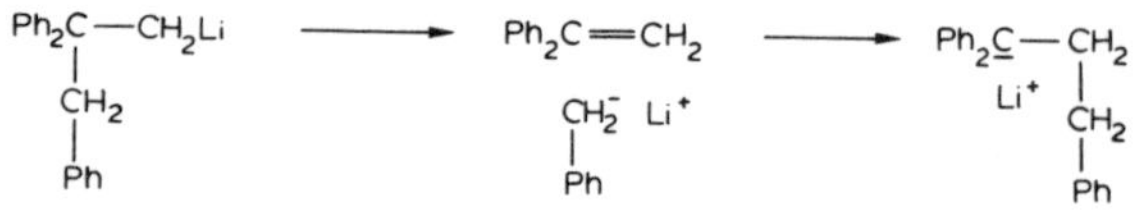

(22)
$$\underset{\substack{|\\ CH_2\\ |\\ Ph}}{Ph_2C-CH_2^-} \longrightarrow \underset{\substack{|\\ \cdot CH_2\\ |\\ Ph}}{Ph_2\bar{C}-\dot{C}H_2} \longrightarrow \underset{\substack{|\\ CH_2\\ |\\ Ph}}{Ph_2\bar{C}-CH_2}$$

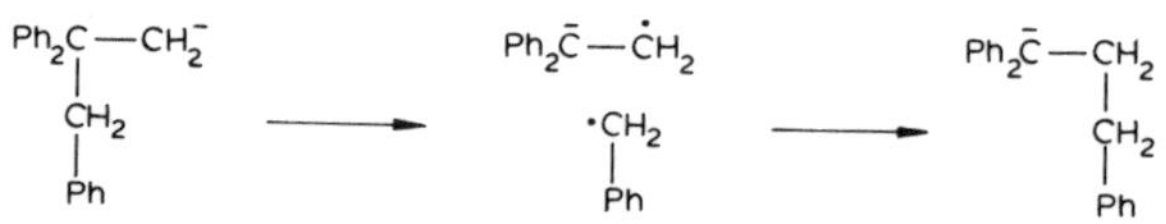

60

The mechanism indicated in eq. (21) is that of elimination-addition of ionic species, whereas eq. (22) shows the dissociation-recombination of radical-ion pair species **60**. The available evidence does not definitively differentiate between these possibilities. However, in either case, the important point is that the 1,2-migration of benzyl in this anionic system is *not a concerted process*, so that it falls outside orbital symmetry considerations. Thus the "rules" for orbital symmetry controlled reactions are selection rules applicable to concerted reactions, and do not exclude the possibility that an alternative, non-concerted, process is energetically feasible and provides a valid alternative reaction pathway.

194

(c) Sigmatropic rearrangements of the order [i, j]

An example of a new type of thermal rearrangement in a carbanion will now be described.

The fluorenyl-methide structure **61**, prepared as the lithium salt from the corresponding bromide and lithium metal, is found to be stable in THF at $-70°$ (e.g. it can be deuterated or carboxylated). However, on warming to $-20°$ (or above) **61** undergoes a facile rearrangement to **62**; on quenching, the fluorene derivative **63** is obtained in 70% yield [63]. The reaction is shown in eq. (23):

23)

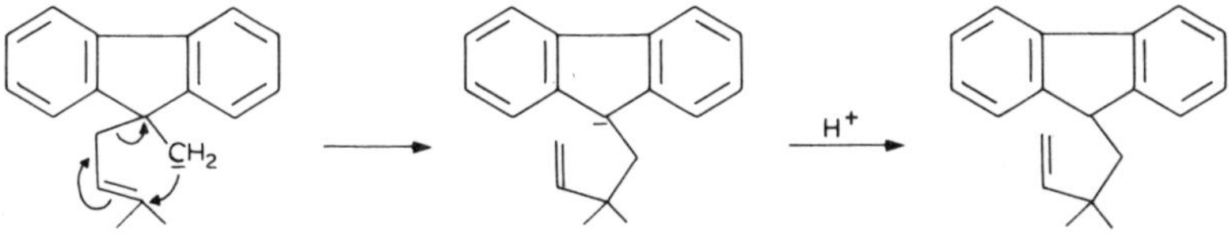

| 61 | 62 | 63 |

The transformation **61→62** involves essentially the migration of a σ-bond from one location in the molecule to another, as seen more clearly in the partial structures **64** and **66**, and proceeds via the transition state configuration **65**. This concerted isomerisation may be described as a sigmatropic rearrangement of the order [2, 3], a terminology which is understood by reference to the relative positions of the migration termini as indicated in **65**.

(24)

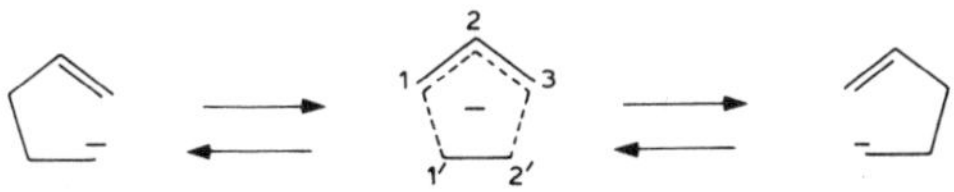

| 64 | 65 | 66 |

The orbital diagram applicable to this rearrangement is shown in Figure 7, as an allylic anion interacting with an ethylene fragment. The suprafacial, suprafacial overlap leads to a Hückel type transition state, and as there are 6 participating electrons, the thermally induced process is favourable.

Although the rearrangement shown in eq. (23) appears to be the only anionic rearrangement of this type reported to date in an all-carbon

195

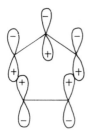

Figure 7. Orbital overlap diagram for the [2,3] sigmatropic rearrangement (eq. (24)).

system, there are a number of heteroatomic systems in which this kind of process obtains. These rearrangements are of the general type shown in eq. (25) where X is commonly oxygen, sulfur, nitrogen, or phosphorus. In some cases two of the five participating atoms may be heteroatomic.

(25)

67   **68**

For example, it is found that the benzyl allyl ether **69**, when treated with *n*-butyllithium in THF followed by quenching, yields the rearranged alcohol **72** [64]. An analogous rearrangement is observed with the corresponding benzyl allyl sulfide [65]. The reaction pathway plausibly involves initial formation of the benzylic carbanion **70**, which then undergoes a concerted [2,3]-sigmatropic rearrangement to yield the "inverted allyl" anion **71**.

(26)

196

The transformation of allylic ether anions is classed under the Wittig rearrangements [66, 67]. Similar migrations are known to occur in benzylic ammonium, and sulfonium ylides. The latter reactions, known as the Sommelet-Hauser rearrangement [68], lead to substitution on the aromatic ring, as for example,

(27)

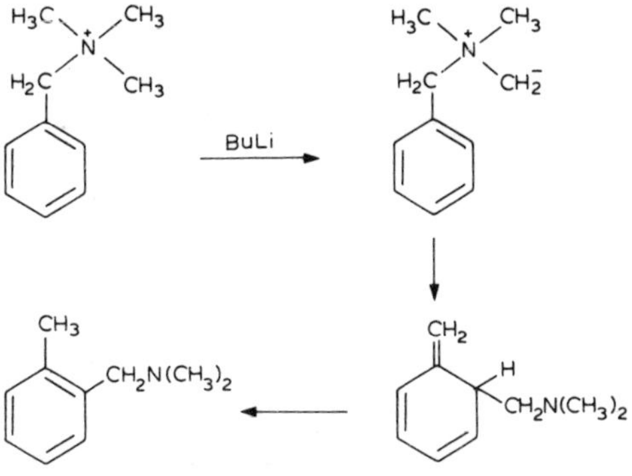

$$ \text{[structure: } H_3C \text{, } CH_3 \text{ on } N^+ \text{, } H_2C \text{, } CH_3 \text{, benzene ring]} \xrightarrow{\text{BuLi}} \text{[structure with } N^+, H_2C, CH_2^- \text{]} $$

$$ \text{[benzene with } CH_3 \text{ and } CH_2N(CH_3)_2] \longleftarrow \text{[structure with } CH_2, H, CH_2N(CH_3)_2] $$

[2,3] Sigmatropic rearrangements of the general type shown in eq. (25) are of considerable value in synthesis, including that of squalene [69–72].

In a number of reactions involving ylid type systems, products have been found which are indicative of a competing dissociation-recombination process involving a radical-ion mechanism. A general equation for this competing reaction is given by

(28)

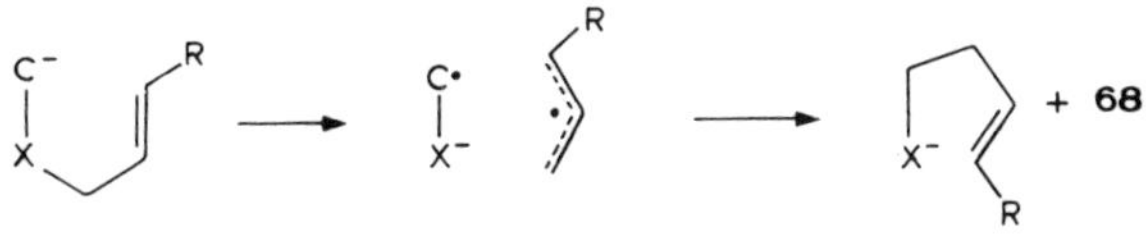

$$ \underset{67}{\overset{C^-}{\underset{X}{\big|}}\!\!\!\diagup\!\!\diagdown^{R}} \longrightarrow \underset{73}{\overset{C^\bullet}{\underset{X^-}{\big|}}\;\;\diagup\!\!\diagdown^{R}} \longrightarrow \underset{74}{\overset{}{\underset{X^-}{}}\diagdown_{R}} + \mathbf{68} $$

A similar mechanism is believed to hold in the case of certain 1,2-anionic migrations which are observed in systems related to those considered above. An appropriate example is given by the Stevens rearrangement [68]:

(29)

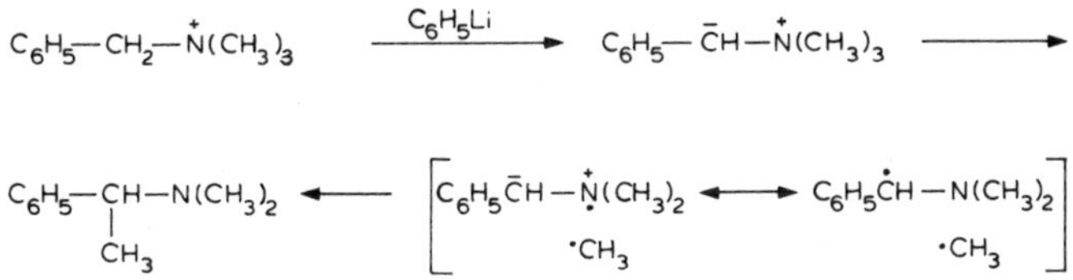

$$ C_6H_5{-}CH_2{-}\overset{+}{N}(CH_3)_3 \xrightarrow{C_6H_5Li} C_6H_5{-}\overset{-}{C}H{-}\overset{+}{N}(CH_3)_3 \longrightarrow $$

$$ \underset{\underset{CH_3}{|}}{C_6H_5{-}CH{-}N(CH_3)_2} \longleftarrow \left[\underset{\bullet CH_3}{C_6H_5\overset{-}{C}H{-}\overset{+}{N}(CH_3)_2} \longleftrightarrow \underset{\bullet CH_3}{C_6H_5\overset{\bullet}{C}H{-}N(CH_3)_2} \right] $$

The radical-ion mechanism for the Stevens rearrangement is indicated by the results of applying the technique of chemically induced dynamic nuclear polarisation (CIDNP) (see, however, ref. 62). Thus, once again, it appears that a choice may exist between concerted and non-concerted processes, and that structural modification can bring about a change from one to the other.

Perhaps the best known rearrangement which falls under the classification of $[i,j]$ sigmatropic migrations is the Cope rearrangement [73]. This is the thermal transformation of 1,5-dienes into the isomeric 1,5-dienes, as shown in eq. (30):

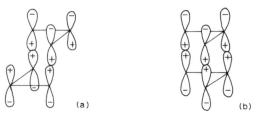

(30)

75   76

Some of the most remarkable discoveries have followed from investigation of this reaction. It is instructive, therefore, to consider some of the characteristics of this process in detail. Subsequently, it will be shown how base-induced hydrogen-deuterium exchange studies have been exploited to give further insight into these systems.

It has been shown that the isomerisations occur intramolecularly, and are uncatalyzed and reversible. In general, elevated temperatures $(150°–300°)$ are required for reaction to proceed; the 2,3-dimethyl derivative, for example, undergoes rearrangement at ca. $200°$. Isomerisation occurs via a six-centered cyclic transition state, analogous to **65**, and may be classified as a $[3,3]$ sigmatropic process. In the aromatic transition state approach one considers orbital overlap in the interaction of two allylic radicals. This leads to two distinct Hückel type transition states, formed by suprafacial, suprafacial interaction of the six p-lobes: a chair (Figure 8a) and a boat configuration (Figure 8b).

Figure 8. Transition states in Cope rearrangement: (a) chair configuration, (b) boat configuration.

198

An alternative approach is to construct correlation diagrams for this process, which leads to the conclusion that the chair-like transition state should be preferred to the boat-like configuration (see Figures 42 and 43 in ref. 29, and also ref. 74).

The Cope rearrangement becomes degenerate (a "null" reaction) in the particular case of the unsubstituted hexadiene, since in the isomerisation process of eq. (30) ($R_1 = R_2 = H$) the original material is regenerated. Nevertheless this reaction has been studied, through isotopic labelling, by following the rearrangement of 1,1-dideuteriohexa-1,5-diene to 3,3-dideuteriohexa-1,5-diene at ca. $300°$ ($E_a = 34$ kcal/mole) [75].

An important case of a degenerate Cope rearrangement which has been well characterized is that of bicyclo[5.1.0]octa-2,5-diene (homotropilidene),77. The process, shown in eq. (31), may be described as a *cis*-divinylcyclopropane isomerisation [76]:

(31)

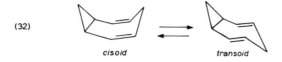

77

As a result of the strain-imparting fused cyclopropyl moiety, the rate of rearrangement is greatly increased. From NMR studies it has been estimated [18] that at $180°$ the isomerisation occurs 1000 times per second, while at $-50°$ it occurs once every second, effectively becoming frozen out on the NMR time scale.

Valence isomerisation of homotropilidene proceeds by way of the *cisoid* conformation shown in eq. (31); however, the thermodynamically favoured conformation is *transoid* [77]:

(32)

cisoid   transoid

The fact that the equilibrium of eq. (32) is on the side of the *transoid* conformation will lead to a decrease in the effective rate of the valence isomerisation. It follows that, if a bridging grouping were introduced so as to fix the *cisoid* geometry of the molecule, then the isomerisation process should become facilitated. Such is indeed the case. Thus, the tricyclic structures **78** (semibullvalene), **79** (barbaralane), **80** (dihydro-

199

bullvalene) and **81** (barbaralone) all undergo rapid degenerate isomerisations of the divinylcyclopropane type. The rearrangement of semibullvalene is the fastest of the known valence bond isomerisation processes.
(From NMR studies the free energy of activation for the rearrangement
of octamethylsemibullvalene is found to be 6.4 kcal/mole [78], i.e.
almost 30 kcal/mole less than for 1,5-hexadiene.) Theoretical treatments
of this system have been presented [79–82].

78   79   80   81

Hydrogen-deuterium exchange in **81** under basic conditions (NaOD/
D_2O) could not be effected [76], which is in agreement with the
discussion in Chapter 3 concerning the difficulty of abstraction of *alpha*
hydrogens from bridgehead positions. Now, if the bridge were extended
by the introduction of a methylene group adjacent to carbonyl, as in
bullvalone (**82**), one would expect that exchange of the two methylene
hydrogens would become feasible. In practice hydrogen-deuterium
exchange in bullvalone has an entirely different character. It is found
that in NaOD/D_2O two hydrogens exchange quickly, but then exchange
of *all* the hydrogens in the molecule follows [76]:

(33)

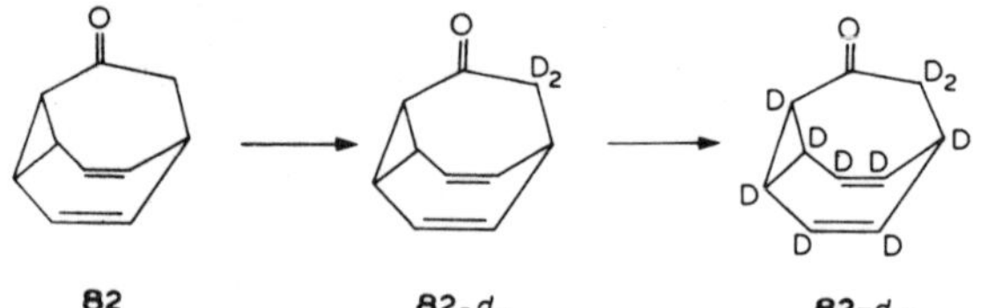

82   82-d_2 82-d_{10}

How can we explain this experimental result when there appear to be
only two enolizable hydrogens in **82**? Let us consider the types of
valence isomerisation which are possible in these systems. We see that
bullvalone itself, like barbaralone, is only capable of the degenerate
divinylcyclopropane rearrangement, analogous to that of the parent
homotropilidene (eq. (31)). On the other hand in basic media bullvalone
is expected to be in equilibrium with a small proportion of the enol

200

tautomer (83), and the opportunity for occurrence of the divinylcyclo-propane rearrangement in the latter is greatly enhanced. Thus, the ring bearing the enol function has now become part of a newly formed homo-tropilidene system, allowing the occurrence of the divinylcyclopropane rearrangement, **83a** ⇌ **83b**. Now enol **83b** will be in equilibrium with ketone **82b**, in which the hydrogen atoms on C10 are enolizable and hence available for isotopic exchange:

(34)

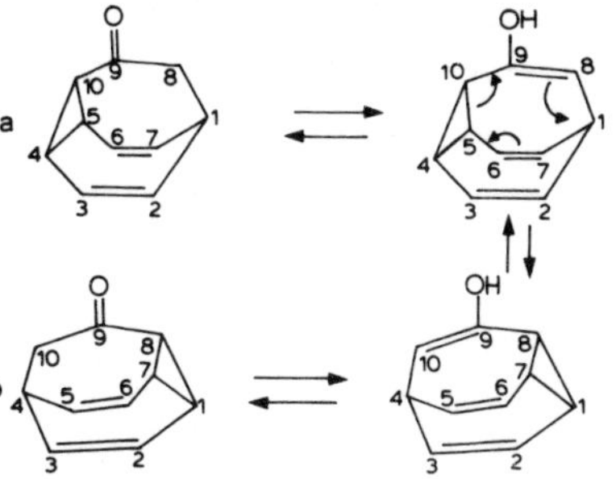

Other sequential divinylcyclopropane rearrangements transform the bonding environment of each of the other carbons, so that each carbon in turn will take the place of C10 in **83b**. In this way all the hydrogens will successively become enolizable and subject to isotopic exchange.

Hydrogen-deuterium exchange in bullvalene (84) takes place under much more stringent conditions, with ROK/ROD (R = tri-*n*-butyl) at 160°, and results in random distribution of the deuterium isotope over the whole molecule [19, 83]. Under comparable conditions the diene (85) exchanges only the olefinic hydrogens.

(35)

84

(36)

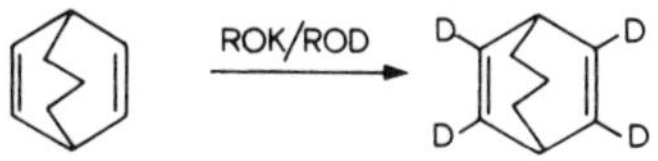

85

Bullvalene provides a fascinating situation of valence isomerism, as the successive divinylcyclopropane rearrangements result in isomers which are all structurally equivalent. Since the molecule has a 3-fold axis of symmetry the number of valence isomers will be $10!/3 = 1,209,600$. The NMR spectrum of bullvalene taken at $120°$ exhibits a single, sharp resonance absorption (τ 5.78), showing that isomerisation is sufficiently fast on the NMR time-scale for the various proton environments (cyclopropyl, olefinic and bridgehead) to become equivalent. On lowering of the temperature to $15°$, the signal broadens and eventually becomes resolved at $-85°$ into a multiplet at τ 4.35 assignable to the olefinic protons and a somewhat broad band at τ 7.42 corresponding to the cyclopropyl and bridgehead hydrogens. Though related degenerate Cope rearrangement systems have been discovered [84–88], bullvalene remains the classical example of a molecule with fluctuating bonds. (The term "constitutional topomerisation" has recently been proposed for this type of ligand interchange occurring in multiple valence bond isomerization [89]). The historical sequence beginning with Cope's initial observations with hexadienes, through Doering's prediction of the fluxional nature of bullvalene*, to its synthesis and the experimental verification of its properties by Schröder, surely forms one of the exciting achievements of organic chemistry.

In an ingeneously devised experiment [90] the barbaryl system has been used to establish the preference of a deuterium atom for the bridgehead (86a) over the cyclopropyl (86b) position, through measurements by NMR spectroscopy:

(37)

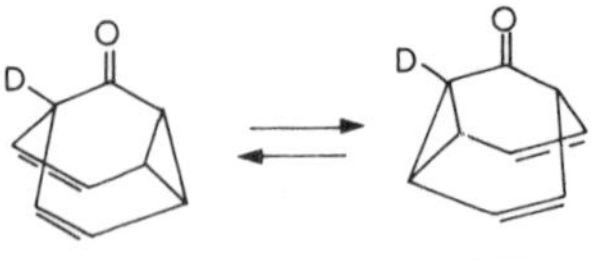

86a
55.5±0 5%

86b
44.5±0.5%

*The author had the good fortune to be present at a seminar given by Professor Doering at Columbia University in 1961 where the properties of the then hypothetical bullvalene were predicted with remarkable foresight before a spellbound audience.

202

A related study focuses on the consequences of generating a carbonium centre adjacent to a homotropilidene system, in the form of the semibullvalenylcarbinyl cation **87b** [91]:

(38)

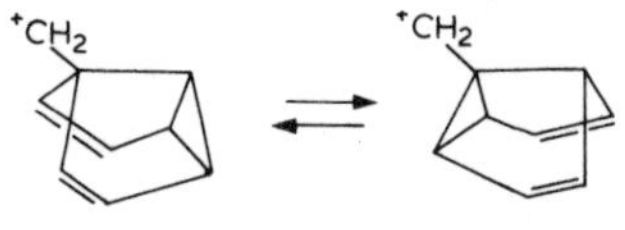

87a   87b

This takes us back to the opening remarks of this chapter, in which the applicability of the principles governing these rearrangements to various charge types was pointed out, and serves to emphasize the wide scope of valence isomerisation processes. Of considerable interest are the theoretical predictions [92, 93] that a suitably substituted semibullvalene might provide the first case of a fluxional organic molecule where the mesovalent intermediate is actually the stable form. Initial experiments to test this hypothesis have already shown some promise [94].

REFERENCES

1 R.B. Woodward and R. Hoffmann, J. Amer. Chem. Soc., 87 (1965) 395, 2511; R. Hoffmann and R.B. Woodward, J. Amer. Chem. Soc., 87 (1965) 2046, 4388.
2 H.C. Longuet-Higgins and E.W. Abrahamson, J. Amer. Chem. Soc., 87 (1965) 2045.
3 K. Fukui, Bull. Chem. Soc. Japan, 39 (1966) 498; Accounts Chem. Res., 4 (1971) 57.
4 M.J.S. Dewar, Tetrahedron Suppl., 8 (1967) 75; Angew. Chem. Intern. Ed., 10 (1971) 761.
5 H.E. Zimmerman, J. Amer. Chem. Soc., 88 (1966) 1564, 1566; Accounts Chem. Res., 4 (1971) 272; ibid., 5 (1972) 393.
6 L. Salem, J. Amer. Chem. Soc., 90 (1968) 543, 553.
7 W.J. van der Hart, J.J.C. Mulder and L.J. Oosterhoff, J. Amer. Chem. Soc., 94 (1972) 5724.
8 C. Kaneko, Tetrahedron, 28 (1972) 4915.
9 J. Langlet and J.-P. Malrieu, J. Amer. Chem. Soc., 94 (1972) 7254.
10 N.D. Epiotis, J. Amer. Chem. Soc., 95 (1973) 1191, 1200, 1206, 1214.
11 H.E. Zimmerman in "Molecular Rearrangements", Vol. 1, edited by P. de Mayo, Interscience, New York, 1963, Chapter 6.
12 D.J. Cram, "Fundamentals of Carbanion Chemistry", Academic Press, New York, 1965, Chapter 6.

13 R.B. Bates and D.A. McCombs, Tetrahedron Lett., No. 12, 977 (1969).
14 H. Kloosterziel and J.A.A. van Drunen, Rec. Trav. Chim. Pays-Bas, 89 (1970) 368.
15a P.R. Stapp and R.F. Kleinschmidt, J. Org. Chem., 30 (1965) 3006.
15b L.H. Slaugh, J. Org. Chem., 32 (1967) 108.
15c M. Tardi, J.-P. Vairon and P. Sigwalt, Bull. Soc. Chim. France, 1791 (1972).
16a J.R. Grunwell and J.F. Sebastian, Tetrahedron, 27 (1971) 4387.
16b R.B. Bates, S. Brenner, C.M. Cole, E.W. Davidson, G.D. Forsythe, D.A. McCombs and A.S. Roth, J. Amer. Chem. Soc., 95 (1973) 926.
17 D.S. Glass, J.W.H. Whatthey and S. Winstein, Tetrahedron Lett., No. 6, 377 (1965).
18 W. v. E. Doering and W.R. Roth, Angew. Chem. Intern. Ed., 2 (1963) 115.
19 G. Schröder, J.F.M. Oth and R. Merényi, Angew. Chem. Intern. Ed., 4 (1965) 752.
20 G. Maier, Angew. Chem. Intern. Ed., 6 (1967) 402.
21 L.A. Paquette, Angew. Chem. Intern. Ed., 10 (1971) 11.
22 E. Ciganek, J. Amer. Chem. Soc., 93 (1971) 2207.
23 G.E. Hall and J.D. Roberts, J. Amer. Chem. Soc., 93 (1971) 2203.
24 W.-D. Stohrer, Ber., 106 (1973) 970.
25 E.N. Marvell, G. Caple and B. Schatz, Tetrahedron Lett., No. 7, 385 (1965).
26 S.I. Miller, Adv. Phys. Org. Chem., 6 (1968) 185.
27 G.B. Gill, Quart. Rev., 22 (1968) 338.
28 J.A. Berson, Accounts Chem. Res., 1 (1968) 152.
29 R.B. Woodward and R. Hoffmann, Angew. Chem. Intern. Ed., 8 (1969) 781; "The Conservation of Orbital Symmetry", Verlag Chemie, 1971.
30a J.J. Gajewski, in "Mechanisms of Molecular Migrations", Vol. 4, edited by B.S. Thyagarajan, Wiley-Interscience, New York, 1971.
30b T.L. Gilchrist and R.C. Storr, "Organic Reactions and Orbital Symmetry", Cambridge University Press, Cambridge, 1972.
31 J.E. Baldwin, A.H. Andrist and R.K. Pinshchmidt, Jr., Accounts Chem. Res., 5 (1972) 402.
32 J.A. Berson, Accounts Chem. Res., 5 (1972) 406.
33 D.J. Atkinson, M.J. Perkins and P. Ward, Chem. Commun., 1390 (1969).
34 R.B. Bates, D.W. Gosselink and J.A. Kaczynski, Tetrahedron Lett., No. 3, 205 (1967).
35 D.H. Hunter and S.K. Sim, Can. J. Chem., 50 (1972) 669, 678.
36a H. Kloosterziel, J.A.A. van Drunen and P. Galama, Chem. Commun., 885 (1969).
36b R.B. Bates, L.M. Kroposki and D.E. Potter, J. Org. Chem., 37 (1972) 560.
36c V. Rautensrauch, Helv. Chim. Acta, 55 (1972) 594.
37a H. Kloosterziel and J.A.A. van Drunen, Rec. Trav. Chim. Pays-Bas, 88 (1969) 1084.
37b R.B. Bates, W.H. Deines, D.A. McCombs and D.E. Potter, J. Amer. Chem. Soc., 91 (1969) 4608.
38 R. Hoffmann and R.A. Olofson, J. Amer. Chem. Soc., 88 (1966) 943.

39   R. Huisgen and P. Eberhard, J. Amer. Chem. Soc., 94 (1972) 1346.
40   H.M. Walborsky and J.M. Motes, J. Amer. Chem. Soc., 92 (1970) 2445.
41   G. Boche and D. Martens, Angew. Chem. Intern. Ed., 11 (1972) 724.
42   J.E. Mulvaney and D. Savage, J. Org. Chem., 36 (1971) 2592.
43   T. Kauffmann, K. Habersaat and E. Koppelmann, Angew. Chem. Intern. Ed., 11 (1972) 291.
44a M.E. Londrigan and J.E. Mulvaney, J. Org. Chem., 37 (1972) 2823.
44b M. Newcomb and W.T. Ford, J. Amer. Chem. Soc., 95 (1973) 7186.
45   G.A. Olah and P.R. Schleyer, Editors, "Carbonium Ions", Vol. 2, Wiley, New York, 1970; Chapters by N.C. Deno and T.S. Sorensen.
46   C.H. DePuy, Accounts Chem. Res., 1 (1968) 33.
47   M.J. Perkins and P. Ward, in "Mechanisms of Molecular Migrations", Vol. 4, edited by B.S. Thyagarajan, Wiley-Interscience, New York, 1971.
48   P.H. Campbell, N.W.K. Chiu, K. Deugau, I.J. Miller and T.S. Sorensen, J. Amer. Chem. Soc., 91 (1969) 6404.
49   R.F. Childs and V. Taguchi, Chem. Commun., 695 (1970); K.E. Hine and R.F. Childs, Chem. Commun., 145 (1972).
50   E.A. van Tamelen, R.H. Greeley and H. Schumacher, J. Amer. Chem. Soc., 93 (1971) 6151.
51   W.R. Roth and J. König, Ann., 699 (1966) 24; W.R. Roth, J. König and K. Stein, Ber., 103 (1970) 426.
52   L.T. Scott and M. Jones, Jr., Chem. Rev., 72 (1972) 181.
53   H. Kloosterziel and E. Zwanenburg, Rec. Trav. Chim. Pays-Bas, 88 (1969) 1373.
54   R.B. Bates, S. Brenner, W.H. Deines, D.A. McCombs and D.E. Potter, J. Amer. Chem. Soc., 92 (1970) 6345.
55a J. Klein and S. Gilly, Tetrahedron, 27 (1971) 3477.
55b R.B. Bates, S. Brenner and C.M. Cole, J. Amer. Chem. Soc., 94 (1972) 2130.
56   M.G. Evans and E. Warhurst, Trans. Far. Soc., 34 (1938) 614; M.G. Evans, Trans. Far. Soc., 35 (1939) 824.
57   N.F. Phelan, H.H. Jaffé and M. Orchin, J. Chem. Educ., 44 (1970) 626.
58   H.E. Zimmerman and A. Zweig, J. Amer. Chem. Soc., 83 (1961) 1196.
59   E. Grovenstein, Jr. and L.P. Williams, J. Amer. Chem. Soc., 83 (1961) 412, 2537.
60   E. Grovenstein, Jr., S. Akabori and J.-U. Rhee, J. Amer. Chem. Soc., 94 (1972) 4734.
61   G. Fraenkel, C.C. Ho, Y. Liang and S. Yu, J. Amer. Chem. Soc., 94 (1972) 4732.
62   E. Grovenstein, Jr. and G. Wentworth, J. Amer. Chem. Soc., 89 (1967) 1852, 2348.
63   J.E. Baldwin and F.J. Urban, Chem. Commun., 165 (1970).
64   J.E. Baldwin, J. DeBernardis and J.E. Patrick, Tetrahedron Lett., No. 5, 353 (1970).
65a V. Rautensrauch, Helv. Chim. Acta, 54 (1971) 739.
65b J.F. Biellmann and J.B. Ducep, Tetrahedron Lett., No. 1, 33 (1971).

66a G. Wittig, Angew. Chem., 66 (1954) 10.
66b U. Schöllkopf, Angew. Chem. Intern. Ed., 10 (1970) 765.
67a D.R. Dimmel and S.B. Gharpure, J. Amer. Chem. Soc., 93 (1971) 3991.
67b J.E. Baldwin and J.E. Patrick, J. Amer. Chem. Soc., 93 (1971) 3556.
68   A.R. Lepley and A.G. Giumanini, in "Mechanisms of Molecular Migrations",
     Vol. 3, edited by B.S. Thyagarajan, Wiley, New York, 1971.
69   J.E. Baldwin, R.E. Hackler and D.P. Kelly, J. Amer. Chem. Soc., 90 (1968)
     4758; J.E. Baldwin and R.E. Hackler, J. Amer. Chem. Soc., 91 (1969) 3646.
70   V. Rautensrauch, Helv. Chim. Acta, 55 (1972) 2233; K. Kondo and I. Ojima,
     Chem. Commun., 62, 860 (1972); P.G. Gassman, G. Gruetzmacher and
     R.H. Smith, Tetrahedron Lett., 497 (1972); S. Julia, C. Huynk and D. Michelot,
     ibid., 3587 (1972).
71   S. Mageswaran, W.D. Ollis, I.O. Sutherland and Y. Thebtaranonth, Chem.
     Commun., 1494 (1971); R.W. Jemison, T. Laird and W.D. Ollis, Chem.
     Commun., 556 (1972); G.M. Blackburn, W.D. Ollis, C. Smith and I.O. Suther-
     land, Chem. Commun., 99 (1969).
72a I.D. Brindle and M.S. Gibson, Chem. Commun., 803 (1969).
72b D.A. Evans, G.C. Andrews, T.T. Fujimoto and D. Wells, Tetrahedron Lett.,
     No. 16, 1385 (1973).
72c B.M. Trost and R.F. Hammen, J. Amer. Chem. Soc., 95 (1973) 963.
73   A.C. Cope and E.M. Hardy, J. Amer. Chem. Soc., 62 (1940) 411; E.G. Foster,
     A.C. Cope and F. Daniels, ibid., 69 (1947) 1893, and other papers in the series.
74   M.J. Goldstein and M.S. Benzon, J. Amer. Chem. Soc., 94 (1972) 7147, 7149.
75   W. v. E. Doering, V.G. Toscano and G.H. Beasley, Tetrahedron, 27 (1971)
     5299.
76   W. v. E. Doering, B.M. Ferrier, E.T. Fossel, J.H. Hartenstein, M. Jones, Jr.,
     G. Klumpp, R.M. Rubin and M. Saunders, Tetrahedron, 23 (1967) 3943.
77   H. Günther, J.B. Pawliczek, J. Ulmen and W. Grimme, Angew. Chem. Intern.
     Ed., 11 (1972) 517.
78   F.A.L. Anet and G.E. Schenck, Tetrahedron Lett., No. 48, 4237 (1970).
79   R. Askani, R. Gleiter, E. Heilbronner, V. Hornung and H. Musso, Tetrahedron
     Lett., No. 46, 4461 (1971).
80   R. Hoffmann and W.-D. Stohrer, J. Amer. Chem. Soc., 93 (1971) 6941.
81   M.J.S. Dewar and W.W. Scholler, J. Amer. Chem. Soc., 93 (1971) 1481.
82   H. Iwamura, K. Morio and T.L. Kunii, Bull. Chem. Soc. Japan, 45 (1972)
     841.
83   G. Schröder, Ber., 97 (1964) 3140.
84   J.S. McKennis, L. Brener, J.S. Ward and R. Pettit, J. Amer. Chem. Soc., 93
     (1971) 4957.
85   R.T. Seidner, N. Nakatsuka and S. Masamune, Can. J. Chem., 48 (1970) 187.
86   L.A. Paquette and M.J. Kukla, J. Amer. Chem. Soc., 94 (1972) 6874.
87   E.E. van Tamelen and B.C.T. Pappas, J. Amer. Chem. Soc., 93 (1971) 6111;
     E.E. van Tamelen, T.L. Burkoth and R.H. Greeley, ibid., 93 (1971) 6120.
88   M.J. Goldstein, R.C. Kraus and S.-H. Dai, J. Amer. Chem. Soc., 94 (1972)
     680; M.J. Goldstein and S.-H. Dai, J. Amer. Chem. Soc., 95 (1973) 933.

89   G. Binsch, E.L. Eliel and H. Kessler, Angew. Chem. Intern. Ed., 10 (1971) 570.
90   J.C. Barborak, S. Chari and P. v. R. Schleyer, J. Amer. Chem. Soc., 93 (1971)
     5275.
91   L.A. Paquette and G.H. Birnberg, J. Amer. Chem. Soc., 94 (1972) 164,
     footnote 8.
92   M.J. Goldstein and R. Hoffmann, J. Amer. Chem. Soc., 93 (1971) 6193.
93   M.J.S. Dewar and D.H. Lo, J. Amer. Chem. Soc., 93 (1971) 7201.
94   L.A. Paquette, R.E. Wingard, Jr. and R.K. Russell, J. Amer. Chem. Soc., 94
     (1972) 4739; G.P. Ceasar, J. Green, L.A. Paquette and R.E. Wingard, Jr.,
     Tetrahedron Lett., No. 20, 1721 (1973).

Chapter 7

CARBANIONS IN REACTIONS OF ORGANOMETALLIC COMPOUNDS

1. METAL CATIONS AS COUNTERIONS IN CARBANIONIC SYSTEMS

In earlier chapters of this monograph we have frequently encountered organometallic compounds in the course of our consideration of carbanions, but such encounters were largely limited in scope to the alkali metal derivatives. We may recall that carbanions which are formed under conditions where they are *stable with respect to their environment* are generally obtained from the parent carbon acid by the action of alkali metal (alkyl) amide in liquid ammonia (amine), or by means of *n*-butyllithium in a saturated hydrocarbon or ether medium. In such cases the pK_a of the carbon acid is smaller than that of the protic solvent medium, and the carbanion can be present in appreciable concentration (see, for example, the valence isomerisation processes in liquid ammonia discussed in Chapter 6, or the equilibrium pK_a investigations in amine or hydrocarbon medium treated in Chapter 1). In the cases cited the metal served merely as a positive counterion to maintain charge neutrality. Thus the metal was actually incidental to the method of formation of the carbanion.

Carbanions may also be generated under conditions in which they are *unstable with respect to their environment*, and in such cases they will be present in very small concentration only. Now, however, the gegenion is not restricted to a metal cation. A pertinent case falling under this category is the typical hydrogen-deuterium exchange process carried out with a weak carbon acid in a proton-donating medium under catalysis of a basic species. The latter may be the lyate ion of the solvent medium or, in general, any Brönsted base whose conjugate acid has a pK_a which is smaller in magnitude than the pK_a of the solvent and of the carbon acid under examination. In these systems the carbanion will be present in very small (steady-state) concentration, being immediately discharged by reaction with the proton- or deuteron-donating solvent molecule. Proton transfer studies of this type commonly involve amine bases, so that the counterion in such cases would be the corresponding alkylammonium ion (see Chapter 1, Sections 3, 5, for examples).

There are of course other systems where carbanions are present as unstable reaction intermediates, in addition to the hydrogen-deuterium exchange case. An example which we have considered in some detail in this monograph is the base-catalyzed bromination of ketones (Chapter 3); another is the Favorskii rearrangement (Chapter 5). Other cases falling under this category include the olefin-forming eliminations proceeding by the E1CB mechanism [1], as well as a variety of synthetic processes which utilize carbanions as unstable intermediates [2, 3]. In such reactions the carbanions undergo rapid transformation through reaction with other reagents or by internal reorganization. In general the counterion in such reactions is again incidental, and is not restricted to metal cations.

For completeness we may consider the proton (suitably solvated) as a possible alternative counterion, in which case we would be dealing with a dissociated carbon acid. However, as was seen in Chapter 1, there are only few carbon acids which dissociate measurably in aqueous solution, notably those with multiple nitro, cyano, or carbonyl substitution in the position *alpha* to the dissociating C–H bond. Of the carbon acids which do not contain such substituents, we are left with fluoradene and related molecules which on dissociation give rise to the aromatic cyclopentadienide type anions. Thus the proton as counterion in carbanionic systems is applicable only to very limited situations.

2. SCOPE OF ORGANOMETALLIC CHEMISTRY AND OF PRESENT COVERAGE

If one were to limit consideration of organometallic compounds to those situations where the metal serves merely as a counterion to the carbanionic species, one would be restricting oneself to a very small aspect of a vast subject. Moreover, our chosen segment would not even be representative of the subject as a whole. Nevertheless it is clearly beyond the scope of this monograph to even attempt to cover in depth representative areas of organometallic chemistry. This is witnessed by the fact that not only are there monographs on the general subject of organometallic chemistry (e.g. refs. 4–8) and on the organic chemistry of individual metals (e.g. refs. 9–16), but there is also a fully-fledged journal of organometallic chemistry as well as series volumes covering the latest advances and developments in various aspects of the field. Under these circumstances the scope of the present treatment must necessarily be limited.

In this chapter no discussion will be given of organometallic compounds of the transition metals and we also exclude from consideration organometallic π-complexes. Thus we limit the present treatment to selected organometallic and organometalloid compounds in which the C–M σ-bond has partial carbanionic character. For convenience the terms "organometallic" and "organometalloid" will, in general, be used synonymously.

We have chosen to discuss in the main some aspects of the chemistry of organosilicon and organomercury compounds, from the viewpoint of the ways in which such compounds may act as precursors of carbanionic species. However, in order to be able to contrast the properties and reactions of organosilicon and organomercury compounds with those of their alkali metal analogs, some knowledge of the factors influencing structure and reactivity in organometallic chemistry is needed. In the following two sections a brief review of such factors is presented.

3. BOND POLARITY IN ORGANOMETALLIC COMPOUNDS AND ELECTRO-NEGATIVITY CONSIDERATIONS

A major characteristic feature through which one may distinguish the properties of organosilicon and organomercury compounds from those of the alkali metals is the polarity of the carbon–metal bond. In alkali metal organic compounds the carbon–metal bond has predominantly ionic character, as seen, for example, in their NMR spectra by an upfield shift of the proton absorptions [17–23]. (This should not be taken to imply that alkali metal alkyls necessarily exist in dissociated form, since the dissociation would be influenced by factors such as the nature of the organic moiety, the metal, and the solvent medium; for further information on ion-pairing phenomena in carbanionic systems, including discussion of contact ion pairs and solvent-separated ion pairs, see refs. 23–25 and Chapter 1, Section 2.) On the other hand, in organosilicon and organomercury compounds the carbon–metal bond has predominantly covalent character. Thus for these metal organic compounds one may exclude ionization as a feasible process under circumstances in

211

which the behaviour of the corresponding saline alkali metal compounds would be governed by considerations based primarily on principles of electrostatic bonding.

One parameter by which bond polarity in organometallic compounds may be estimated is the electronegativity difference between the two elements forming the C—M bond. In Table I are presented electronegativity values for carbon and some representative metallic and metalloid elements belonging to Groups Ia, Ib, IIa, IIb, IIIa, IVa, Va, and VIa of the Periodic Table. The data given are the Pauling electronegativity values (see ref. 26 for comparison with electronegativity values calculated by the Allred-Rochow and Mulliken methods). Bond polarity, $C^{\delta-}$—$M^{\delta+}$, will increase as the magnitude of the difference in electronegativity for the two elements increases.

Table 1

ELECTRONEGATIVITIES (PAULING) OF SOME METALLIC AND METALLOID NONTRANSITION ELEMENTS

Ia		IIa		IIIa		IVa	
Li	0.98	Be	1.57	B	2.04	C	2.55
Na	0.93	Mg	1.31	Al	1.61	Si	1.90
K	0.82	Ca	1.00	Ga	1.81	Ge	2.01
Rb	0.82	Sr	0.95	In	1.78	Sn	1.96
Cs	0.79	Ba	0.89	Tl	2.04	Pb	2.33
Ib		**IIb**		**Va**		**VIa**	
Cu	1.90	Zn	1.65	As	2.18	Se	2.55
Ag	1.93	Cd	1.69	Sb	2.05	Te	2.1
Au	2.54	Hg	2.00	Bi	2.02	Po	1.9

Use of Pauling's relationship [27] between electronegativity differences and degree of ionic character leads to the following estimates for the percent ionic character in various C—M bonds: C—Li, 42%; C—Cs, 55%; C—Si, 12%; C—Hg, 9%. These estimates should be regarded as qualitative only, in part because electronegativity values differ when derived by different methods [26]. More importantly, it is now recognized that the simplifying assumptions used by Pauling to derive the relationship between electronegativity difference and ionic character are

212

open to doubt [26]. Nevertheless, when used as a rough guide, the data serve well to contrast bond polarity in organosilicon and organomercury compounds from their alkali metal analogs. The placing at virtually opposite ends of the polarity scale of the C–Cs and C–Hg bonds must certainly be significant in determining the properties of the respective organometallic compounds.

Another point to which attention may be drawn is that the polarity of a given carbon–metal bond will be dependent on the molecular environment. For example, the *alpha* carbon in benzyllithium will have a greater ability to attract electrons to itself than the carbon in methyllithium, and this will lead to a greater carbon–lithium bond polarity in the former compound. Thus bond polarity, $C^{\delta-}$—$M^{\delta+}$, in a given metal alkyl R_3CM is actually related to the carbanion stability of R_3C^-, which takes us back to the subject matter of Chapter 1.

Bond polarity, though an important factor, does not provide a sufficient basis for full understanding of the structure and reactivity of organometallic compounds. The availability of low-lying unoccupied electronic orbitals on the metal centre is clearly a matter of consequence. Moreover, structural studies with organometallic compounds have brought to light certain aspects of chemical bonding which are not apparent in the typical organic compounds. This point is elaborated in the next section.

4. MOLECULAR STRUCTURE OF ORGANOMETALLIC COMPOUNDS

Whereas most organic functional compounds are monomeric, metal alkyls often exist as dimers and higher aggregates in solution and in the gaseous, liquid or solid state. Study of these association phenomena has revealed a number of interesting facets relating to structure and bonding. Some representative metal alkyls will be considered in this context.

Lithium alkyls derived from the lower alkanes are known to be associated in ether or in hydrocarbon solution as well as in the vapour phase. Aggregation into tetramers and hexamers is established by means of a variety of measurements of the colligative properties of such solutions [28a, b] and of the vapour phase spectral characteristics [28a, c]. Elucidation of the structure of the methyllithium tetramer derives largely from NMR studies (^{1}H, ^{7}Li, and ^{13}C) [29] and from X-ray crystallographic analysis [30, 31]. The structure of the (MeLi)$_4$ unit is

213

shown below. In this structure, 1, the lithium atoms are placed at the apixes of a tetrahedron, while the methyl groups are situated over each face and are bridged to three lithium atoms. The ethyllithium tetramer has an analogous structure. A somewhat similar, octahedral, structure has been advanced for the methyllithium hexamer [28a].

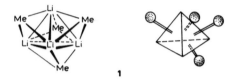

1

The nature of the chemical bonding in electron deficient structures of this type, in which there are insufficient valence electrons for normal covalent bonding to occur, has been discussed by a number of authors (with particular reference to the boron hydrides) on the basis of multi-centre molecular orbitals [31–34]. The methyl group in 1 is said to be "bridged" and its bonding described by a four-centre two-electron bond.

A bridged structure (2) is also indicated for dimethylberyllium, which is found to be polymeric in the crystalline state [35]. Dimethylmagnesium has an analogous polymeric structure [36].

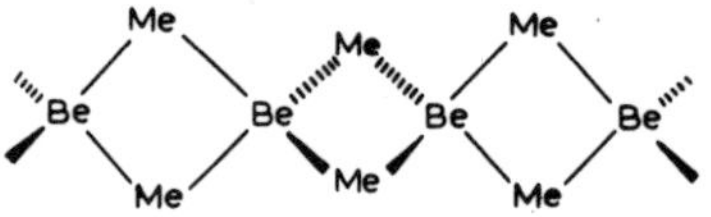

2

Trimethylaluminium has been shown by X-ray crystallography to be dimeric, with symmetrical Al–C–Al bridges as indicated in structure 3 [33]:

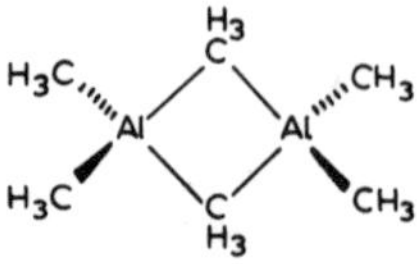

3

An important consequence of the equidistant Al–C–Al bridging is that this rules out the possibility of a dynamic equilibrium between two

214

unsymmetrical structures (4) which, if present, would negate the idea of electron deficient bonding in this compound:

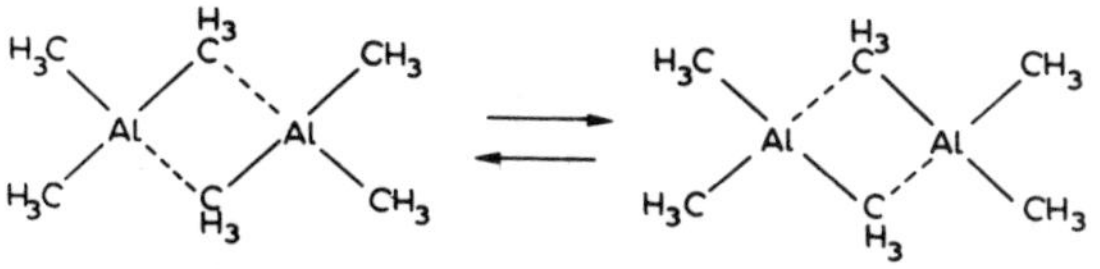

4a   **4b**

Trimethylaluminium is also dimeric in solution (hydrocarbon solvents); however, NMR studies point to alkyl exchange via solvated $AlMe_3$ monomer [37–39]. Tri-isobutylaluminium is largely monomeric [40], indicating a decreasing tendency towards association as the size of the alkyl group increases.

Though diborane (B_2H_6) is probably the archetypal case of an electron deficient molecule (the bridging hydrogens may be described by means of three-centre, two-electron bonds), trimethylboron does not form stable bridged structures [10a]. The compound exists as a monomer both in the solid state and in solution. However, the occurrence of re-distribution reactions of the type shown in eq. (1) [41] suggests that dimer structures (5) may be formed as transition states or as metastable intermediates:

$$(1) \quad R_2BR' \; + \; R_2BR' \; \rightleftharpoons$$

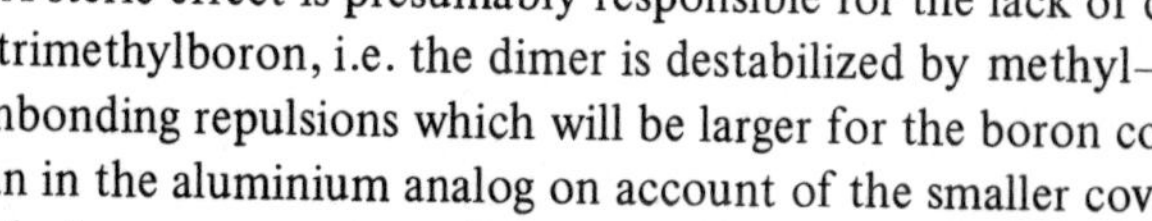

$$\rightleftharpoons \; R_3B \; + \; RBR'_2$$

5

A steric effect is presumably responsible for the lack of dimerisation of trimethylboron, i.e. the dimer is destabilized by methyl–methyl nonbonding repulsions which will be larger for the boron compound than in the aluminium analog on account of the smaller covalent radius of the boron atom. An unfavourable metal–metal interaction may contribute to destabilization of the boron dimer. A MO analysis of the problem has been made [34a].

The metal alkyls of cadmium, zinc and mercury are monomeric. However, compounds such as dimethylcadmium, dimethylzinc and dimethylmercury undergo alkyl exchange, presumably by the mechanism

215

of eq. (1). An elegant example of such an exchange is provided by the following isotope exchange reaction in THF solution [42]:

(2) $\qquad (CH_3)_2Hg + (CD_3)_2Hg \rightleftharpoons 2 CD_3HgCH_3$

A similar exchange is observed between $(CH_3)_2Mg$ and $(CD_3)_2Hg$ [42]. As an alternative approach, exchange between R_2Hg and R'_2Hg can be studied via the mercury isotope [43, 44]. The $R_2Mg + Ph_2Hg \rightleftharpoons Ph_2Mg + R_2Hg$ system, as studied by NMR in dimethoxyethane medium, has been used to establish a scale of carbanion stabilities [45]; it is apparent that equilibrium in this metal—metal interchange process is attained such that the more electropositive element becomes preferentially linked to the more stable incipient carbanion. (Analogous results are obtained in study [46] of the metal—halogen exchange, $RLi + PhI \rightleftharpoons RI + PhLi$.)

5. HETEROLYTIC BOND SCISSION PROCESSES OF ORGANOMETALLIC COMPOUNDS

On the basis of bond polarity considerations, as discussed in Section 3 of this chapter, one would predict that heterolytic scission of the C—M bond in the typical organometallic compound would result in the formation of a carbanion and the metal cation:

(3) $\qquad \overset{\delta- \ \delta+}{R:M} \longrightarrow R:^- + M^+$

In the general case, the carbanion thus generated would not be expected to be stable under typical reaction conditions (see Section 1) and would partake in reaction(s) with some incipient electrophilic species resulting in formation of a stable entity. It follows that in these circumstances the overall reaction becomes one of substitution of M^+ by electrophile E^+ at the carbon centre of the organometallic:

(4) $\qquad R—M + E^+ \longrightarrow R—E + M^+$

Among the electrophilic species which may partake in the reaction (4) the simplest would be the proton, in which case the overall reaction would be one of protolysis or acidolysis. (We may recall that a commonly used procedure for the preparation of deuterated compounds makes use of the sequence, $R—X \xrightarrow{M} RM \xrightarrow{D^+} RD$.) Another possible

216

electrophilic reagent is the halonium ion, corresponding to the halogenation of organometallics. However, of particular interest to us is the situation in which another metal cation constitutes the electrophilic species. Such metal–metal interchange reactions have been among the most informative of the transformations undergone by organometallics. Clearly the metal–metal interchange reaction (eq. (5)) may be classified as an electrophilic substitution at carbon.

$$(5) \qquad R-M_1 + M_2^+ \longrightarrow R-M_2 + M_1^+$$

In the historical sense, as well as pedagogically, one may look upon electrophilic substitution at saturated carbon (S_E processes) as the counterpart of nucleophilic substitution (S_N processes), with the same challenges arising in understanding of reaction mechanisms [47, 48]. Studies with organometallics have contributed largely toward our current understanding of electrophilic substitution.

For completeness, we should note that organomercurials may react also by another type of heterolytic process, leading to the formation of carbonium ions. Thus, under solvolytic conditions we may have the following processes taking place in sequence:

$$(6) \qquad R-Hg-X \longrightarrow R-Hg^+ + X^-$$
$$(7) \qquad R-Hg^+ \longrightarrow R^+ + Hg^0$$
$$(8) \qquad R^+ \longrightarrow \text{solvolysis products}$$

This type of *demercuration* reaction would be favoured by strongly electronegative ligands (X), and should be highly sensitive to the nature of the alkyl moiety (R), and the solvent medium. Thus, in general, the organomercury halides undergo demercuration extremely slowly, if at all, the acetates react at moderate rates, while the perchlorates do so extremely rapidly. *t*-Butylmercurials solvolyse many orders of magnitude faster than the methyl analogues (e.g. $k(t\text{-BuHgX})/k(\text{MeHgX}) \sim 10^{13}$ for acetolysis at $25°$) [49]. *Exo* norbornyl mercuric perchlorate is more reactive in acetolysis than the *endo* isomer, by a factor of 1600 [50]; evidence for the formation of nonclassical carbonium ions in these processes has been presented [51]. Rearrangements, typical of carbonium ion processes, are normally observed in demercurations. Similar considerations apply also to demetallation reactions involving lead, thallium, and some other organometallic derivatives [52].

In contrast, organosilicon derivatives, R_3SiX, do not undergo ioniza-

tion to positive "siliconium" ion species, $R_3SiX \xrightarrow{\ \ \ } R_3Si^+ + X^-$, an
observation that has been the subject of considerable discussion [11].

In conclusion of this section, mention is made of a number of
reactions which proceed via silyl anions or silylcarbanions, among which
are found a number of interesting rearrangements. These intriguing
systems are outside our scope of coverage and the interested reader is
referred to the original literature [53–56, 74].

6. CLASSIFICATION OF S_E MECHANISMS

The generalized case of the S_E reaction could perhaps be considered,
on the basis of the limited discussion given in the previous section, to
proceed by rate-determining ionization of RM according to eq. (3),
followed by rapid reaction of R^- with an electrophilic reagent E^+. It is
immediately seen, however, that the carbanion mechanism of electrophilic substitution need not be the only possible pathway in the conversion of RM to RE. A plausible alternative pathway would be a concerted
displacement of M^+ by E^+. These two possibilities may be designated as
the S_E1 mechanism (unimolecular electrophilic substitution) and the
S_E2 mechanism (bimolecular electrophilic substitution), by analogy with
the dichotomy of mechanistic pathways applicable to nucleophilic
substitution processes at saturated carbon.

The S_E1 and S_E2 mechanisms are differentiated fundamentally by
the number of species which are present in the transition state of the
rate-limiting step, structures 6 and 7, respectively:

S_E1:

(9a) $\quad R-M \xrightarrow{\text{slow}} \left[R \cdots\cdots M\right]^{\ddagger} \longrightarrow R^- + M^+$
$\qquad\qquad\qquad\qquad\qquad\qquad 6$

(9b) $\qquad\qquad\quad R^- + E^+ \xrightarrow{\text{fast}} R-E$

S_E2:

(10) $\quad E^+ + R-M \xrightarrow{\text{slow}} \left[E\cdots R\cdots M\right]^{\ddagger} \longrightarrow E-R + M^+$
$\qquad\qquad\qquad\qquad\qquad\qquad 7$

The kinetic distinction between the S_E1 and S_E2 mechanisms is quite
clear in principle. The former process (eq. (9)) will be first order in
substrate and zero order in electrophilic reagent, provided that the
ionization step (eq. (9a)) is not appreciably reversible. The latter
process (eq. (10)) will obey second order kinetics, first order in substrate

218

and first order in electrophilic reagent. These rate laws are directly analogous to established kinetic behaviour in S_N1 and S_N2 processes.

The stereochemical criteria of S_E1 and S_E2 reactions are subject to greater uncertainty than in corresponding S_N reactions. One would predict that the S_E1 mechanism would lead to racemization, provided that the carbanion is not configurationally stable and that there is no shielding effect from the departing metal cation (cf. Chapter 2, Section 3).

In the case of the S_E2 reaction, by analogy with the S_N2 process it was initially anticipated that the stereochemical result would be inversion of configuration, through rearward displacement as shown in the alternative formulations of transition state 8:

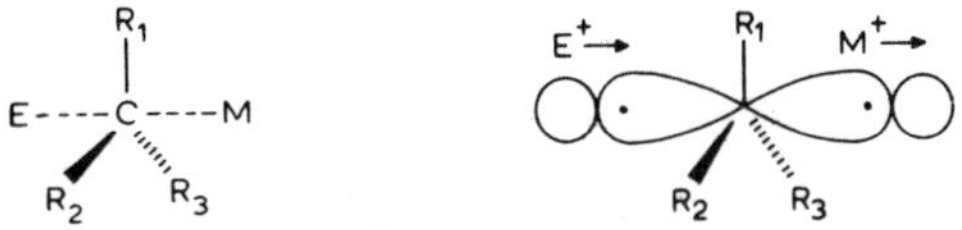

$$E\text{----}\underset{\underset{R_2}{|}\;\;\underset{R_3}{}}{\overset{R_1}{C}}\text{----}M$$

8

However, this presumed analogy discounts the fact that in the S_E2 transition state there are only four electron pairs surrounding the central atom, so that electrostatic repulsion between the incoming and outgoing ligands no longer has the dominant importance which is assigned to such interaction in the S_N2 transition state. Thus an alternative mode of attack by electrophile, adjacent to the departing group (i.e. front side attack), leads to the transition state configuration shown in 9. (The arrows indicating directionality in 8 and 9 are for illustrative purposes only.)

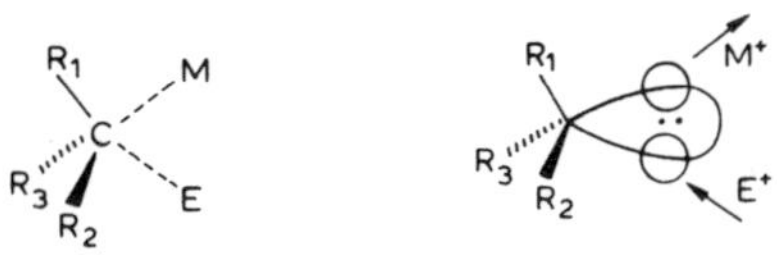

9

Clearly the stereochemical result of 9 will be retention of configuration. In fact it is now known that retention is the normal result in electrophilic substitution. (An exception has been found [57] in the bromine cleavage of organotin compounds.) It has been pointed out [58] that a modification of 9 to include orbital overlap between E and M should be energetically favourable, by analogy with the 2π-electron system of the cyclopropenium cation.

A further point of difference arises between electrophilic and nucleophilic processes as a result of constitutive differences in the structure of reagents. Thus, in practice, electrophilic reagents rarely consist of an unattached positive atomic species. Rather, electrophiles generally come firmly attached to nucleophilic ligands. Pertinent examples are the mercuric salts, HgX_2, and the monoalkyl mercury derivatives, RHgX, which are commonly used reagents in electrophilic substitution, shedding the ligand X in the process. Thus, the following reaction types have all been well characterized:

(11) X_2Hg ⌢R—HgX ⇌ XHg—R⌒ HgX_2

(12) X_2Hg ⌢R—HgR ⇌ XHg—R⌒ HgRX

(13) XRHg ⌢R—HgR ⇌ RHg—R⌒ HgRX

These "redistribution" reactions may be called one-alkyl (eq. (11)), two-alkyl (eq. (12)) and three-alkyl (eq. (13)) exchanges [47, 59].

As a result of this structural characteristic in the electrophilic reagent, one may view the reacting entity as consisting of an electrophilic portion (E) and a nucleophilic portion (N), covalently bonded. This now raises the possibility of coordination between the electronegative pole of E–N and the (electropositive) metal atom of R–M, as an integral part of the rate process. If this nucleophilic participation occurs simultaneously with electrophilic attack then a four-centred transition state would obtain:

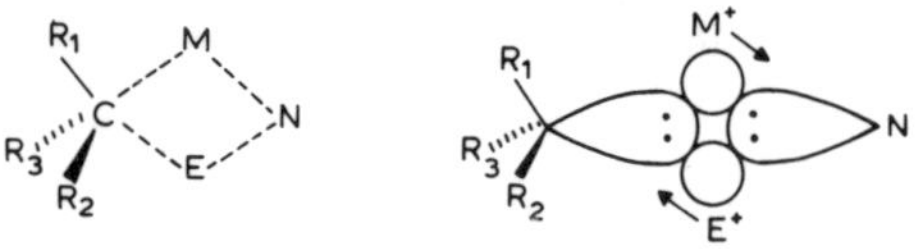

10

This cyclic process, which is named as the S_Ei mechanism (internal electrophilic substitution, also called on occasion the four-centre S_F2 mechanism), would be subject to second order kinetics and should *a priori* proceed with retention of configuration. Thus, the kinetic and stereochemical criteria will apply alike to the S_E2 and S_Ei pathways.

As an alternative to simultaneous attack by E and N, leading to the cyclic mechanism, one may visualize nucleophilic attack on the metal atom to *precede* electrophilic attack on carbon. This new mechanism may be denoted as S_EC (electrophilic substitution via coordination) [60].

Finally, the S_E1 mechanism may be modified so as to include nucleophilic coordination to the metal, called the S_E1 (N) mechanism [61]. The various reaction types may be represented as follows, with the nature of bonding changes occurring in the rate-determining transition state indicated by the arrows:

Some clarification of these diagramatic representations is needed, however, particularly for the S_EC and S_E1 (N) cases. First, in both these cases it has been assumed here that coordination between M and N occurs in an initial, fast step. Second, in actual S_E1 (N) processes (see Sections 7, 8) the nucleophilic coordination is typically by halide ions or alkoxide ions (e.g. added as the alkali metal salts), which are not associated with the electrophilic species that subsequently reacts with the carbanions. Thus the E—N representation for the S_E1 (N) case is not strictly valid and is used only to keep uniformity throughout the set of diagramatic representations.

One can envisage a gradation of mechanistic types, covering the broad spectrum of bonding changes and transition state configurations. Differentiation between such mechanisms may well be difficult in practice. Mechanistic criteria which have been suggested for this purpose include the polar and steric requirements of the alkyl moiety, the nature of E and N and of the ligand (X) coordinated to M (not shown in the above formulations for simplicity), as well as the nature of the reaction medium. Considerable progress in this direction has already been made [8, 16, 47, 48].

A reaction shown to proceed by the S_Ei mechanism is the following [62]. The one-alkyl exchange reaction (eq. (11), R = Me or s-Bu), which proceeds with retention of configuration, is strongly catalyzed by added anions, in ethanol solution, the observed catalytic order being $I^- > Br^- > Cl^- > OAc^- > NO_3^-$. From the rate data it was possible to identify a one-anion as well as a two-anion kinetic process. These results point to two separate catalytic pathways proceeding by cyclic (S_Ei) mechanisms, as shown by the transition state structures 11 and 12, respectively. The anion which is added first assists via a bridging function, while the second anion serves by coordinating with the mercury atom of RHgX. The relative order of anion efficacy parallels the ability of these ions to partake in complex formation.

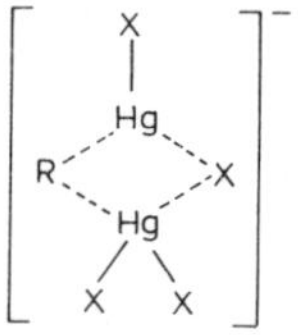

11 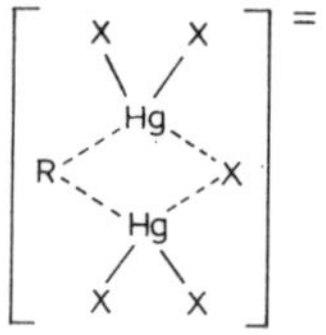   12

The S_E2 mechanism has been established for all three types of alkyl exchange processes (eq. (11)–(13)). Only one example will be given here, which will demonstrate the degree of elegance with which various criteria may be combined to bear upon the problem of the elucidation of reaction mechanisms. The reaction concerned [63] is the three-alkyl exchange process, eq. (14) with R = s-Bu, which was studied in ethanol solution both polarimetrically and by means of radiomercury-203:

(14a) s-BuHgBu-s + s-Bu$\overset{*}{\text{Hg}}$Br $\longrightarrow$ s-Bu$\overset{*}{\text{Hg}}$Bu-s + s-BuHgBr

(14b) s-BuHgBu-s + s-Bu$\overset{\circ}{\text{Hg}}$Br $\longrightarrow$ s-Bu$\overset{\circ}{\text{Hg}}$Bu-$s$ + s-BuHgBr

The reaction is first order in each reactant (second order overall) and the rate of transfer of the radio label (indicated by the asterisk in eq. (14a)) is equal to the rate of transfer of optical label (indicated by the degree sign in eq. (14b)). There is a complete retention of configuration. The rate of reaction increases along the series of electrophilic reagents, s-BuHgBr $<$ s-BuHgOAc $<$ s-BuHgNO$_3$, which follows the order of increasing ease of ionization of the coordinated ligand. All these relationships are characteristic of the S_E2 mechanism of reaction.

Having dwelt on metal–metal interchange as representative of one type of electrophilic substitution process at carbon, we now return to the other reaction types introduced in Section 5, that is protolysis and halogenation of metal-organic compounds. Also, the illustrative examples will now be chosen from the chemistry of organo-silicon and other Group IV metal derivatives, in addition to organomercurials.

The acid cleavage of dialkylmercurials, resulting in rupture of the carbon-mercury bond, has been known for over 100 years [64]. Two relatively recent examples are given in equations (15) [65] and (16) [66].

(15)

(16)

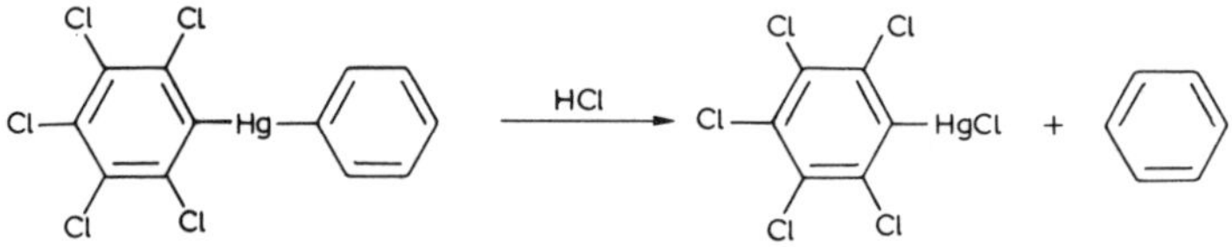

The direction of cleavage in eq. (16) is in accord with an electrophilic process, since it is noted that electron density will be lower at the phenyl group bearing the halogen substituents. Also in accord with electrophilic attack in such processes is the general experience that alkylmercuric salts RHgX usually react slower than dialkylmercurials. However, some alkyl-mercuric salts are sufficiently reactive and have been studied extensively; notable cases are methyl-, vinyl-, and allylmercuric iodides [67–69]. A rate-determining proton transfer (S_E2) mechanism is proposed in the allyl case on the basis of isotope effect studies in aqueous acids:

(17) $H^+ + CH_2{=}CH{-}CH_2{-}HgI \xrightarrow{\text{slow}} CH_3{-}CH{=}CH_2 + HgI^+$

Allylgermanes are cleaved by HBr in similar fashion [70], though the detailed mechanism has not been investigated:

(18) $HBr + CH_2{=}CH{-}CH_2GeBu_3\text{-}n \longrightarrow CH_3{-}CH{=}CH_2 + n{-}Bu_3GeBr$

Perhaps the most intensively investigated protic cleavage of organo-metallics is that of the aryl carbon–metal bond in group IV metal derivatives [71–77], that is eq. (19) with M = Si, Ge, Sn, Pb:

(19) $ArMR_3 + HX \longrightarrow ArH + R_3MX$

These *protodemetallation* reactions are typically performed in aqueous acid media (with the hydronium ion as HX), which allows for investiga-tion of the kinetic form of the acid dependence, of substituent effects, and the effect of varying the arene moiety. The mechanisms favoured for these reactions are centered on formation of sigma complex inter-mediates 13 and 14, though in some cases the cyclic mechanism (transition state structure 15) cannot be excluded. (These mechanisms could be described as $A-S_E2$, $A-S_EC$ and $A-S_Ei$, denoting that substitution occurs here at aromatic carbon, with consequent possibility for delocalization of positive charge over the benzene ring.)

223

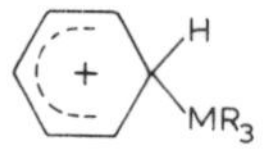

13 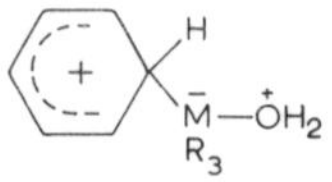   **14** 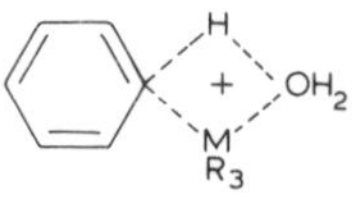   **15**

Similar considerations apply to aryl derivatives of boron [78], mercury [79] and so on. Not surprisingly, such reactions are often discussed under the heading of electrophilic aromatic substitution [80].

Finally, we consider, very briefly, the halogen cleavage of the carbon—metal bond, reactions which are of wide importance synthetically [74—77]. The mechanism is often complex. For example, the cleavage of dialkylmercurials by halogens

$$(20) \qquad R-Hg-R' + X_2 \longrightarrow RHgX + R'X$$

is often complicated by the occurrence of competing homolytic and heterolytic processes [81, see also 82 for recent work on homolytic displacements in organometallic compounds]. Conditions favouring heterolytic scission may be attained by use of pyridine complexes of the halogens, in which halogen is more polarized ($C_5H_5N:Br\text{-}Br \leftrightarrow C_6H_5N\text{-}Br^+ Br^-$), or by use of tri-iodide ion (I_3^-) as halogenating agent.

Extensive mechanistic studies of the halogen cleavage of tetraalkylstannanes have been reported and discussion has been given on the basis of S_E2 and S_Ei mechanisms [83, 84]. Aromatic bromo- and chloro-desilylation has also been studied mechanistically [85, 86]. The remarkable observation [57] of inversion of configuration at carbon in the bromine cleavage of an alkyltin compound has already been referred to.

Having reviewed the background of the various modes of electrophilic substitution, we now proceed to examine the carbanionic pathways in more detail. Organomercurials are treated first, followed by organosilicon compounds and, more briefly, the germanium, tin and lead analogues.

7. CARBANION MECHANISMS IN ELECTROPHILIC SUBSTITUTION REACTIONS OF ORGANOMERCURY COMPOUNDS

The dissociative, unimolecular, carbanion mechanism for electrophilic substitution at mercury has been established only during the past decade. The time lag between the discovery of the S_E1 and the other mechanistic types of processes is due, in part, to certain recent developments in the chemistry of organomercurials, such as the synthesis of some key optically active substrates. An equally important contributing factor, however, is the rapid progress which has been made during the last decade toward the understanding of the underlying principles which govern carbanionic systems, their formation, properties, and reactivity.

(a) Isotopic exchange in α-carbethoxybenzylmercuric bromide

The key experiments were performed in the early 1960's in the schools of Reutov and Ingold [87, 88]. The studies were undertaken independently using the same organomercury derivative though by somewhat different, complementary, approaches. The substrate concerned is α-carbethoxybenzylmercuric bromide, and the combined approaches involve kinetic, stereochemical and isotopic evidence.

The isotopic exchange reaction of α-carbethoxybenzylmercuric bromide with radiomercuric bromide in anhydrous dimethyl sulfoxide (DMSO) medium is first-order in substrate but zeroth-order in Hg^*Br_2. When the isotopic exchange is carried out with optically active α-carbethoxybenzylmercuric bromide one observes complete racemization of the substrate. Moreover, the rate of incorporation of isotopic label into the substrate is equal to the rate of racemization. These observations uniquely define the reaction as proceeding by the unimolecular, S_E1, mechanism:

$$(21) \quad Ph\!-\!\underset{\underset{HgBr}{|}}{CH}\!-\!CO_2Et \xrightarrow{\text{slow}} Ph\!-\!\overset{(-)}{CH}\!-\!CO_2Et \;+\; {}^+HgBr$$

$$(22) \quad Ph\!-\!\overset{(-)}{CH}\!-\!CO_2Et \;+\; \overset{*}{Hg}Br_2 \xrightarrow{\text{fast}} Ph\!-\!\underset{\underset{\overset{*}{Hg}\,Br}{|}}{CH}\!-\!CO_2Et \;+\; Br^-$$

We may note that the carbanion obtained is actually an enolate ion, which is known to be planar (see Chapter 3, Section 1e); hence the observation of racemization.

It should also be pointed out that it is not necessary to implicate free carbanions in the reaction, since an ion-pair mechanism can satisfactorily account for the kinetic and stereochemical observations. A more complete scheme would then include the following steps:

$$\text{Ph}-\underset{\underset{\displaystyle \text{HgBr}}{|}}{\text{CH}}-\text{CO}_2\text{Et} \quad \xrightleftharpoons[\text{}]{\text{slow}} \quad \text{Ph}-\overset{(-)}{\underset{\underset{\displaystyle {}^+\text{HgBr}}{:}}{\text{CH}}}-\text{CO}_2\text{Et}$$

(23)

$$\text{Ph}-\underset{\underset{\displaystyle \text{Hg}^*\text{Br}}{|}}{\text{CH}}-\text{CO}_2\text{Et} \quad \xleftarrow[\text{slow}]{\text{}} \quad \text{Ph}-\overset{(-)}{\underset{\underset{\displaystyle {}^+\text{Hg}^*\text{Br}}{:}}{\text{CH}}}-\text{CO}_2\text{Et}$$

An interesting argument has developed over the question of the applicability of the principle of microscopic reversibility to such schemes for isotopic exchange; the reader is referred to the original sources [89a, 90].

The reaction medium has a vital role on mechanism. Thus, when the reaction of α-carbethoxybenzylmercuric bromide with Hg^*Br_2 is carried out in aqueous acetone, aqueous ethanol, or in pyridine, second order kinetics are followed (first order in substrate and first order in reagent) and reaction occurs with retention of configuration. Presumably the reaction pathway which is followed under these conditions is S_E2. Stabilization of carbanionic species in dipolar aprotic solvents such as DMSO, relative to the common protic solvents, is well established [91].

Of considerable interest is the observation of bromide ion nucleophilic catalysis in the isotopic exchange reaction of α-carbethoxybenzylmercuric bromide with Hg^*Br_2 [88]. The rate of reaction in DMSO medium is dependent on the *square* of the concentration of added bromide ion. This result is in accord with occurrence of successive equilibria leading to formation of complex, $PhCH(CO_2Et)HgBr_3^{=}$, which then undergoes rate-determining fission to two mutually repelling fragments, $HgBr_3^-$ and the carbanion. The overall sequence of reactions is:

(24) $\quad PhCH(CO_2Et)HgBr + Br^- \xrightleftharpoons[\text{}]{\text{fast}} PhCH(CO_2Et)HgBr_2^-$

(25) $\quad PhCH(CO_2Et)HgBr_2^- + Br^- \xrightleftharpoons[\text{}]{\text{fast}} PhCH(CO_2Et)HgBr_3^=$

(26) $\quad PhCH(CO_2Et)HgBr_3^= \xrightarrow{\text{slow}} \overset{(-)}{Ph}CHCO_2Et + HgBr_3^-$

(22) $\quad \overset{(-)}{Ph}CHCO_2Et + Hg^*Br_2 \xrightarrow{\text{fast}} PhCH(CO_2Et)Hg^*Br + Br^-$

This catalytic mechanism may be denoted as S_E1-2Br^- and is essentially an example of the $S_E1(N)$ mechanism discussed in Section 6.

226

(b) Isotopic exchange in p-nitrobenzylmercuric bromide

The next case of the $S_E 1$ mechanism in organomercury compounds to be considered involves p-nitrobenzylmercuric bromide [92]. It is found that this compound undergoes isotopic exchange with Hg*Br$_2$ in DMSO solution by a first-order process, in accord with rate-determining ionization to yield the resonance-stabilized p-nitrobenzyl anion 17.

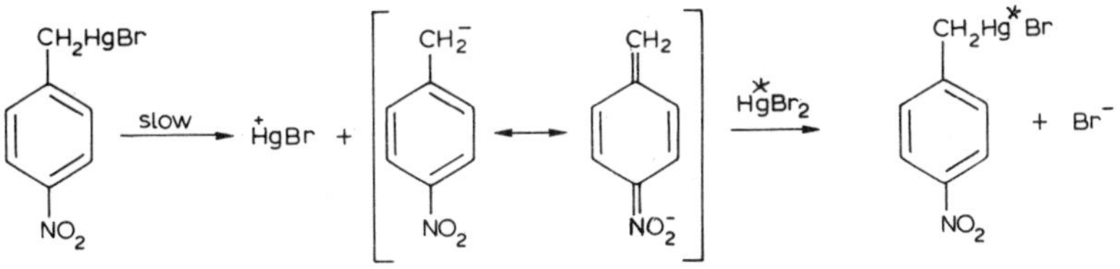

16   17

It is noteworthy that in the case of ring substituents, such as p-methyl or p-chloro, a bimolecular mechanism is followed. The p-nitrobenzyl-mercuric bromide isotope exchange reaction is also catalyzed by added bromide ion, though the rate dependence on bromide ion concentration is not a simple one. It has been pointed out [89b] that such catalytic effects of the added bromide ion could be due to the intervention of free radical processes; possible mechanisms for radical-catalyzed exchange have been given [89b].

(c) Protolysis of 4-pyridiomethylmercuric chloride

The last example of the $S_E 1$ mechanism discussed concerns 4-pyridio-methylmercuric chloride (18), reacting in aqueous solution to yield the 4-methylpyridinium ion (20) [93]. The reaction type involved here is somewhat different, in that it is an overall hydrogen-for-metal substitution (protolysis), in contrast to the metal–metal substitution considered in the previous two cases. Though the starting pyridinium salt is stable in dilute aqueous perchloric acid solution, protolysis proceeds readily in the presence of chloride ions. Catalytic pathways can be identified corresponding to $S_E 1 - Cl^-$ (path a) and $S_E 1 - 2Cl^-$ (path b) mechanisms. In these pathways the complexed ions formed by pre-equilibria undergo slow cleavage with loss of HgCl$_2$ (path a) and HgCl$_3^-$ (path b) to yield the picolinium betaine 19 which is then rapidly protonated.

(28)

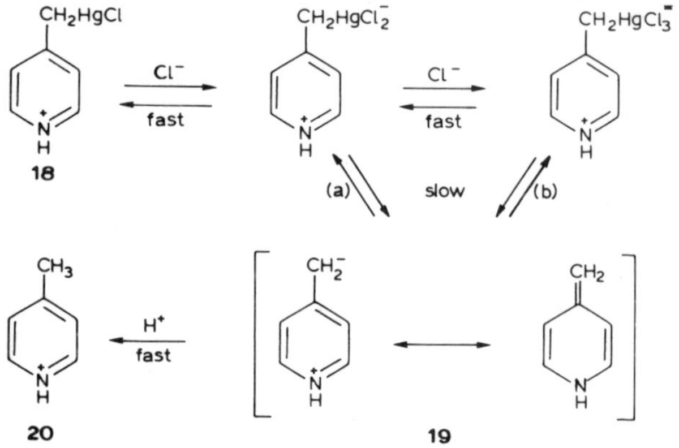

It is noteworthy that 3-pyridiomethylmercuric chloride is unreactive under these conditions. This lack of reactivity readily follows from the absence of conjugative stabilization in the 3-pyridiomethide anion.

If the mechanism of eq. (28) is correct, it might be anticipated that the 4-pyridiomethide anion might also be captured by electrophilic species other than the solvated proton. In accord with this expectation it was shown that added $HgCl_2$ represses the rate of acidolysis of 4-pyridiomethylmercuric chloride [94]. Similarly, rate measurements in presence of added $HgCl_2$ and Cl^- showed that the $HgCl_3^-$ and $HgCl_4^=$ species are also capable of intercepting the 4-pyridiomethide anion. The last mentioned electrophile probably suffers loss of chloride ion in the process:

(29)

These observations are analogous to the well known common ion effect in S_N1 processes [95].

8. CARBANIONS IN ELECTROPHILIC SUBSTITUTION REACTIONS OF ORGANOSILICON AND ORGANOTIN COMPOUNDS

As one proceeds from organomercury derivatives to those of silicon and tin, one finds a change in emphasis in the pathways available for carbanion formation. The dissociative unimolecular (S_E1) mechanism is

228

not observed, while nucleophilic participation becomes a requirement, in these carbanion-forming electrophilic substitution processes.

The types of reactions where the possibility of carbanion formation arises is between mixed tetraalkyl (or aryl) derivatives R'_3MR and alkoxide (or hydroxide) ion in alcoholic medium, resulting in formation of RH. A typical example [96] of such a reaction is given in eq. (30), which is seen to be an overall electrophilic substitution at the benzylic carbon center of the organometallic.

(30) $Et_3SiCH_2Ph + MeOH \xrightarrow{MeO^-} Et_3SiOMe + PhCH_3$

Trialkyl (or aryl) organosilicon derivatives R'_3SiX, where X is a leaving group such as halogen, acetate, or alkoxy, also undergo facile *nucleophilic* displacement; e.g.,

(31) $Ph_3SiCl + MeOH \longrightarrow Ph_3SiOMe + HCl$

Such reactions have been extensively studied [e.g., 11, 97, 98], with particular interest attaching to the stereochemistry of the process when the reaction is performed with chiral compounds [99–102].

Current interpretations of reactions of the type given in eq. (30) recognize several possible mechanistic pathways. We will now examine these pathways, illustrating with an organosilicon derivative of the general structure R'_3SiR, it being understood that our discussion extends to corresponding derivatives of germanium and tin. For convenience one may subdivide into six mechanistic cases [cf. 129].

Case I. Formation of a pentacoordinate intermediate in a rapidly attained pre-equilibrium, followed by slow breakdown of the complex with expulsion of a carbanion (eq. (32a)). The latter is then rapidly protonated by interaction with a solvent molecule (eq. (32b)). This case would be an analogue of the $S_E1(N)$ mechanism considered previously.

(32a) $R''O^- + R'_3Si{-}R \underset{\longleftarrow}{\overset{fast}{\longrightarrow}} R''O{-}\underset{R'}{\overset{R'}{\underset{|}{Si}}}{-}R \xrightarrow{slow} R''O{-}SiR'_3 + R^-$

(32b) $R''OH + R^- \xrightarrow{fast} RH + R''O^-$

Case II. The pentacoordinate intermediate is formed in the slow stage, while its breakdown to yield carbanion is fast. Reaction is completed as in Case I.

229

$$\text{(33)} \quad R''O^- + R'_3Si{-}R \xrightarrow{\text{slow}} R''O{-}\underset{\underset{R'}{\overset{|}{\diagdown}}}{\overset{\overset{R'}{\overset{|}{\diagup}}}{Si}}{-}R \underset{\text{fast}}{\rightleftharpoons} R''O{-}SiR'_3 + R^-$$

Case III. The complex is formed in a pre-equilibrium, as in Case I, but it does not break down spontaneously to a free carbanion. Instead, a solvent molecule acting as a proton-donating electrophile provides the driving force for the removal of carbanion as the conjugate acid, RH. Thus we have:

$$\text{(34)} \quad R''O^- + R'_3Si{-}R \underset{\text{fast}}{\rightleftharpoons} R''O{-}\underset{\underset{R'}{\overset{|}{\diagdown}}}{\overset{\overset{R'}{\overset{|}{\diagup}}}{Si}}{-}R \xrightarrow[\text{slow}]{R''OH} R''O{-}SiR'_3 + RH + R''O^-$$

Case IV. Here we have a slow formation of complex, analogous to Case II, but followed by reaction with a solvent molecule so that a free carbanion is not formed:

$$\text{(35)} \quad R''O^- + R'_3Si{-}R \xrightarrow{\text{slow}} R''O{-}\underset{\underset{R'}{\overset{|}{\diagdown}}}{\overset{\overset{R'}{\overset{|}{\diagup}}}{Si}}{-}R \xrightarrow[\text{fast}]{R''OH} R''O{-}SiR'_3 + RH + R''O$$

Case V. This is a direct, concerted, displacement of carbanion by $R''O^-$, without prior formation of an intermediate complex. Subsequent discharge of carbanion would occur rapidly as in eq. (32b).

$$\text{(36)} \quad R''O^- + R'_3Si{-}R \longrightarrow \left[R''O\cdots\underset{\underset{R'}{\overset{|}{\diagdown}}}{\overset{\overset{R'}{\overset{|}{\diagup}}}{Si}}\cdots R \right]^{\ddagger} \longrightarrow R''O{-}SiR'_3 + R^-$$

Case VI. This is also a concerted process but it assumes participation by solvent, so that a free carbanion is not present.

$$\text{(37)} \quad R''O^- + R'_3Si{-}R + HOR'' \longrightarrow \left[R''O\cdots\underset{\underset{R'}{\overset{|}{\diagdown}}}{\overset{\overset{R'}{\overset{|}{\diagup}}}{Si}}\cdots R\cdots H\cdots OR'' \right]^{\ddagger} \longrightarrow R''O{-}SiR'_3 + RH + R$$

The various mechanistic pathways may be illustrated by means of potential-energy reaction-coordinate diagrams, and Figures 1 and 2 show two of these. Figure 1 corresponds to the mechanism given by Case 1

230

and it is seen that there are two potential energy minima, corresponding to the pentacoordinate intermediate and the carbanion. In Figure 2, which corresponds to Case III, there is only one minimum since a free carbanion is not present. The diagram for Case VI (not shown) would simply have one maximum corresponding to the single activation process of eq. (37)).

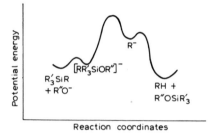

Figure 1. Potential energy-reaction coordinates diagram for substitution proceeding by the pentacoordinate complex and carbanion intermediates.

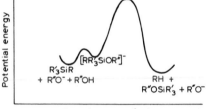

Figure 2. Potential energy-reaction coordinates diagram for solvent participation in breakdown of pentacoordinate intermediate.

To summarise, we have here an overall electrophilic displacement at the alkyl carbon of the organometallic derivative $R_3'MR$ which requires nucleophilic coordination at metal, and may proceed via pentacoordinate intermediates or by direct displacement, and may, or may not, involve free carbanion entities. In the examples to be discussed we will weigh some of the evidence for and against one or more of these possible pathways.

(a) Alkali cleavage of (phenylethynyl)-silanes and -germanes

The alkaline cleavage of (phenylethynyl)-silanes and -germanes [103] is our first reaction to be examined. The reaction, as given by eq. (38) (where M = Si or Ge; R' = alkyl or aryl; and R" = H or Me), is generally performed in aqueous methanol containing NaOH and is readily followed spectrophotometrically, with the rate of reaction being directly proportional to the concentration of substrate and of added base.

$$(38) \quad XC_6H_4C \equiv CMR'_3 + R''OH \xrightarrow{R'O^-} XC_6H_4C \equiv CH + R'_3MOR''$$

Useful information is derived through study of the effect of structural modification on the rate of reaction [103a]. In the case at hand the rate is sensitive to the nature of M, R' and X. When R' = alkyl, the rate of reaction decreases markedly on going from Me to Et, indicating that a steric effect is important, either directly in interfering with approach of nucleophile at M, or in providing hindrance to the orientation of solvent molecules in the transition state. For R' = aryl, it is found that electron withdrawing substituents accelerate the rate of reaction. A similar effect is obtained with substituents (X) on the phenylethynyl residue; clearly there is a facile transmission of electronic effects through the phenylethynyl system [104]. The observed substituent effects indicate that in the transition state for reaction there is a build-up of negative charge, both on the metal atom and on the *alpha* acetylenic carbon. One may best account for these observations on the basis of nucleophilic attack on metal in a pre-equilibrium process, followed by rate-determining ejection of the phenylethynyl carbanion, $XC_6H_4C \equiv C^-$. In terms of the classifications listed above, this reaction pathway would hence correspond to Case I.

The rate of the alkaline cleavage is also markedly dependent on the nature of the metal. Thus $PhC \equiv CSiMe_3$ reacts faster than $PhC \equiv CGeMe_3$, by a factor of 30 in 20% aqueous methanol at 30°. Qualitative data obtained with $PhC \equiv CSnMe_3$ indicate this compound to be more reactive than the corresponding silane. Thus the general order of reactivity for $ArC \equiv CMR_3$ is (M=) Ge < Si < Sn. Factors which may contribute to this reactivity order include the coordinating ability of the metal atoms, steric and solvation effects, and the electronic nature of the C—M bond, in particular the relative importance of $p_\pi - d_\pi$ bonding interactions in the C—M bonds along such a series. Many interesting discussions of the

232

relative contributions of these various factors have been given [105–110] though at the present time the problem is by no means resolved.

(b) Alkali cleavage of (phenylallyl)-silanes and -stannanes

The system which we examine next concerns the alkaline cleavage of 3-phenylallyl derivatives of silicon and tin. The processes under observation are [111]:

(39) $Me_3SiCH_2CH=CHPh + H_2O \xrightarrow{OH^-} Me_3SiOH + CH_3CH=CHPh$

(40) $Et_3SnCH_2CH=CHPh + H_2O \xrightarrow{OH^-} Et_3SnOH + CH_3CH=CHPh$

These reactions are also followed spectrophotometrically, since the reactants have characteristic absorption in the ultraviolet which is significantly different from that of the *trans-β*-methylstyrene product. In each case the rate of reaction is linearly dependent on the substrate concentration and on base concentration.

The reaction mechanism suggested by the authors [111] is a concerted, rate-determining, displacement of the phenylallyl carbanion, followed by its rapid discharge by reaction with solvent (i.e. Case V). The main evidence on which this conclusion is based is the kinetic solvent isotope effect (KSIE) [112]. It is found that the reaction is characterized by an inverse isotope effect, $k_{H_2O}/k_{D_2O} = 0.50$. This value of the KSIE is consistent with theoretical calculations on the basis of the model proposed by Bunton and Shiner [113] for a concerted displacement of carbanion by OH⁻. It is interesting that the authors consider the possibility of several transition state configurations: a "normal" SN2 type configuration (21) and two types of octahedral configurations, corresponding to a transannular displacement mode (22) or to edge displacement (23). The present evidence does not allow one to distinguish between these alternatives.

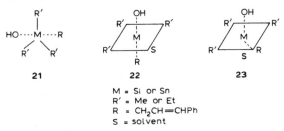

M = Si or Sn
R' = Me or Et
R = CH₂CH=CHPh
S = solvent

233

(c) Alkali cleavage of (benzyltrimethyl)-silanes, -germanes and -stannanes

The next reaction we consider is the alkaline cleavage of benzyltrimethylsilanes, and of the germane and stannane analogs, in aqueous methanol [114]:

(41) 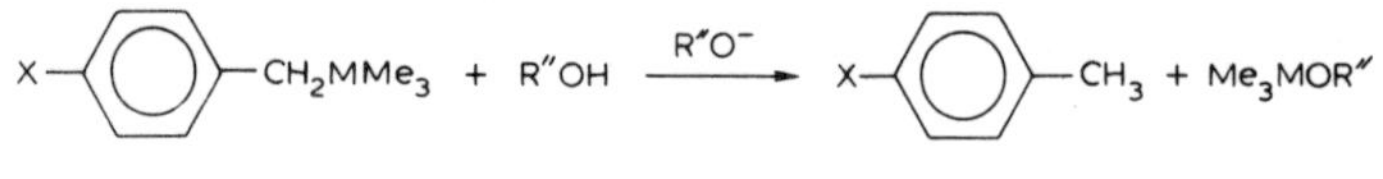 $X-\langle\bigcirc\rangle-CH_2MMe_3 + R''OH \xrightarrow{R'O^-} X-\langle\bigcirc\rangle-CH_3 + Me_3MOR''$

The relative reactivities at $50°$ for $X = m\text{-}Cl$ are: (M=) Ge, 1×10^{-3}; Si, 1.0; Sn, 17, i.e. the same relative order as in reaction (a). Comparing reactivities for the reaction series (a), (b) and (c), we have $PhC{\equiv}CSiMe_3$ > $PhCH{=}CHCH_2SiMe_3$ > $PhCH_2SiMe_3$, though no linear relationship exists between $\log k$ and pK_a of RH, where R is the displaced moiety.

The degree of charge development on the benzylic carbon in such a system can be probed by determining the effect of substituents X (in o-, m-, or p-positions) on the rate of reaction. It is found in this case that electron withdrawing substituents greatly accelerate the reaction. For the benzyltrimethylsilane system, a detailed analysis of the kinetic data in terms of Hammett σ-ρ correlations [115] leads to a positive ρ value (+4.9), which indicates a substantial development of negative charge on the benzylic carbon in the rate-determining transition state. Moreover, substituents such as p-COMe, p-NO$_2$, and p-SO$_2$Ph require the use of σ^- constants [116], indicating that a conjugative type of interaction operates in delocalizing the negative charge developed at the reaction site.

The above results are strongly suggestive of a pathway involving carbanion formation in the rate-limiting step, either through direct displacement (Case V) or via pre-equilibrium formation of a pentacoordinate complex (Case I). However, evidence has been put forth [117, 118] in support of an alternative thesis, requiring participation by solvent in the rate-determining transition state. Thus, when the reaction is carried out in alkaline CH_3OH–CH_3OD one observes a product isotope effect ($XC_6H_4CH_3/XC_6H_4CH_2D$) ranging over 1.4–1.6 for the $XC_6H_4CH_2SiMe_3$ series and 2.0–2.8 for the $XC_6H_4CH_2SnMe_3$ series [117]. A value of the product isotope effect appreciably greater than unity implies that there is discrimination between ROH and ROD in the course of reaction, which is consistent with Case III or with Case VI. If free carbanions were formed, their reaction with solvent would be expected to be rapid, and no significant discrimination between ROH and ROD could be anticipated [20]. In fact, when an ethereal solution of phenyllithium was quenched with CH_3OH-CH_3OD, the benzene which was produced had a ratio $C_6H_6 : C_6H_5D$ equal, within experimental error, to that of the reaction medium.

(d) Alkali cleavage of (phenyltrimethyl)-silanes and -stannanes

The last main reaction we consider is the alkaline cleavage of phenyl-trimethylsilanes and phenyltrimethylstannanes [119–121]. As in the case of the benzyl series, the germane analog reacts the slowest and quantitative data for this case are not yet available. Even the aryltri-methylsilanes react too slowly to allow their study in aqueous methanol, though reaction with OH^- occurs at a convenient rate in aqueous DMSO [119]. The reaction type can be represented by:

$$(42) \quad X-\!\!\left\langle\!\!\bigcirc\!\!\right\rangle\!\!-SiMe_3 \ + \ R''OH \ \xrightarrow{R''O^-} \ X-\!\!\left\langle\!\!\bigcirc\!\!\right\rangle \ + \ Me_3SiOR''$$

At a first approximation, the alkaline cleavage in the $XC_6H_4MMe_3$ series and in the $XC_6H_4CH_2MMe_3$ series appear to have similar characteristics. In both series there is a rate acceleration by electron withdrawing substituents (indicating carbanion character) and a product isotope effect (indicating electrophilic participation by solvent). The latter criterion is even more pronounced in the aryl series, with product isotope effect ratios ranging between 2.2 and 5.5 for the $XC_6H_4SnMe_3$ series [117, 118].

However, an intensive analysis of the rate data, in terms of the electrical effects of substituents, reveals some unexpected facets with interesting consequences. Thus it is found that whereas *meta* substituents exert a normal effect, with an excellent linear relationship between log k and σ_m, *para* substituents which can interact conjugatively with the reaction center deviate from that line. Electron withdrawing substituents such as p-NO_2 *activate markedly less* than predicted on the basis of the σ-constant, and electron releasing substituents such as p-OMe and p-NMe_2 *deactivate less* (or activate more) than anticipated. In aqueous methanol, reaction of p-$Me_2NC_6H_4SnMe_3$ actually occurs *faster* than of the un-substituted derivative so that p-NMe_2 apparently acts as an electron withdrawing group! (In aqueous DMSO, where the ρ value is larger,

235

p-NMe$_2$ and p-OMe become deactivating though these points still deviate markedly from the log $k-\sigma_m$ line; the qualitative argument hence remains unchanged but the importance of solvent effects is to be noted.)

The results for the ArMMe$_3$ series, including observation of the abnormal substituent effects and of the product isotope effect, can be accommodated by the postulate of a transition state structure **24**:

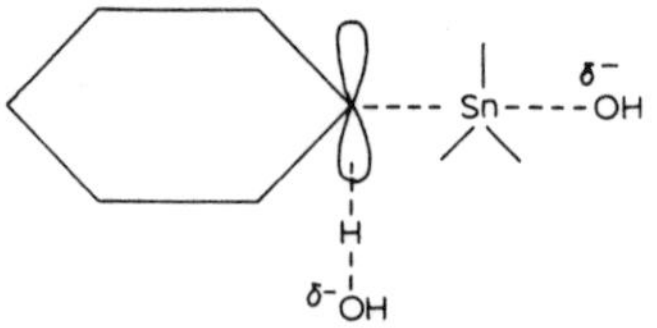

The consequences of this transition state may be seen through its contributing valence bond structures **25–28**[*]:

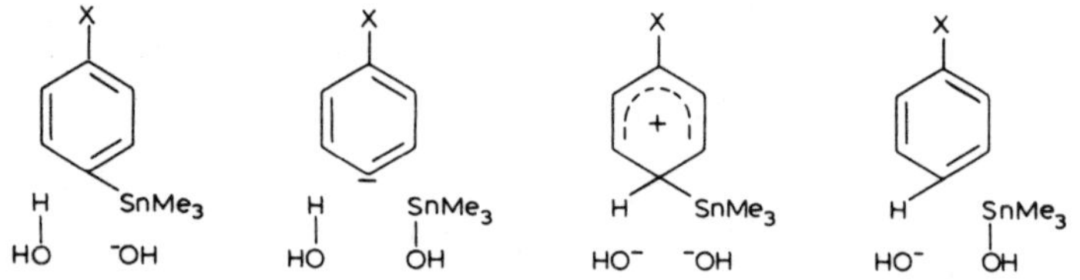

Now structure **26** reflects the carbanion character of the hybrid and accounts for the observation of rate enhancement by electron with-drawing substituents. However, structure **27** has essentially the character of an electrophilic attack by H$^+$ (derived from a solvent molecule) on XC$_6$H$_4$MMe$_3$. Thus, to the extent that **27** contributes to the overall hybrid structure **24**, substituents such as p-OMe or p-NMe$_2$ which are conjugatively electron releasing will have an accelerating effect, as in normal electrophilic aromatic substitution [80]. In this way the reacting system can call forth a response from both electron-withdrawing and electron-supplying substituents.

These conclusions also have a bearing on base-catalyzed aromatic hydrogen exchange. Thus a scrutiny of the rate data for proton exchange,

---

[*]Nothing is meant to be implied in transition state **24** and structures **25–28** about the stereochemistry of the reaction at the tin centre. There are some arguments in favour of the probability of flank attack of the $^-$OH ion at tin.

under catalysis by potassium amide in liquid ammonia [122], reveals a significant (positive) deviation for substituents such as p-OMe and p-NMe$_2$, using as standard the line defined by *meta* substituents. Thus it has been proposed [121] that base-catalyzed aromatic proton exchange, similarly to the alkaline cleavage of ArMR$_3$, essentially involves the transfer of a proton from a solvent molecule to aromatic carbon, concertedly with cleavage of the C—H bond. Under these circumstances free carbanions would not be generated.

One may represent the transition state for the amide ion catalyzed proton exchange by the contributing structures 29—32. The analogy between these and 25—28 is obvious and leads to an analogous argument on substituent effects.

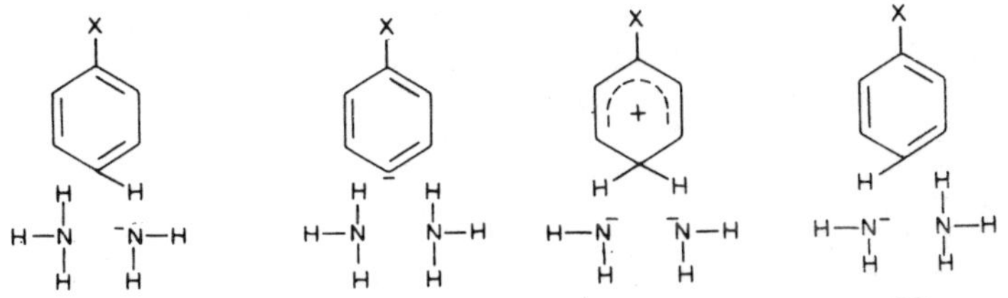

This new interpretation raises questions concerning the true mechanism of other base-catalyzed aromatic hydrogen exchange processes which have hitherto been considered to proceed via the free aryl carbanions [122—125]. In these systems, is there an electrophilic attack by solvent on carbon occurring concertedly with nucleophilic attack by base on hydrogen? Apart from the general question of the effect of aryl substituents on carbanion stability, is it possible that certain substituents such as *ortho*-halogeno [126, 127] or *ortho*-nitro [128] serve a special function in stabilizing the resulting aryl carbanions? Clearly answers to these questions will have a direct bearing on the requirement for electrophilic participation by solvent in these processes.

It seems fitting to conclude this chapter on this note of uncertainty so that future research in this area may continue to gain in stimulus. However, it is clear in any case that study of the basic cleavage of the carbon—metal bond has provided valuable information concerning carbon—hydrogen cleavage and that a novel viewpoint of carbanion chemistry has thereby been uncovered.

237

REFERENCES

1a J. Sicher, Angew. Chem. Intern. Ed., 11 (1972) 200.
1b A. Fry, Chem. Soc. Revs., 1 (1972) 163.
1c D.J. McLennan, Quart. Rev., 21 (1967) 490.
1d J.F. Bunnett, Surv. Prog. Chem., 5 (1969) 53.
1e D.V. Banthorpe, "Elimination Reactions", Elsevier, Amsterdam, 1963.
2   D.C. Ayres, "Carbanions in Synthesis", Oldbourne Press, London, 1966.
3   H.O. House, "Modern Synthetic Reactions", 2nd Ed., Benjamin, New York, 1972.
4   E.G. Rochow, "Organometallic Chemistry", Reinhold, New York, 1964.
5   G.E. Coates, M.L.H. Green and K. Wade, "Organometallic Compounds", Methuen, London, 1967.
6   J.J. Eisch, "The Chemistry of Organometallic Compounds. The Main Elements", MacMillan, New York, 1967.
7   P.L. Pauson, "Organometallic Chemistry", Edward Arnold, London, 1967.
8   O.A. Reutov and I.P. Beletskaya, "Reaction Mechanisms of Organometallic Compounds", North-Holland Publishing Co., Amsterdam, 1968.
9   D.A. Everest, "The Chemistry of Beryllium", Elsevier, Amsterdam, 1964.
10a W. Gerrard, "Organic Chemistry of Boron", Academic Press, New York, 1961.
10b H. Steinberg, "Organoboron Chemistry", Vol. 1, Wiley, New York, 1964, and following volumes in the series.
11a C. Eaborn, "Organosilicon Compounds", Academic Press, New York, 1960.
11b L.H. Sommer, "Stereochemistry, Mechanism, and Silicon", McGraw-Hill, New York, 1965.
12 M. Lesbre, P. Mazerolles and J. Satgé, "The Organic Compounds of Germanium", Wiley, New York, 1971.
13a R.C. Poller, "The Chemistry of Organotin Compounds", Academic Press, New York, 1970.
13b "Organotin Compounds", Vol. 3, edited by A.K. Sawyer, M. Dekker, New York, 1972.
14 H. Shapiro and F.W. Frey, "The Organic Compounds of Lead", Wiley, New York, 1968.
15 G.O. Doak and L.D. Freedman, "Organometallic Compounds of Arsenic, Antimony, and Bismuth", Wiley, New York, 1970.
16 F.R. Jensen and B. Rickborn, "Electrophilic Substitution of Organomercurials", McGraw-Hill, New York, 1968.
17 C.S. Johnson, Jr., M.A. Weiner, J.S. Waugh and D. Seyferth, J. Amer. Chem. Soc., 83 (1961) 1306.
18 F.J. Kronzer and V.R. Sandel, J. Amer. Chem. Soc., 94 (1972) 5750; V.R. Sandel, S.V. McKinley and H.H. Freedman, J. Amer. Chem. Soc., 90 (1968) 495.
19 R. Waack, M.A. Doran, E.B. Baker and G.A. Olah, J. Amer. Chem. Soc., 88 (1966) 1272.
20 Y. Pocker and J.H. Exner, J. Amer. Chem. Soc., 90 (1968) 6764.

21 R.B. Bates, W.H. Deines, D.A. McCombs and D.E. Potter, J. Amer. Chem. Soc., 91 (1969) 4608.
22 S. Brownstein and D.J. Worsfold, Can. J. Chem., 50 (1972) 1246.
23 J.B. Grutzner, J.M. Lawlor and L.M. Jackman, J. Amer. Chem. Soc., 94 (1972) 2306.
24a M. Szwarc, "Carbanions, Living Polymers and Electron Transfer Processes", Interscience, New York, 1968.
24b "Ions and Ion-Pairs in Organic Reactions", edited by M. Szwarc, Interscience, New York, 1972.
25 J. Smid, Angew. Chem. Intern. Ed., 11 (1972) 112.
26 F.A. Cotton and G. Wilkinson, "Advanced Inorganic Chemistry", 2nd Ed., Interscience, New York, 1966, Chapter 4.
27 L. Pauling, "The Nature of the Chemical Bond, 3rd Ed., Cornell University Press, Ithaca, N.Y., 1960, p. 98.
28a T.L. Brown, Advan. Organometal. Chem., 3 (1965) 365.
28b S. Bywater and D.J. Worsfold, J. Organometal. Chem., 10 (1967) 1.
28c J.B. Swart, R. Hogan, P.A. Scherr, L. Ferrier and J.P. Oliver, J. Amer. Chem. Soc., 94 (1972) 8371.
29a T.L. Brown, L.M. Seitz and B.Y. Kimura, J. Amer. Chem. Soc., 90 (1968) 3245.
29b D.P. Novak and T.L. Brown, J. Amer. Chem. Soc., 94 (1972) 3793.
29c L.D. McKeever, R. Waack, M.A. Doran and E.B. Baker, J. Amer. Chem. Soc., 90 (1968) 3244.
30a H. Dietrich, Acta Cryst., 16 (1963) 681.
30b E. Weiss and G. Henchen, J. Organometal. Chem., 21 (1970) 265.
31 E. Weiss and E.A.C. Lucken, J. Organometal. Chem., 2 (1964) 197.
32 W.N. Lipscomb, "Boron Hydrides", Benjamin, New York, 1963; W.E. Palke and W.N. Lipscomb, J. Chem. Phys., 45 (1966) 3948; W.N. Lipscomb and I.R. Epstein, Inorg. Chem., 10 (1971) 1921; W.N. Lipscomb, Accounts Chem. Res., 6 (1973) 257.
33 R.G. Vranka and E.L. Amma, J. Amer. Chem. Soc., 89 (1967) 3121.
34a A.H. Cowley and W.D. White, J. Amer. Chem. Soc., 91 (1969) 34.
34b R.N. Grimes, "Carboranes", Academic Press, New York, 1970.
34c M.F. Guest, I.H. Hillier and V.R. Saunders, J. Organometal. Chem., 44 (1972) 59.
35 A.I. Snow and R.E. Rundle, Acta Cryst., 4 (1951) 348.
36 E. Weiss, J. Organometal. Chem., 2 (1964) 314.
37 N. Muller and D.E. Pritchard, J. Amer. Chem. Soc., 82 (1960) 248.
38 K.C. Ramey, J.F. O'Brien, I. Hasegawa and A.E. Borchert, J. Phys. Chem., 69 (1965) 3418.
39 K.C. Williams and T.L. Brown, J. Amer. Chem. Soc., 88 (1966) 5460.
40a E.G. Hoffmann, Ann., 629 (1960) 104.
40b M.B. Smith, J. Organometal. Chem., 22 (1970) 273.
41 P.A. McCusker, J.V. Marra and G.F. Hennion, J. Amer. Chem. Soc., 83 (1961) 1924.

42 R.E. Dessy, F. Kaplan, G.R. Coe and R.M. Salinger, J. Amer. Chem. Soc., 85 (1963) 1191.
43 O.A. Reutov, Rec. Chem. Prog., 22 (1961) 1.
44 D.R. Pollard and M.H. Thompson, J. Organometal. Chem., 36 (1972) 13.
45 R.E. Dessy, W. Kitching, T. Psarras, R. Salinger, A. Chen and T. Chivers, J. Amer. Chem. Soc., 88 (1966) 460.
46 D.E. Applequist and D.F. O'Brien, J. Amer. Chem. Soc., 85 (1963) 743.
47 C.K. Ingold, "Structure and Mechanism in Organic Chemistry", 2nd Ed., Cornell University Press, Ithaca, N.Y., 1969.
48 M.H. Abraham, "Electrophilic Substitution at Saturated Carbon Atom", in "Comprehensive Chemical Kinetics", Vol. 12, edited by C.H. Bamford and C.F.H. Tipper, Elsevier, Amsterdam, 1973.
49a F.R. Jensen and R.J. Quellette, J. Amer. Chem. Soc., 83 (1961) 4477, 4478.
49b F.R. Jensen and R.J. Quellette, J. Amer. Chem. Soc., 85 (1963) 363.
50 F.R. Jensen, R.J. Quellette, G. Knutson and D.A. Babbe, Tetrahedron Lett., 339 (1963).
51a S. Winstein, E. Vogelfanger, K.C. Pande and H.F. Ebel, J. Amer. Chem. Soc., 84 (1962) 4993.
51b J.P. Hardy, A.F. Diaz and S. Winstein, J. Amer. Chem. Soc., 94 (1972) 2363.
52a R.E. Dessy and W. Kitching, Adv. Organometal. Chem., 4 (1966) 267.
52b W. Kitching, Rev. Pure and Appl. Chem. (Australia), 19 (1969) 1.
53 A.G. Brook, Pure Appl. Chem., 13 (1966) 215.
54a R.A. Benkeser, Pure Appl. Chem., 19 (1969) 389.
54b R. West, Pure Appl. Chem., 19 (1969) 291.
55a T.H. Chan and E. Vinokur, Tetrahedron Lett., 75 (1972).
55b A.G. Evans, M.L. Jones and N.Y. Rees, J. Chem. Soc. Perkin II, 389 (1972).
56a F.A. Carey and A.S. Court, J. Org. Chem., 37 (1972) 939.
56b J.J. Eisch and M.-R. Tsai, J. Amer. Chem. Soc., 95 (1973) 4065.
57 F.R. Jensen and D.D. Davis, J. Amer. Chem. Soc., 93 (1971) 4048.
58 D.S. Matteson, Organometal. Chem. Rev. A, 4 (1969) 263.
59 F.G. Thorpe, in "Studies on Chemical Structure and Reactivity", edited by J.H. Ridd, Wiley, New York, 1966.
60 M.H. Abraham and J.A. Hill, J. Organometal. Chem., 7 (1967) 11.
61 I.P. Beletskaya, K.P. Butin and O.A. Reutov, Organometal. Chem. Rev. A, 7 (1971) 51.
62 H.B. Charman, E.D. Hughes, C.K. Ingold and H.C. Volger, J. Chem. Soc., 1142 (1961).
63 H.B. Charman, E.D. Hughes, C.K. Ingold and F.G. Thorpe, J. Chem. Soc., 1121 (1961).
64 C. Schorlemmer, Ann. Chem., 132 (1864) 243; R. Otto, Ann. Chem., 154 (1870) 188.
65 M.S. Kharasch, R.R. Legault and W.R. Sprowls, J. Org. Chem., 3 (1938) 409, and earlier papers in the series.
66 F.E. Paulik, S.I.E. Green and R.E. Dessy, J. Organometal. Chem., 3 (1965) 229.

67 M.M. Kreevoy, J. Amer. Chem. Soc., 79 (1957) 5927.
68 M.M. Kreevoy and R.A. Kretchmer, J. Amer. Chem. Soc., 86 (1964) 2435.
69 M.M. Kreevoy, P.J. Steinwand and W.V. Kayser, J. Amer. Chem. Soc., 88 (1966) 124.
70 P. Mazerolles and M. Lesbre, Compt. Rend., 248 (1959) 2018.
71 R.A. Benkeser, D.I. Hoke and R.A. Hickner, J. Amer. Chem. Soc., 80 (1958) 5294.
72 C. Eaborn and K.C. Pande, J. Chem. Soc., 1566 (1960).
73 C. Eaborn, Z. Lasocki and J.A. Sperry, J. Organometal. Chem., 35 (1972) 245.
74 C. Eaborn and R.W. Bott, "Synthesis and Reactions of the Silicon–Carbon Bond", in "Organometallic Compounds of the Group IV Elements", Vol. 1, edited by A.G. MacDiarmid, Marcel Dekker, New York, 1968.
75 F. Glockling and K.A. Hooton, "Synthesis and Properties of the Germanium-Carbon Bond", in Ref. 74.
76 J.G.A. Luijten and G.J.M. van der Kerk, "Synthesis and Properties of the Tin-Carbon Bond", in Ref. 74.
77 L.C. Willemsens and G.J.M. van der Kerk, "Synthesis and Properties of the Lead-Carbon Bond", in Ref. 74.
78 H.G. Kuivila and K.V. Nahabedian, J. Amer. Chem. Soc., 83 (1961) 2159, 2164; K.V. Nahabedian and H.G. Kuivila, ibid., 83 (1961) 2167.
79 R.D. Brown, A.S. Buchanan and A.A. Humffray, Australian J. Chem., 18 (1965) 1507, 1513.
80 R.O.C. Norman and R. Taylor, "Electrophilic Substitution in Benzenoid Compounds", Elsevier, Amsterdam, 1965.
81a S. Winstein and T.G. Traylor, J. Amer. Chem. Soc., 78 (1956) 2597.
81b F.R. Jensen and L.H. Gale, J. Amer. Chem. Soc., 82 (1960) 145, 148.
82a K.U. Ingold and B.P. Roberts, "Free Radical Substitution Reactions", Wiley, New York, 1971.
82b A.G. Davies and B.P. Roberts, Accounts Chem. Res., 5 (1972) 387.
83 M. Gielen and J. Nasielski, Rec. Trav. Chim. Pays-Bas, 82 (1963) 228; J. Organometal. Chem., 1 (1963) 173.
84 M. Gielen, P. Baekelmans and J. Nasielski, J. Organometal. Chem., 34 (1972) 329.
85 C. Eaborn and D.E. Webster, J. Chem. Soc., 179 (1960).
86 C. Eaborn, Pure Appl. Chem., 19 (1969) 375.
87 O.A. Reutov, B. Praisnar, I.P. Beletskaya and V.I. Sokolov, Izvest. Akad. Nauk. S.S.S.R., Otdel. Khim. Nauk., 970 (1963).
88 E.D. Hughes, C.K. Ingold and R.M.G. Roberts, J. Chem. Soc., 3900 (1964).
89 Ref. 16, (a) p. 164, (b) p. 176.
90 M.H. Abraham, D. Dodd, M.D. Johnson, E.S. Lewis and R.A. More O'Ferrall, J. Chem. Soc. (B), 762 (1971).
91 D.J. Cram, "Fundamentals of Carbanion Chemistry", Academic Press, New York, 1965.
92 V.A. Kalyavin, T.A. Smolina and O.A. Reutov, Doklady Akad. Nauk. S.S.S.R., 156 (1964) 95.

93 J.R. Coad and M.D. Johnson, J. Chem. Soc. (B), 633 (1967).
94 D. Dodd and M.D. Johnson, J. Chem. Soc. (B), 1071 (1969).
95 C.A. Bunton, "Nucleophilic Substitution at a Saturated Carbon Atom", Elsevier, Amsterdam, 1963.
96 H. Gilman, A.G. Brook and L.S. Miller, J. Amer. Chem. Soc., 75 (1953) 4531.
97 A.D. Allen and G. Modena, J. Chem. Soc., 3671 (1957).
98 E. Buncel and A.G. Davies, J. Chem. Soc., 1550 (1958).
99 A.G. Brook and G.J.D. Peddle, J. Amer. Chem. Soc., 85 (1963) 2338.
100 C. Eaborn, R.E.E. Hill and P. Simpson, J. Organometal. Chem., 37 (1972) 251.
101 R.J.P. Corriu and B.J.L. Henner, Chem. Commun., 116 (1973); R.J.P. Corriu and G. Royo, Bull. Soc. Chim. France, 1490, 1497 (1972).
102 L.H. Sommer, W.D. Korte and C.L. Frye, J. Amer. Chem. Soc., 94 (1972) 3463.
103a C. Eaborn and D.R.M. Walton, J. Organometal. Chem., 4 (1965) 217.
103b C.S. Kraihanzel and J.E. Poist, J. Organometal. Chem., 8 (1967) 239.
104a J.D. Roberts and R.A. Carboni, J. Amer. Chem. Soc., 77 (1955) 5554.
104b C. Eaborn, A.R. Thompson and D.R.M. Walton, J. Chem. Soc. (B), 859 (1969).
105 J.R. Chipperfield and R.H. Prince, J. Chem. Soc., 3567 (1963).
106 J. Nagy and J. Réffy, J. Organometal. Chem., 23 (1970) 71.
107 V. Vaisarova, J. Hetflejs and V. Chavalovsky, J. Organometal. Chem., 22 (1970) 395.
108 J.M. Angelelli, J.C. Maire and Y. Vignollet, J. Organometal. Chem., 22 (1970) 313.
109 C.G. Pitt, J. Organometal. Chem., 22 (1970) C35.
110 G.M. Whitesides, J.G. Selgestad, S.P. Thomas, D.W. Andrews, B.A. Morrison, E.J. Panek and J.S. Filippo, Jr., J. Organometal. Chem., 22 (1970) 365.
111 R.M.G. Roberts and F.E. Kaissi, J. Organometal. Chem., 12 (1968) 79.
112 P.M. Laughton and R.E. Robertson, in "Solute-Solvent Interactions", edited by J.F. Coetzee and C.D. Ritchie, Marcel Dekker, New York, 1969.
113 C.A. Bunton and V.J. Shiner, J. Amer. Chem. Soc., 83 (1961) 3207, 3214.
114a R.W. Bott, C. Eaborn and T.W. Swaddle, J. Chem. Soc., 2342 (1963).
114b R.W. Bott, C. Eaborn and B.M. Rushton, J. Organometal. Chem., 3 (1965) 448.
115 L.P. Hammett, "Physical Organic Chemistry", McGraw-Hill, New York, 1970.
116 P.R. Wells, "Linear Free Energy Relationships", Academic Press, London, 1968.
117 R. Alexander, C. Eaborn and T.G. Traylor, J. Organometal. Chem., 21 (1970) P65.
118 C. Eaborn and I.D. Jenkins, unpublished work.
119 J. Cretney and G.J. Wright, J. Organometal. Chem., 28 (1971) 49.
120 C. Eaborn, H.L. Hornfeld and D.R.M. Walton, J. Chem. Soc. (B), 1036 (1967).

121 A.R. Bassindale, C. Eaborn, R. Taylor, A.R. Thompson, D.R.M. Walton, J. Cretney and G.J. Wright, J. Chem. Soc. (B), 1155 (1971).
122 A.I. Shatenshtein, Adv. Phys. Org. Chem., 1 (1963) 155.
123 E. Buncel and E.A. Symons, J. Org. Chem., 38 (1973) 1201; E. Buncel, A.R. Norris and K.E. Russell, Quart. Rev., 22 (1968) 123.
124 A. Streitwieser, Jr., J.A. Hudson and F. Mares, J. Amer. Chem. Soc., 90 (1968) 648.
125 R.D. Guthrie and D.P. Wesley, J. Amer. Chem. Soc., 92 (1970) 4057.
126 J.D. Roberts, D.A. Semenov, H.E. Simmons, Jr. and L.A. Carsmith, J. Amer. Chem. Soc., 78 (1956) 601.
127 J.F. Bunnett, Accounts Chem. Res., 5 (1972) 139.
128 E. Buncel, unpublished work.
129 C. Eaborn, Intra Science Reports, Vol. 7, in the press.

AUTHOR INDEX

Numbers of pages on which the full references appear are given in *italics*

Abraham, M.H., 217, 220, 221, *240*, 226, *241*
Abrahamson, E.W., 171, 175, 177, 178, *203*
Abrams, G.D., 122, *140*, 164, *169*
Achmad, S.A., 157, *168*
Adams, R., 110, *138*
Adolph, H.G., 51, *67*
Ahlberg, P., 98, *107*, 114, *139*
Akabori, S., 193, *205*
Akhrem, A.A., 144, *167*
Albagli, A., 17, *31*
Albin, J., 132, *141*
Alcais, P., 72, *102*, 80, *104*
Alexander, R., 234, 235, *242*
Allen, A.D., 229, *242*
Allen, L.C., 35, 36, 37, 39, 42, 48, *66*
Allinger, J., 14, *31*, 35, *66*, 72, *102*
Allred, A.L., 72, *102*
Almy, J., 58, *67*, 148, *167*
Amburn, H.W., 18, *32*, 80, 90, *103*, 96, *106*
Amis, E.S., 78, *103*
Amma, E.L., 214, *239*
Anderson, K.K., 64, *68*
Ando, T., 100, *107*
Andose, J.D., 37, 42, *66*
Andreades, S., 19, *32*, 48, *66*
Andrews, D.W., 233, *242*
Andrews, G.C., 197, *206*
Andrist, A.H., 179, *204*
Anet, F.A.L., 200, *206*

Angelelli, J.M., 233, *242*
Applequist, D.E., 216, *240*
Arai, K., 59, *67*
Arigoni, D., 120, *139*
Armstrong, D.R., 131, 132, *140*
Arnett, J.F., 44, *66*
Askani, R., 200, *206*
Atkinson, D.J., 182, *204*
Atkinson, J.G., 92, *105*
Auerbach, R.A., 78, 83, *103*
Averbeck, H., 71, *102*
Ayres, D.C., 91, *105*, 210, *238*

Baer, H.H., 95, *106*
Badoz-Lambling, J., 11, *31*
Baechler, R.D., 42, *66*
Baekelmans, P., 224, *241*
Baker, E.B., 211, *238*, 213, *239*
Baker, W., 157, *168*
Bakuzis, P., 166, *169*
Balasubramanian, K., 135, *141*
Baldeschweiler, J.D., 21, *32*
Baldwin, J.E., 151, *168*, 165, *169*, 179, *204*, 195, 196, *205*, 197, *206*
Bamford, C.H., 217, *240*
Bank, K.C., 64, *68*
Bank, S., 98, *106*
Banthorpe, D.V., 210, *238*
Barbarella, G., 59, *67*
Barborak, J.C., 202, *207*
Barnes, D.J., 92, *105*
Barnes, R.K., 132, *141*

Barnett, C., 18, *32*
Bartlett, P.D., 75, *102*, 86, 88, *105*, 113, *138*
Barton, T.J., 114, *139*
Bassindale, A.R., 235, *243*
Bates, R.B., 98, *107*, 172, 182, 183, 184, *204*, 189, 191, *205*, 211, *239*
Bates, R.G., 22, *33*
Battiste, M., 132, *141*
Baumgarten, H.E., 158, 159, *168*
Baumgarten, R.J., 166, *169*
Beasley, G.H., 199, *206*
Beauchamp, J.L., 21, *33*
Beeson, J.H., 158, 160, *168*
Beib, K.H., 18, *32*
Belanger, P., 92, *105*
Beletskaya, I.P., 78, *103*, 210, 221, *238*, 221, *240*, 225, *241*
Bell, R.P., 2, 5, 18, *30*, 27, *33*, 71, 72, 75, *102*, 77, 78, 79, *103*, 85, *104*, 91, *105*
Bender, M.L., 79, *103*
Benkeser, R.A., 218, *240*, 223, *241*
Benkovic, S.J., 79, *103*
Benson, R.E., 7, *30*
Bent, H.A., 6, *30*
Benzon, M.S., 199, *206*
Bergander, H., 86, *104*
Bergman, N.-A., 27, *33*, *34*
Bergmann, E.D., 130, *140*
Bergon, M., 72, *102*
Berrigan, P.J., 93, *105*
Berson, J.A., 175, 179, 187, *204*
Bethell, D., 15, *31*, 58, *67*, 113, *138*
Betts, M.J., 122, *140*, 164, 165, *169*
Bieber, J.B., 165, *169*
Biellmann, J.F., 196, *205*
Bigeleisen, J., 23, *33*, 101, *107*
Bigelow, W.B., 138, *141*
Billman, J.H., 72, *102*
Binsch, G., 202, *207*
Birnberg, G.H., 203, *206*
Bischof, P., 116, *139*
Bissell, R., 4, *30*
Blackburn, G.M., 197, *206*

Blair, L.K., 21, *33*
Boche, G., 186, *205*
Bodor, N., 151, *167*
Bohme, D.K., 21, *33*
Bonhoeffer, K.F., 18, *32*, 59, *67*, 95, *106*
Borchert, A.E., 215, *239*
Bordwell, F.G., 18, *32*, 59, *67*, 96, *106*, 144, 147, 148, 151, 152, *167*, 152, *168*
Borowitz, I.J., 152, *168*
Bory, S., 64, *68*
Bothner-By, A.A., 82, *104*
Bott, R.W., 218, 223, *241*, 234, *242*
Bourns, A.N., 113, *138*, 148, *167*
Bowden, K., 9, 22, *30*, 23, *33*, 98, *106*
Bowman, N.S., 23, *33*, 85, *104*
Boyd, R.H., 22, *33*
Boyle, Jr., W.J., 18, *32*, 96, *106*
Brauman, J.I., 17, *31*, 19, *32*, 21, *33*, 114, *139*
Bredt, J., 88, *105*
Brener, L., 7, *30*, 202, *206*
Brenner, S., 98, *107*, 172, *204*, 189, 191, *205*
Breslow, R., 129, 130, 131, 133, 135, *140*, 132, 133, 135, 136, 137, 138, *141*, 147, 148, *167*
Brindle, I.D., 197, *206*
Broaddus, C.D., 98, 99, *107*
Broadhurst, M.J., 114, *139*
Brönsted, J.N., 17, *31*
Brook, A.G., 218, *240*, 229, *242*
Brooks, J.J., 4, *30*
Brouillard, R., 72, *102*, 80, *140*
Brown, H.C., 113, *138*
Brown, J., 130, 135, *140*
Brown, J.M., 124, 125, 126, *140*
Brown, M.D., 61, *68*
Brown, R.D., 224, *241*
Brown, T.L., 11, *31*, 40, *66*, 213, 214, 215, *239*
Brownstein, S., 211, *239*
Bruice, T.C., 18, *32*, 79, *103*, 96, *106*
Bryson, J.A., 17, *31*
Buchanan, A.S., 224, *241*

Buchanan, G.L., 88, *105*
Bugg, C., 5, *30*
Bullock, E., 61, *68*
Buncel, E., 15, 16, 17, *31*, 19, *32*, 59, 67, 61, 64, *68*, 80, 90, *103*, 96, *106*, 113, *138*, 148, *167*, 229, *242*, 237, *243*
Bunnett, J.F., 210, *238*, 237, *243*
Bunton, C.A., 228, 233, *242*
Burkoth, J.L., 202, *206*
Burr, Jr., J.G., 151, *167*
Burton, D.J., 19, *32*, 49, 51, *67*
Burwell, R.L., 14, 15, *31*
Butin, K.P., 221, *240*
Bywater, S., 213, *239*

Cain, E.N., 126, *140*
Caldin, E.F., 26, *33*, 85, 96, *104*, 96, *106*
Caldwell, R.A., 4, *30*, 19, *32*, 29, *34*
Calmon, J.-P., 72, *102*, 85, *104*
Calmon, M., 85, *104*
Camp, R.L., 151, *168*
Campbell, P.H., 187, *205*
Caple, G., 174, 176, *204*
Capon, B., 111, *138*
Carboni, R.A., 232, *242*
Carey, F.A., 218, *240*
Carlson, M.W., 148, *167*
Carnighan, R.H., 98, *107*
Carsmith, L.A., 237, *243*
Carter, J.S., 75, *102*
Casanova, J., 153, *168*
Catchpole, A.G., 83, *104*
Cava, M.P., 135, *141*
Cavill, G.W.K., 157, *168*
Ceasar, G.P., 203, *206*
Chan, T.H., 218, *240*
Chang, C.J., 12, 13, 14, 23, *31*
Chang, H.W., 131, *140*, 137, *141*
Chantooni, Jr., M.K., 22, *33*
Chari, S., 202, *207*
Charman, H.B., 221, 222, *240*
Chavalovsky, V., 233, *242*
Chen, A., 18, *32*, 80, 90, *103*, 216, *240*
Chen, Y.H., 3, *30*

Cheng, Y.-M., 164, *169*
Chevalier, M., 83, *104*
Chiang, J.F., 115, *139*
Chiang, Y., 27, *33*
Childs, R.F., 114, *139*, 187, *205*
Chipperfield, J.R., 233, *242*
Chirinko, Jr., J.M., 92, *105*
Chiu, N.W.K., 187, *205*
Chivers, T., 216, *240*
Cholod, M.S., 79, *103*
Chong, J.A., 89, *105*
Chow, L.W., 72, *102*
Chu, K.C., 58, *67*
Chu, W., 137, *141*
Ciganek, E., 174, *204*
Ciuffarin, E., 12, 22, *31*
Clare, B.W., 22, *33*
Clark, D.T., 131, 132, *140*
Clark, R.A., 115, *139*
Clark, R.D., 158, 159, *168*
Clementi, E., 36, 37, *66*
Clifford, P.R., 114, *138*
Coad, J.R., 227, *242*
Coates, G.E., 210, *238*
Cockerill, A.F., 9, 22, *30*, 15, *31*, 23, 27, *33*
Coe, G.R., 216, *240*
Coetzee, J.F., 22, *33*, 78, *103*, 233, *242*
Coffen, D.L., 61, 64, *68*
Cole, C.M., 172, *204*, 189, *205*
Collins, C.J., 23, *33*, 85, *104*
Collis, M.J., 80, 90, *103*
Conant, J.B., 12, *31*, 93, *105*
Conia, J.M., 145, *167*
Cook, D., 22, *33*, 114, *139*
Cook, M.J., 61, *68*
Cook, R.S., 98, *106*
Cookson, R.C., 151, *168*
Cope, A.C., 198, *206*
Cordner, J.P., 77, *103*
Corey, E.J., 61, 62, *68*
Corriu, R.J.P., 229, *242*
Cortegiano, H., 88, *105*
Corver, H.A., 114, *139*
Cotton, F.A., 212, *239*

247

Coulson, C.A., 6, *30*, 61, *68*
Court, A.S., 218, *240*
Cover, R.E., 88, *105*
Covey, D.F., 119, *139*, 164, *169*
Coward, J.K., 79, *103*
Cowley, A.H., 214, 215, *239*
Cox, B.G., 27, *33*, 79, *103*
Cox, R.A., 80, 81, 82, *104*
Cram. D.J., 2, 23, *30*, 13, 14, *31*, 19, *32*,
    28, *34*, 35, 64, *66*, 55, 57, 58, *67*,
    61, 62, 65, *68*, 87, 91, 100, *105*, 97,
    98, *106*, 98, 101, *107*, 143, *167*, 171,
    203, 226, *241*
Crandall, J.K., 151, *168*
Crasnier, F., 130, *140*
Cretney, J., 235, *242*, 235, *243*
Csizmadia, I.G., 36, 37, *66*, 61, 62, 63,
    68

Dai, S.-H., 202, *206*
Dalsin, P.D., 51, *67*
Daniels, F., 198, *206*
Darwish, D., 9, *30*
Dauben, Jr., H.J., 132, 137, *141*
Davidson, E.S., 172, *204*
Davies, A.G., 224, *241*, 229, *242*
Davis, D.D., 138, *141*, 219, 224, *240*
Dawson, H.M., 75, *102*
DeBernardis, J., 165, *169*, 196, *205*
De Boer, Th.J., 90, *105*, 134, *141*
Deines, W.H., 184, *204*, 189, 191, *205*,
    211, *239*
Deno, N.C., 187, 188, *205*
DePuy, C.H., 187, 188, *205*
DeSalas, E., 99, *107*
Desiderato, R., 5, *30*
Dessy, R.E., 18, *32*, 80, 90, *103*, 216,
    217, 222, *240*
Deugau, K., 187, *205*
Dewar, M.J.S., 37, *66*, 137, *141*, 151,
    167, 171, 175, 179, 190, *203*, 200,
    203, *206*
Diaz, A., 114, *139*, 127, *140*, 217, *240*
Dietrich, H., 213, *239*
Digiorgio, J.B., 161, *168*

Dillon, R.L., 18, *32*, 96, *106*
Di Milo, A.J., 77, *103*
Dimmel, D.R., 197, *206*
Dirlam, J.P., 114, *139*, 153, *168*
Dittmar, B., 61, *68*
Dixon, J.E., 18, *32*, 96, *106*
Doak, G.O., 210, *238*
Dodd, D., 226, *241*, 228, *242*
Doddrell, D., 166, *169*
Doering, W. von E., 59, 60, *67*, 98, *106*,
    136, *141*, 174, 199, *204*, 199, *206*
Dolman, D., 15, 16, *31*
Doran, M.A., 211, *238*, 213, *239*
Dorko, E.A., 132, *141*
Douek, M., 133, *141*
Draus, R.C., 202, *206*
Dubois, J.E., 71, 72, *102*, 75, *103*, 80,
    104
Ducep, J.B., 196, *205*
Dunlap, R.P., 80, 81, *104*
Dupuy, Jr., A.E., 160, 161, *168*
Durst, T., 64, *68*

Eaborn, C., 210, 218, 229, *238*, 218,
    223, 224, *241*, 229, 232, 234, 235,
    242, 235, *243*
Earls, D.W., 85, *104*
Eastham, J.F., 145, 155, *167*
Ebel, H.F., 2, *30*, 217, *240*
Eberhard, P., 185, *205*
Eberson, L., 153, *168*
Edelson, S.S., 151, 154, *168*
Eggers, Jr., D., 134, *141*
Ehret, A., 111, *138*
Eicher, T., 147, *167*
Eigen, M., 18, *31*, 78, 83, 100, *103*
Eisch, J.J., 210, *238*, 218, *240*
Ela, S.W., 98, *106*
Elia, V.J., 159, *168*
Eliel, E.L., 59, *67*, 202, *207*
Endres, L.S., 159, *168*
Epiotis, N.D., 171, 175, *203*
Epstein, I.R., 214, *239*
Evans, A.G., 218, *240*
Evans, D.A., 197, *206*

248

Evans, M.G., 190, *205*
Everest, D.A., 210, *238*
Exner, J.H., 211, *238*

Farnum, D., 4, *30*
Fava, A., 59, *67*
Favorskii, A., 144, *167*
Feather, J.A., 77, *103*
Feld, W.A., 120, 126, *139*
Feldman, M., 114, 137, *139*
Fendley, J.A., 85, *104*
Ferrier, B.M., 199, *206*
Ferrier, L., 213, *239*
Feuer, H., 94, *105*
Fiato, R.A., 115, *139*
Figuera, J.M., 98, *106*
Filippo, Jr., J.S., 233, *242*
Fischer, H., 2, *30*
Fischer, H.P., 58, *67*, 97, *106*, 98, *107*
Flachskam, N.W., 98, 99, *107*
Flanagan, P.W.K., 18, *32*, 96, *106*
Fluendy, M.A.D., 79, *103*
Flygare, W.H., 151, *168*
Flythe, W.C., 114, 137, *139*
Fong, W.C., 145, *167*
Fonzes, L., 86, *104*
Ford, R.A., 72, *102*
Ford, W.T., 58, *67*, 187, *205*
Forsén, S., 72, 73, 93, *102*
Forsythe, G.D., 172, *204*
Fort, A.W., 151, *167*
Fossel, E.T., 199, *206*
Foster, E.G., 198, *206*
Fraenkel, G., 3, *30*, 193, *205*
Frainier, L., 18, *32*, 96, *106*
Frame, R.R., 148, *167*
Frank, G.A., 152, *168*
Frank, R., 147, *167*
Fraser, R.R., 61, 64, *68*
Freeburger, M.E., 120, 126, *139*
Freedman, H.H., 211, *238*
Freedman, L.D., 210, *238*
Freeman, J.P., 59, *68*, 122, *140*
Freeman, P.K., 127, *140*
Frey, F.W., 210, *238*

Frick, W.G., 37, *66*
Fritzberg, A.R., 147, *167*
Fry, A., 210, *238*
Fry, A.J., 154, *168*
Frye, C.L., 229, *242*
Fuerholzer, J.F., 158, *168*
Fujimoto, T.T., 197, *206*
Fukui, K., 171, 175, 179, 190, *203*
Fukunaga, M., 59, *67*
Fukunaga, T., 162, *169*
Fukuyama, M., 18, *32*, 96, *106*
Funderburk, L.H., 26, *33*, 96, *106*
Fuson, R.C., 91, *105*

Gagosian, R.B., 146, 147, *167*
Gajewski, J.J., 130, 135, *140*, 175, 188, *204*
Galama, P., 183, *204*
Gale, L.H., 224, *241*
Gall, M., 78, 83, *103*
Gamboa, J.M., 98, *106*
Garbesi, A., 59, *67*
Garbisch, Jr., E.W., 152, *168*
Garratt, P.J., 7, 8, *30*, 130, *140*
Gaspar, P.E., 136, *141*
Gaspar, P.P., 98, *106*
Gassman, P.G., 89, *105*, 197, *206*
Geib, K.H., 95, *106*
Genkina, N.K., 78, *103*
Gero, A., 71, *102*
Gerrard, W., 210, 215, *238*
Gharpure, S.B., 197, *206*
Ghirardelli, R.G., 2, 28, *30*
Gibson, M.S., 197, *206*
Gielen, M., 224, *241*
Gilbert, J.M., 15, *31*
Gilbert, J.R., 23, *33*
Gilchrist, T.L., 175, *204*
Gill, G.B., 175, 187, *204*
Gillespie, R.J., 36, *66*
Gilly, S., 189, *205*
Gilman, H., 229, *242*
Gilmore, W.F., 151, *167*, 157, *168*
Gitter, A., 93, *105*
Giumanini, A.G., 165, *169*, 197, *206*

249

Glass, D.S., 174, *204*
Gleicher, G.J., 88, *105*
Gleiter, R., 115, 116, *139*, 200, *206*
Glockling, F., 223, *241*
Goering, H.L., 61, *68*
Gold, E.H., 114, *138*
Gold, V., 58, *67*, 113, *138*, 77, *103*, 85, 86, *104*, 93, *105*
Golding, P.D., 61, *68*
Goldstein, M.J., 115, *139*, 128, *140*, 199, 202, 203, *206*
Goldstein, S., 122, *140*
Goldstein, S.M., 164, *169*
Gordon, A.W., 92, *105*
Gordon, M., 92, *105*
Gorodetsky, M., 72, *102*
Gosselink, D.W., 182, *204*
Gosser, L., 55, 57, 58, *67*, 97, *106*
Gould, S.J., 120, *139*
Graham, E.W., 58, *67*
Granger, M.R., 4, *30*
Greeley, R.H., 187, *205*, 202, *206*
Green, J., 203, *206*
Green, M.L.H., 210, *238*
Green, S.I.E., 222, *240*
Greenberg, A., 151, *168*
Greene, F.D., 151, *168*
Gregory, M.J., 18, *32*, 96, *106*
Grieco, P.A., 9, *30*
Grimes, R.N., 214, *239*
Grimme, W., 199, *206*
Grist, S., 86, *104*
Grovenstein, Jr., E., 164, *169*, 193, 194, 198, *205*
Grubbs, R., 135, *141*
Gruetzmacher, G., 197, *206*
Grunwald, E., 47, 50, *66*
Grunwell, J.R., 83, *104*, 172, *204*
Grutzner, J.B., 128, 129, *140*, 211, *239*
Guest, M.F., 214, *239*
Guether, A., 71, *102*
Gunther, H., 199, *206*
Guthrie, R.D., 101, *107*, 237, *243*
Gutsche, C.D., 144, *167*

Haberfield, P., 57, *67*, 91, *105*
Habersaat, K., 186, *205*
Hackler, R.E., 197, *206*
Hadzi, D., 72, *102*
Hagemeier, L.D., 159, *168*
Haines, W.J., 36, *66*
Hall, G.E., 174, *204*
Halpern, Y., 114, *138*
Hamberger, H., 114, *139*
Hammen, R.F., 197, *206*
Hammett, L.P., 15, *31*, 47, *66*, 234, *242*
Hammond, G.S., 72, 73, *102*, 83, *104*
Hammond, W.B., 145, 147, *167*
Hammons, J.H., 2, 12, 29, *30*, 12, 22, *31*, 19, *32*
Hamon, D.P.G., 121, *140*
Hampton, K.G., 80, 83, *103*
Hantzsch, A., 94, *105*
Harding, D.R.K., 59, *67*
Hardy, E.M., 198, *206*
Hardy, J.P., 217, *240*
Hardy, T.A., 127, *140*
Harget, A., 151, *167*
Hargreaves, M.K., 47, *66*
Harmony, M.D., 26, *33*
Harper, E.T., 79, *103*
Harris, D.L., 114, *139*, 127, *140*
Hartenstein, J.H., 199, *206*
Hartmann, A.A., 59, *67*
Hasegawa, I., 215, *239*
Haselbach, E., 151, *167*
Hauck, F., 13, 14, *31*
Hautala, J.A., 18, *32*, 96, *106*
Hegazi, M., 79, *103*
Hehre, W.J., 49, *67*, 73, *102*
Heilbronner, E., 115, 116, *139*, 200, *206*
Henchen, G., 213, *239*
Henderson, J.W., 42, *66*
Hendon, J.E., 92, *105*
Henner, B.J.L., 229, *242*
Hennion, G.F., 215, *239*
Henshall, J.B., 79, *103*
Herlem, M., 11, *31*
Hess, Jr., B.A., 136, *141*

250

Hetflejs, J., 233, *242*
Hibbert, F., 18, *32*, 97, *106*
Hickner, R.A., 223, *241*
Hill, J.A., 220, *240*
Hill, R., 131, *140*
Hill, R.E.E., 229, *242*
Hillier, I.H., 214, *239*
Hine, J., 2, 28, *30*, 19, *32*, 47, *66*, 49,
    51, *67*, 79, 80, 83, *103*, 83, *104*, 92,
    105, 98, 99, *107*
Hine, K.E., 187, *205*
Ho, C.C., 193, *205*
Hochberg, J., 59, *67*
Hoffman, D.H., 58, *67*
Hoffmann, A.K., 59, 60, *67*
Hoffmann, E.G., 215, *239*
Hoffmann, H.M.R., 74, *102*
Hoffmann, R., 4, *30*, 49, *67*, 115, *139*,
    151, *168*, 164, *169*, 171, 175, 190,
    203, 175, 184, 187, 189, 199, *204*,
    200, 203, *206*
Hofmann, J.E., 19, *32*
Hogan, R., 213, *239*
Hoke, D.I., 223, *241*
Holland, W.V., 137, *141*
Hollyhead, W.B., 13, *31*
Holtz, D., 49, 51, *67*
Hooton, K.A., 223, *241*
Hornfeld, H.L., 235, *242*
Hornung, V., 200, *206*
Hoskins, C.R., 75, *102*
House, H.O., 78, 83, 91, *103*, 151, *167*,
    152, 157, *168*, 210, *238*
Houston, J.G., 92, *105*
Howe, R., 162, *169*
Hsü, S.K., 86, *105*, 101, *107*, 119, *139*
Hubner, G., 82, *104*
Hückel, E., 7, *30*, 109, *138*
Hudson, J.A., 19, 28, *32*, 237, *243*
Hughes, E.D., 83, *104*, 101, *107*, 221,
    222, *240*, 225, 226, *241*
Huisgen, R., 185, *205*
Hulett, J.R., 85, *104*
Humffray, A.A., 224, *241*
Hunter, D.H., 119, *139*, 164, *169*, 182,
    204

Hunter, F.R., 137, *141*
Hutchinson, B.J., 61, 64, *68*
Huynk, C., 197, *206*

Ingold, C.K., 47, 52, *66*, 72, 73, *102*, 83,
    86, *104*, 86, 93, *105*, 99, 101, *107*,
    119, *139*, 130, *140*, 217, 220, 221,
    222, *240*, 225, 226, *241*
Ingold, K.U., 224, *241*
Iwamura, H., 59, *67*, 200, *206*

Jackman, L.M., 211, *239*
Jaeger, D.A., 101, *107*
Jaffé, H.H., 193, *205*
Jamieson, G., 88, *105*
Jarczewski, A., 96, *106*
Jeffs, P.W., 166, *169*
Jemison, R.W., 197, *206*
Jencks, W.P., 79, *103*
Jenkins, I.D., 234, 235, *242*
Jensen, F.R., 210, 221, 226, 227, *238*,
    217, 219, 224, *240*, 224, *241*
Jensen, J.H., 79, *103*, 92, *105*
Jindal, S.P., 119, 120, 126, *139*
Johanson, R.G., 114, *138*
Johnson, A.L., 119, *139*, 164, *169*
Johnson, A.W., 10, *30*
Johnson, Jr., C.S., 211, *238*
Johnson, M.D., 226, *241*, 227, 228, *242*
Jones, J.R., 2, *30*, 15, 17, *31*, 18, *32*,
    80, 85, *104*
Jones, Jr., M., 188, *205*, 199, *206*
Jones, M.L., 218, *240*
Joshi, G.C., 120, *139*
Julia, S., 197, *206*
Jullien, J., 83, *104*

Kaczynski, J.A., 182, *204*
Kahl, D.C., 17, *31*
Kaiser, E.M., 2, 10, *30*
Kaiser, E.T., 61, *68*
Kaissi, F.E., 233, *242*
Kalyavin, V.A., 227, *241*
Kamlet, M.J., 51, *67*
Kaneko, C., 171, 175, *203*
Kang, S., 151, *168*

251

Kaplan, F., 216, *240*
Kaplan, L.A., 95, *106*
Kari, R.E., 36, 37, *66*
Kasai, P., 134, *141*
Katritzky, A.R., 61, 64, *68*
Katz, T.J., 7, *30*, 114, *138*, 126, *140*
Kauffman, K.C., 18, *32*, 80, 90, *103*
Kauffmann, T., 186, *205*
Kayser, W.V., 223, *241*
Keese, R., 89, *105*
Keller, C.E., 114, *139*
Kelly, D.P., 197, *206*
Kelly, J.F., 116, *139*
Kende, A.S., 144, *167*
Kessler, H., 202, *207*
Kharasch, M.S., 222, *240*
Kheifets, G.M., 72, *102*
Kiehlmann, E., 80, *103*
Kimura, B.Y., 213, *239*
King, J.F., 59, *67*
Kingsbury, C.A., 28, *34*, 55, 57, *67*, 91,
   105
Kirby, G.W., 93, *105*
Kistiakowsky, G.B., 93, *105*
Kitching, W., 216, 217, *240*
Klabunde, K.J., 19, *32*, 49, 51, *67*
Klein, J., 98, *107*, 189, *205*
Kleinschmidt, R.F., 172, *204*
Kloosterziel, H., 127, *140*, 172, 182,
   183, *204*, 189, *205*
Kluger, R., 59, *68*
Klumpp, G., 199, *206*
Klunder, A.J.H., 163, *169*
Knorr, L., 71, *102*
Knox, G.R., 14, *31*
Knutson, G., 217, *240*
Knutsson, L., 146, 147, *167*
Ko, C.E.F., 22, *33*
Kobrich, G., 88, *105*
Koch, H.F., 19, *32*
Koeppl, G.W., 38, 40, *66*
Kohn, M.C., 136, *141*
Kollmeyer, W.D., 19, *32*
Kolthoff, I.M., 22, *33*
Koltsov, A.I., 72, *102*

Kondo, K., 197, *206*
König, H., 61, *68*
König, J., 188, 189, *205*
Kopecky, K.R., 13, 14, *31*
Kopp, J., 85, *104*
Koppelmann, E., 186, *205*
Kornblum, N., 95, *106*
Kornblum, P.J., 93, *105*
Korte, W.D., 229, *242*
Kouba, J., 27, *34*
Kovacs, J., 87, 88, *105*
Kraihanzel, C.S., 232, *242*
Krebs, A., 147, *167*
Krebs, E.-P., 89, *105*
Kreevoy, M.M., 223, *241*
Kreiter, C.G., 114, *139*
Kresge, A.J., 18, *32*, 27, *33*
Kretchmer, R.A., 223, *241*
Krishnamurthy, G.S., 38, 40, *66*
Krolls, U., 158, *168*
Kronzer, F.J., 211, *238*
Kroposki, L.M., 183, *204*
Kuhn, R., 10, *30*
Kuivila, H.G., 224, *241*
Kukla, M.J., 202, *206*
Kunii, T.L., 200, *206*
Kurts, A.L., 78, *103*
Kwasnik, H., 161, *168*

Labarre, J.-F., 130, *140*
Ladhar, F., 98, *107*
Lagowski, J.J., 12, *31*
Laird, T., 197, *206*
Laity, J.L., 132, *141*
LaLancette, E.A., 7, *30*
Lamaty, G., 86, *104*
Lambert, J.B., 35, 39, 40, 42, 48, *66*
Lambert, J.L., 118, 119, *139*, 163, 164,
   169
La Mer, V.K., 96, *106*
Langemann, A., 14, *31*, 35, *66*
Langford, C.H., 14, 15, *31*
Langlet, J., 171, 175, *203*
Langworthy, W.C., 29, *34*
Lansbury, P.T., 165, *169*

Lapworth, A., 75, *102*
Lasocki, Z., 223, *241*
Lathan, W.A., 73, *102*
Laughton, P.M., 233, *242*
Lawler, R.G., 19, *32*
Lawlor, J.M., 211, *239*
Lee, J.R., 93, *105*
Lee, K.C., 18, *32*
Lee, W.G., 43, *66*
Lee-Ruff, E., 21, *33*
Leffek, K.T., 96, *106*
Leffler, J.E., 47, 50, *66*
Legault, R.R., 222, *240*
Lehn, J.M., 35, 39, 42, 48, *66*
Leicht, C.L., 166, *169*
Leinhard, G.E., 77, *103*
Lengyel, I., 158, *168*
Le Noble, W.J., 93, *105*
Lepley, A.R., 165, *169*, 197, *206*
Lesbre, M., 210, *238*, 223, *241*
Lesko, P.M., 88, *105*
Lett, R., 64, *68*
Levin, J.-O., 90, *105*
Lewis, E.S., 26, *33*, 96, *106*, 226, *241*
Liang, G., 114, *139*
Liang, Y., 193, *205*
Liberles, A., 151, *168*
Lieder, C.A., 21, *33*
Lii, R.-R., 80, 81, *104*
Lin, Y.T., 82, *104*
Liotta, C.L., 19, *32*, 51, *67*
Lipscomb, W.N., 214, *239*
Lloyd, D., 130, *140*
Lo, D.H., 203, *206*
Londrigan, M.E., 187, *205*
Long, F.A., 2, 28, *30*, 18, *32*, 27, *33*, 85, *104*, 97, *106*
Longridge, J.L., 27, *33*
Longuet-Higgins, H.C., 75, *102*, 132, *141*, 171, 175, 177, 178, *203*
Lotfield, R.B., 146, *167*
Lowry, T.H., 61, 62, *68*
Lucken, E.A.C., 213, *239*
Ludwig, U., 165, *169*
Luijten, J.G.A., 223, *241*

Luinstra, E.A., 59, *67*
Luz, Z., 72, *102*
Lynch, R.A., 82, *104*
Lynne, Jr., J.L., 79, *103*

Mac, Y.C., 22, *33*
Macais, A., 78, *103*
MacDiarmid, A.G., 223, *241*
Machleder, W.H., 151, *168*
Madden, J.P., 161, *168*
Mageswaran, S., 197, *206*
Mahone, L.G., 19, *32*, 51, *67*
Maier, G., 174, *204*
Maire, J.C., 233, *242*
Malpass, J.R., 114, *139*
Malrieu, J.-P., 171, 175, *203*
Marchand, A.P., 49, *67*
Mares, F., 19, 28, *32*, 51, *67*, 237, *243*
Margolin, Z., 2, 28, *30*
Marks, R.E., 80, 85, *104*
Maron, S.H., 96, *106*
Marquet, A., 64, *68*
Marra, J.V., 215, *239*
Marshall, D.R., 130, *140*
Martens, D., 186, *205*
Marvell, E.N., 88, *105*, 174, 176, *204*
Masamune, S., 7, *30*, 202, *206*
Maskornick, M.J., 19, 20, *32*
Massicot, J., 93, *105*
Mateo, S., 26, *33*
Mateos, J.L., 14, *31*
Matteson, D.S., 219, *240*
Mayer, M.G., 101, *107*
Mayers, G.L., 87, 88, *105*
Mayo, P. de, 143, *167*, 171, *203*
Mazerolles, P., 210, *238*, 223, *241*
Mazur, Y., 72, *102*
McClory, M.R., 64, *68*
McCombs, D.A., 172, 184, *204*, 189, 191, *205*, 211, *239*
McCusker, P.A., 215, *239*
McElhill, E.A., 53, *67*
McEwen, W.K., 12, 23, *31*
McIver, Jr., R.T., 21, *33*
McIvor, M.C., 126, *140*

253

McKeever, L.D., 213, *239*
McKennis, J.S., 7, *30*, 202, *206*
McKinley, S.V., 211, *238*
McLean, S., 98, *107*
McLennan, D.J., 210, *238*
McManus, S.P., 2, 10, *30*
Meany, J.E., 79, *103*
Mecke, R., 72, *102*
Meijere, A. de, 115, *139*
Meister, W., 101, *107*
Melander, L., 23, 27, *33*, 27, *34*, 85, *104*
Menon, B.C., 80, 83, *103*
Merenyi, R., 174, 201, *204*
Meyer, K.H., 71, *102*
Meyerson, S., 148, *167*
Michelott, D., 197, *206*
Miller, B., 120, *140*
Miller, I.J., 187, *205*
Miller, J.A., 78, *103*
Miller, L.S., 229, *242*
Miller, S.A., 18, *32*, 96, *106*
Miller, S.I., 38, 40, 43, *66*, 80, 81, *104*, 175, 187, *204*
Mislow, K., 35, 37, 39, 42, 48, *66*
Mitchell, K.A.R., 61, *68*
Mitchell, M.J., 135, *141*
Mitchell, R.W., 132, *141*
Modarai, B., 47, *66*
Modena, G., 229, *242*
Moffitt, W., 6, *30*
Moir, R.Y., 61, 64, *68*
Moreau, B., 64, *68*
Morio, K., 200, *206*
Morrison, B.A., 233, *242*
Motes, J.M., 45, 46, *66*, 90, *105*, 186, *205*
Mowery, P.C., 13, *31*
Mueller, R.A., 166, *169*
Mulder, J.J.C., 171, 175, *203*
Mulders, J., 92, *105*
Muller, N., 215, *239*
Muller, R.J., 19, *32*
Mulvaney, J.E., 186, 187, *205*
Murahashi, S.-I., 135, *141*
Murdoch, J.R., 13, *31*

Musso, H., 200, *206*
Myers, R., 134, *141*

Nace, H.R., 148, *167*
Nachod, F.C., 21, *33*
Nagy, J., 233, *242*
Nahabedian, K.V., 224, *241*
Nakatsuka, N., 202, *206*
Nasielski, J., 224, *241*
Nelson, N.J., 17, *31*
Newcomb, M., 187, *205*
Newman, M.S., 47, 50, *66*, 72, 73, *102*
Nickols, R.E., 19, *32*
Nickolson, J.M., 125, *140*
Nickon, A., 118, 119, *139*, 161, *168*, 163, 164, *169*
Nielsen, A.T., 94, *105*
Nielsen, W.D., 13, 14, *31*, 35, *66*, 61, *68*
Niemeyer, H.M., 19, 20, *32*
Nilsson, M., 72, 73, 93, *102*
Nishihata, K., 61, *68*
Nishio, M., 61, *68*
Nordmann, B., 95, *106*
Norman, R.O.C., 224, 236, *241*
Norris, A.R., 96, *106*, 237, *243*
Novak, D.P., 213, *239*
Novikov, S.S., 94, *106*
Nye, M.J., 151, *168*
Nyholm, R.S., 36, *66*

Oae, S., 60, 61, *68*
O'Brien, D.F., 216, *240*
O'Brien, J.F., 215, *239*
Occolowitz, J.L., 124, 125, *140*
O'Ferrall, R.A. More, 27, *34*, 85, *104*, 226, *241*
Ogliaruso, M., 125, 126, *140*
Ogunkoya, L., 93, *105*
Ohno, A., 60, *68*
Ojima, I., 197, *206*
Oki, M., 59, *67*
Okuzumi, Y., 18, *32*, 80, 90, *103*
Olah, G.A., 113, 114, *138*, 178, *205*, 211, *238*
Oleneva, G.I., 94, *106*

254

Oliver, J.E., 119, *139*
Oliver, J.P., 213, *239*
Ollis, W.D., 197, *206*
Olmo, V.S. Del, 101, *107*
Olmstead, H.D., 78, 83, *103*
Olofson, R.A., 155, *168*, 184, *204*
Olsen, B.A., 148, *167*
Oosterhoff, L.J., 171, 175, *203*
Orchin, M., 193, *205*
Ordronneau, C., 114, *138*
Osipov, V.G., 94, *106*
Ossorio, R.P., 101, *107*
Ostermann, G., 165, *169*
Oth, J.F.M., 8, *30*, 114, *139*, 174, 201, *204*
Otsubo, T., 100, *107*
Otto, R., 222, *240*
Overdal, A.W., 137, *141*
Owens, P.H., 36, *66*

Palke, W.E., 214, *239*
Pande, K.C., 217, *240*, 223, *241*
Panek, E.J., 233, *242*
Pappas, B.C.T., 202, *206*
Paquette, L.A., 59, *68*, 114, 116, *139*, 174, *204*, 202, 203, *206*
Parker, A.J., 14, *31*, 22, *33*
Patrick, J.E., 165, *169*, 196, *205*, 197, *206*
Patsch, M., 165, *169*
Pattison, V.A., 165, *169*
Patton, J.T., 95, *106*
Paul, I.C., 161, *168*
Paulik, F.E., 222, *240*
Pauling, L., 49, *67*, 109, *138*, 212, *239*
Pauson, P.L , 210, *238*
Pawliczek, J.B., 199, *206*
Pazos, J.F., 151, *168*
Pearson, R.G., 18, *32*, 96, *106*
Peddle, G.J.D., 229, *242*
Pedersen, K.J., 17, *31*
Peet, N.P., 78, 83, *103*
Perkins, M.J., 182, *204*, 187, 188, *205*
Perrin, C., 19, *32*
Peterson, R.A., 147, *167*

Pettit, R., 7, *30*, 114, *139*, 202, *206*
Pfeffer, P.E., 92, *105*
Phelan, N.F., 193, *205*
Phillips, D.D., 59, *67*
Piccolini, R.J., 112, *138*
Pinshchmidt, Jr., R.K., 179, *204*
Pitt, C.G., 233, *242*
Plonka, J.H., 122, *140*
Pochan, J.M., 151, *168*
Pocker, Y., 85, *104*, 211, *238*
Pogonowski, C.S., 9, *30*
Poist, J.E., 232, *242*
Pollard, D.R., 216, *240*
Poller, R.C., 210, *238*
Pople, J.A., 49, *67*, 132, *141*
Posner, J., 147, *167*
Potter, D.E., 183, 184, *204*, 189, 191, *205*, 211, *239*
Praisnar, B., 225, *241*
Prange, U., 114, *139*
Price, C.C., 61, *68*
Prince, R.H., 233, *242*
Prinzbach, H., 116, *139*
Pritchard, D.E., 215, *239*
Psarras, T., 216, *240*
Pudjaatmaka, A.H., 19, *32*, 49, *67*
Pullman, B., 130, *140*
Putze, B., 114, *139*

Quast, H., 147, *167*
Quellette, R.J., 217, *240*

Radlick, P., 7, *30*, 116, *139*
Radom, L., 49, *67*
Raisin, C.G., 93, *105*
Ramey, K.C., 215, *239*
Rao, S.C., Subba, 80, 82, 85, *104*
Rapoport, H., 8, *30*
Rappe, C., 80, *103*, 82, 86, *104*, 90, *105*, 146, 147, *167*
Rauk, A., 35, 36, 37, 39, 42, 48, *66*, 61, 62, 63, 64, *68*
Rautensrauch, V., 183, *204*, 196, *205*, 197, *206*
Redmore, D., 144, *167*

255

Rees, N.Y., 218, *240*
Réffy, J., 233, *242*
Regan, C.M., 109, *138*
Reichard, D.W., 128, *140*
Reif, L., 55, *67*
Reitz, O., 18, *32*, 80, *103*, 85, *104*, 95, *106*
Relles, H.M., 98, *107*
Reuben, D.M.E., 12, 14, 23, *31*
Reutov, O.A., 78, *103*, 210, 221, *238*, 216, 221, *240*, 225, 227, *241*
Reuwer, Jr., J.F., 25, *33*, 85, *104*, 97, *106*
Rewicki, D., 2, 10, 21, *30*
Rhee, J.-U., 193, *205*
Richey, Jr., F.A., 152, *168*
Rickborn, B., 14, *31*, 28, *34*, 35, *66*, 55, 57, *67*, 91, *105*, 210, 221, 226, 227, *238*
Ridd, J.H., 220, *240*
Rieke, R., 126, *140*
Rifi, M.R., 137, *141*
Riley, T., 85, *104*
Ritchie, C.D., 11, *31*, 22, *33*, 78, *103*, 233, *242*
Riveros, J.M., 21, *33*
Robb, M.A., 36, *66*
Roberts, B.P., 224, *241*
Roberts, J.D., 53, *67*, 109, *138*, 174, *204*, 232, *242*, 237, *243*
Roberts, M., 114, *139*
Roberts, R.M.G., 225, 226, *241*, 233, *242*
Robertson, R.E., 233, *242*
Robinson, G.C., 13, *31*
Robinson, J.K., 26, *33*
Robinson, R., 157, *168*
Rochester, C.H., 15, *31*
Rochow, E.G., 210, *238*
Rogers, L.C., 164, *169*
Roitman, J.N., 65, *68*
Roques, A., 86, *104*
Rosen, W., 7, *30*
Rosenberg, A.S., 86, *104*
Roth, A.S., 172, *204*

Roth, W.R., 174, 199, *204*, 188, 189, *205*
Rothe, O., 71, *102*
Rottschaefer, S., 27, *33*
Rowe, Jr., C A., 98, *106*
Rowland, N.E., 8, *30*
Royo, G., 229, *242*
Rubin, R.M., 199, *206*
Rumney, T.G., 85, *104*
Rundle, R.E., 214, *239*
Rushton, B.M., 234, *242*
Russell, G.A., 46, *66*
Russell, J.G., 3, *30*
Russell, K.E., 96, *106*, 237, *243*
Russell, R.K., 116, *139*, 203, *206*
Rutherford, R.J.D., 98, *107*
Rys, P., 115, *139*

Sachs, W.H., 27, *33*, 82, *104*, 90, *105*
Sackur, O., 145, *167*
Sagatys, D.S., 38, 40, *66*
St. Janiak, P., 61, 62, *68*
Sakai, M., 114, *139*, 125, 127, *140*
Salaun, J.R., 145, *167*
Salem, L., 49, *67*, 130, *140*, 171, 175, *203*
Salinger, R.M., 216, *240*
Salisbury, K., 88, *105*
Salmond, W.G., 61, *68*
Sandel, V.R., 211, *238*
Sannes, K.N., 123, *140*, 138, *141*
Santos, J., 98, *106*
Sarel, S., 158, *168*
Sass, R.L., 5, *30*
Satgé, J., 210, *238*
Saunders, M., 199, *206*
Saunders, V.R., 214, *239*
Saunders, Jr., W.H., 23, 27, *33*, 27, *34*
Savage, D., 186, *205*
Sawyer, A.K., 210, *238*
Scamehorn, R.G., 148, 152, *167*, 152, *168*
Scannon, P.J., 19, 20, *32*
Schaad, L.J., 25, *33*, 85, *104*, 97, *106*, 136, *141*

Schatz, B., 174, 176, *204*
Schechter, H., 80, 90, *103*, 96, *106*
Schellenberger, A., 82, *104*
Schenck, G.E., 200, *206*
Schenk, H.P., 166, *169*
Scherer, K.V., 145, *167*
Scherr, P.A., 213, *239*
Schipperijn, A.J., 132, *141*
Schleyer, P.v.R., 49, *67*, 113, *138*, 187, *205*, 202, *207*
Schmalstieg, F.C., 79, *103*
Schmidt, E.A., 74, *102*
Schmitt, E., 147, *167*
Scholler, W.W., 200, *206*
Schollköpf, U., 165, *169*, 197, *206*
Schorlemmer, C., 222, *240*
Schreck, R., 71, *102*
Schriesheim, A., 19, *32*, 98, *106*
Schröder, G., 8, *30*, 114, *139*, 174, 201, *204*, 201, *206*
Schuber, F.J., 61, 64, *68*
Schultze, O.W., 94, *105*
Schumacher, H., 187, *205*
Schwarzenbach, G., 71, 72, *102*
Schweiger, J.R., 7, *30*
Sclove, D.B., 151, *168*
Scoggins, R., 154, *168*
Scott, D.A., 61, *68*, 98, *107*
Scott, J.A., 21, *33*
Scott, J.M.W., 61, *68*, 92, *105*
Scott, L.T., 188, *205*
Sebastian, J.F., 83, *104*, 172, *204*
Seidl, P., 114, *139*
Seidner, R.T., 7, *30*, 202, *206*
Seitz, L.M., 213, *239*
Selgestad, J.G., 233, *242*
Selman, S., 145, 155, *167*
Semenov, D.A., 237, *243*
Seyferth, D., 211, *238*
Shanshal, M., 37, *66*
Shapiro, H., 210, *238*
Shapiro, I.O., 2, *30*
Shatenshtein, A.I., 2, *30*, 19, 29, *32*, 237, *243*
Shechter, H., 18, *32*

Sheehan, J.C., 158, 160, *168*
Sheppard, W.A., 53, *67*
Shigorin, D.N., 72, *102*
Shiner, V.J., 233, *242*
Shlyapochnikov, V.A., 94, *106*
Shoppee, C.W., 110, *138*
Short, M.R., 116, *139* /
Sicher, J., 210, *238*
Sidler, J.D., 165, *169*
Sigwalt, P., 172, *204*
Silbert, L.S., 92, *105*
Sim, S.K., 182, *204*
Simmons, Jr., H.E., 237, *243*
Simonetta, M., 112, *138*
Simpson, P., 229, *242*
Sinclair, R.W., 121, *140*
Skrabal, P., 115, 116, *139*
Skrobek, A., 152, *168*
Slaugh, L.H., 172, *204*
Sloacum, D.W., 2, 10, *30*
Smid, J., 13, *31*, 211, *239*
Smith, C., 197, *206*
Smith, M.B., 215, *239*
Smith, P.W., 71, 72, *102*
Smith, R.H., 197, *206*
Smolina, T.A., 227, *241*
Smolinsky, G., 8, *30*
Smucker, L.D., 82, *104*
Snow, A.I., 214, *239*
Snyder, J.P., 130, *140*, 132, *141*
Sohoni, S.S., 119, 126, *139*
Sojka, S.A., 72, *102*, 151, *168*
Sokolov, V.I., 225, *241*
Sommer, L.H., 210, 218, 229, *238*, 229, *242*
Sondheimer, F., 8, *30*, 132, *141*
Sonnenberg, J., 113, 114, *138*
Sorensen, T.S., 187, 188, *205*
Sperry, J.A., 223, *241*
Springer, W.R., 148, 152, *167*
Sprowls, W.R., 222, *240*
Stackhouse, J., 42, *66*
Staley, S.W., 128, *140*, 137, *141*
Stange, H., 155, *168*
Staples, C.E., 98, *107*

257

Stapp, P.R., 172, *204*
Starkey, J.D., 15, *31*
Stauffer, C.H., 86, *105*
Stein, K., 188, 189, *205*
Stein, Y., 158, *168*
Steinberg, H., 90, *105*, 98, *106*, 134, *141*, 210, *238*
Steiner, E.C., 15, *31*
Steinwand, P.J., 223, *241*
Stekoll, L.H., 161, *168*
Stewart, R., 9, 22, *30*, 15, 16, 17, *31*, 18, *32*, 80, *104*
Stille, J.K., 120, 126, *139*, 123, *140*, 138, *141*
Stivers, E.C., 25, *33*, 85, *104*, 97, *106*
Stohrer, W.-D., 174, 200, *206*
Stone, H.W., 96, *106*
Stork, G., 152, *168*
Storr, R.C., 175, *204*
Stothers, J.B., 119, *139*, 164, *169*
Streitwieser, Jr., A., 2, 4, 12, 29, *30*, 12, 13, 14, 22, 23, *31*, 19, 20, 28, *32*, 29, *34*, 36, *66*, 49, 51, 55, *67*, 109, *138*, 130, 131, 136, 137, *140*, 237, *243*
Strong, J.G., 148, *167*
Stuart, R.S., 92, *105*
Stucky, G.D., 4, *30*
Sturmer, D., 88, *105*
Subrahmanyam, G., 151, *168*
Sun, C., 82, *104*
Sutherland, I.O., 197, *206*
Swaddle, T.W., 234, *242*
Swain, C.G., 25, *33*, 77, *103*, 80, 81, 85, 86, *104*, 97, *106*
Swart, J.B., 213, *239*
Swartz, T., 161, *168*
Swingle, R.B., 64, *68*
Symons, E.A., 15, 16, 17, *31*, 19, *32*, 59, *67*, 237, *243*
Sytilin, M.S., 80, *103*
Szwarc, M., 13, *31*, 211, *239*

Taft, Jr., R.W., 47, 50, *66*
Tagaki, W., 60, *68*
Taguchi, V., 187, *205*

Tai, W.T., 145, *167*
Taillefer, R., 119, 122, *139*
Takemoto, J.H., 12, *31*
Talaty, E.R., 159, 160, 161, *168*
Talcott, C., 126, *140*
Tan, T.-L., 91, *105*
Tang, R., 37, *66*
Tardi, M., 172, *204*
Taylor, D.R., 19, *32*
Taylor, J.D., 9, *30*
Taylor, P.R., 72, *102*
Taylor, R., 224, 236, *241*, 235, *243*
Tchoubar, B., 145, *167*, 152, *168*
Tee, O.S., 82, *104*
Tefertiller, B.A., 78, 83, *103*
Tel, L.M., 36, *66*, 61, *68*
Thebtaranonth, Y., 197, *206*
Thiebault, A., 11, *31*
Thio, J., 114, *139*
Thoi-Lai, Nguyen, 83, *104*
Thomas, R., 145, *167*
Thomas, S.P., 233, *242*
Thompson, A.R., 232, *242*, 235, *243*
Thompson, D.W., 72, *102*
Thompson, H.W., 72, *102*
Thompson, M.H., 216, *240*
Thompson, R.D., 158, *168*
Thornton, E.K., 23, *33*
Thornton, E.R., 23, 26, *33*
Thorpe, F.G., 220, 222, *240*
Thorpe, J.W., 80, 81, 82, *104*
Thyagarajan, B.S., 165, *169*, 175, *204*, 187, *205*, 197, *206*
Tidwell, T.T., 119, 120, 126, *139*
Tipper, C.F.H., 217, *240*
Titov, Yu.A., 144, *167*
Tomilson, R.L., 9, *30*
Toscano, V.G., 199, *206*
Toullec, J., 75, *103*, 80, *104*
Towns, D.L., 61, *68*
Tranter, R.L., 27, *33*
Traylor, T.G., 224, *241*, 234, 235, *242*
Traynham, J.G., 96, *106*
Trepka, R.D., 61, 62, *68*
Treves, G.R., 155, *168*

Trinajstic, N., 136, *141*
Trost, B.M., 197, *206*
Tsai, M.-R., 218, *240*
Tucker, R., 96, *106*
Turnbull, K.W., 120, *139*
Turner, L.M., 43, 45, *66*
Turner, R.B., 88, *105*
Turro, N.J., 144, 145, 146, 147, 151, *167*, 151, 154, *168*

Ullman, E.F., 80, 90, *103*
Ulmen, J., 199, *206*
Untch, K.G., 116, *139*, 132, *141*
Urban, F.J., 195, *205*
Urbas, L., 95, *106*
Uschold, R.E., 11, *31*
Ustynyuk, T.K., 144, *167*
Utermoelen, C.M., 159, *168*

Vairon, J.-P., 172, *204*
Vaisarova, V., 233, *242*
Van der Hart, W.J., 171, 175, *203*
Van der Kerk, G.J.M., 223, *241*
Van Drunen, J.A.A., 127, *140*, 172, 182, 183, 184, *204*
Van Hoboken, N.J., 98, *106*
Van Hook, W.A., 23, *33*, 85, *104*
Van Sickle, D.E., 29, *34*, 55, *67*
Van Tamelen, E.A., 187, *205*, 202, *206*
Van Wijnen, W.Th., 90, *105*, 134, *141*
Viau, R., 64, *68*
Vignollet, Y., 233, *242*
Vincenti, S.P., 82, *104*
Vincow, G., 137, *141*
Vinokur, E., 218, *240*
Vogelfanger, E., 217, *240*
Volgelsanger, R., 115, *139*
Volger, H.C., 221, *240*
Vranka, R.G., 214, *239*

Waack, R., 211, 213, *239*
Wade, K., 210, *238*
Walborsky, H.M., 43, 44, 45, 46, *66*, 90, *105*, 186, *205*
Walisch, W., 71, 72, *102*

Walters, E.A., 97, *106*
Walton, D.R.M., 232, 235, *242*, 235, *243*
Wang, A.H.-J ,^s161, *168*
Wang, T.-C., 77, *103*
Ward, J.S., 202, *206*
Ward, P., 182, *204*, 187, 188, *205*
Warhurst, E., 190, *205*
Warkentin, J., 18, *32*, 80, 81, 82, *104*
Warner, P., 114, *139*
Warnhoff, E.W., 120, *139*, 145, *167*
Washburn, W., 136, *141*
Wasserman, E., 131, *140*
Wasserstein, P., 59, *68*
Watson, D., 85, *104*
Waugh, J.S., 211, *238*
Webb, R.L., 53, *67*
Webster, C.J., 98, *107*
Webster, D.E., 224, *241*
Webster, O.W., 9, *30*
Weiner, M.A., 211, *238*
Weiss, E., 213, 214, *239*
Weissman, B.A., 158, *168*
Wells, D., 197, *206*
Wells, P.R., 234, *242*
Wenkert, E., 166, *169*
Wentworth, G., 194, 198, *205*
Werstiuk, N.H., 119, 122, *139*, 163, *169*
Wesley, D.P., 237, *243*
West, R., 218, *240*
Westheimer, F.H., 23, *33*
Wharton, P.S., 147, *167*
Whatthey, J.W.H., 174, *204*
Wheland, G.W., 12, *31*, 72, 73, *102*, 109, *138*
White, M.J., 21, *33*
White, W.D., 214, 215, *239*
Whitesides, G.M., 233, *242*
Wiberg, K.B., 132, 134, *141*
Wiesbock, R., 2, 28, *30*
Wigfield, Y.Y., 64, *68*
Wilcox, Jr., C.F., 115, *139*
Wilkinson, G., 212, *239*
Willemsens, L.C., 223, *241*
Willey, F., 98, *107*
Williams, A., 79, *103*

Williams, Jr., F.J., 18, *32*
Williams, Jr., F.T., 96, *106*
Williams, Jr., J.M., 59, *67*
Williams, K.C., 215, *239*
Williams, L.P., 193, *205*
Williams, R.O., 161, *168*, 163, *169*
Wilson, C.L., 86, *104*, 86, 93, *105*, 99,
    101, *107*, 119, *139*
Wilson, H., 26, *33*, 96, *106*
Wilson, J.D., 132, *141*
Wingard, Jr., R.E., 116, *139*, 203, *206*
Wingrove, A.S., 55, *67*
Winstein, S., 12, *31*, 110, 111, 112, 113,
    114, 124, 138, *138*, 114, 116, *139*,
    125, 126, 127, 128, 129, *140*, 162,
    169, 174, *204*, 217, *240*, 224, *241*
Wiseman, J.R., 89, *105*
Wittig, G., 197, *206*
Witwer, C., 71, 72, *102*
Wohllebe, J., 152, *168*
Wolf, J.F., 9, *30*
Wolfe, S., 36, *66*, 61, 62, 63, 64, *68*
Wolfsberg, M., 23, *33*, 85, *104*
Wong, C.M., 145, *167*
Wong, S.M., 58, *67*, 97, *106*
Wood, J., 96, *106*
Woodgate, S.S., 21, *32*
Woods, G.F., 88, *105*
Woodward, R.B., 171, 175, 190, *203*,
    175, 187, 189, 199, *204*

Worsfold, D.J., 211, 213, *239*
Wragg, R.T., 64, *68*
Wright, G.J., 235, *242*, 235, *243*
Wright, W.V., 75, *102*
Wynne-Jones, W.F.K., 96, *106*
Wyvratt, M.J., 59, *68*

Yates, K., 75, *102*
Yates, P., 122, *140*, 164, 165, *169*
Yee, K.C., 96, *106*
Yoneda, S., 114, *138*
Yoshida, Z., 114, *138*
Yoshikoshi, A., 166, *169*
Young, L.B., 21, *33*
Young, W.R., 19, *32*
Yu, S., 193, *205*
Yukawa, Y., 100, *107*

Zabel, A.W., 17, *31*, 19, *32*, 59, *67*
Zalar, F.V., 89, *105*
Zey, R.L., 158, *168*
Ziegler, G.R., 4, *30*
Zimmerman, H.E., 143, *167*, 164, *169*,
    171, 175, 179, 190, *203*, 193, *205*
Zollinger, H., 115, 116, *139*
Zook, H.D., 78, *103*
Zuckerman, J.J., 21, *33*
Zwanenburg, B., 163, *169*, 189, *205*
Zweig, A., 164, *169*, 193, *205*

SUBJECT INDEX

Ab initio MO calculations on methyl
    anion, 36, 37
Acetylide anion, 1
Acidity constant, absolute, 21
Acidity function, 15
Acidity in gas phase, 21
Acidity scales, 20
Acidolysis of organometallic compounds,
    216
Aci-nitro compounds, 94
Activities, thermodynamic, 22
Activity coefficients, 22
Acyl-nitrogen bond fission, 158
1-Adamantyl-3-bisnorcholanyl
    aziridone, 158
Addition-elimination of alkenes, 43
Aggregation of metal alkyls, 214
Aldol condensation, 91
Alkoxy ketone, from α-halogeno ketone,
    152
Alkylation, 93, 185
Alkyl exchange reactions of organo-
    metallics, 220, 222
Alkyl migration, 191
Alkyl-nitrogen bond fission, 158
Allene oxide, 151
Allyl anion, in isomerisation of cyclo-
    propyl anion, 185
    structure, 3
Allylbenzene, 98
Allyl ethers, isomerisation, 99
Allylmercuric iodide, 223
Amarine, 183
Amides, tautomerism, 91
Amine catalysis, 79, 87

Ammonia, inversion barrier, 37
    liquid, 11, 172, 209
Ammonium-carbanide ion pair, 56
Amphiprotic solvent, 11
Anchimeric assistance, 111
Angle strain, 44, 90, 134
Anhydrides, tautomerism, 92
Anionic rearrangements, 171
Anionotropy, 69
[12]Annulene, 8
Antarafacial migration, 189
Antiaromaticity, 129, 190
Antibonding molecular orbital, 181
Anti-Hückel transition state, 190
Aromatic transition state approach, 179
Aryl migration, 191
Asymmetric solvation, 57
Autoprotolysis, 10
Axis of symmetry, 176
Aziridine, N-lithio-2,3 (*cis*) diphenyl-, 186
Aziridone, 1-*t*-butyl-3,3-dimethyl-, 158

Barbaralane, 199
Barbaralone, 200
Barrier to inversion, 36, 131, 161
Base-catalyzed hydrogen exchange, 18,
    237
Benzene, delocalization energy, 129
Benzo[a]spiro[2,5]octa-1,4-diene-3-one,
    115
9-Benzylfluorene, ionization of, 14
Benzylic acid rearrangement, 155
Benzylic anion, deuteration of, 46
    rearrangement of, 196
    structure, 3

Benzyllithium, 213
Benzyl migration, 194
Bicycloaromaticity, 128
Bicycloheptenyl anion, 182
Bicyclohexenyl anion, 182
cis-Bicyclo[3,3,0]oct-2-ene, 172
Bifunctional catalysis, 79
Bimolecular electrophilic substitution, 218
Bimolecular nucleophilic substitution, 219
Bishomoantiaromatic, 138
Bishomocyclopentadienide anion, 124
Bishomotropylium ion, 114
Boat configuration, in Cope rearrangement, 198
Boron hydrides, 215
Bredt's rule, 88
Bridged ions, 111
Bridgehead alkene, 89
Bridgehead hydrogens, 4, 49, 89, 118, 162, 202
Bromination of enols, 71
Bromochlorofluoromethane, 47
Bromodesilylation, 224
Brönsted base, 69
Brönsted catalysis law, 17
Brönsted exponent, 17, 97
Bullvalene, 201
Bullvalone, 200
Butadienolate anion, 183
2-Butanone, 81, 82
t-Butyl hypochlorite, 158
t-Butyl isocyanide, 160
Butyllithium, 184, 196
t-Butylmalononitrile, 97

Calicene, 116
Camphenilone, racemization, 118
Camphor, specific deuteration, 120
Carbanion electrocyclizations, 181
Carbenes, 192
α-Carbethoxybenzylmercuric bromide, 225

Carbonation, of fluorenyl-methide, 195
of sulfinyl carbanions, 64
Carbon-fluorine resonance, 49
Carbon-14, isotope effect, 101
Carbonium ion electrocyclization, 187
Carbonium ion rearrangements, 192
Carbon-13, NMR, 72, 119
tracer, 165
Carbon-14, tracer in Favorskii reaction, 146
Catalysis, of enolization, 77
Cesium cyclohexylamide, 14, 29
Chair configuration in Cope rearrangement, 198
Charge in delocalized anions, 172
Chirality of carbanions, 47, 59
Chlorocyclohexanone, 144
Chlorodesilylation, 224
Cholesteryl chloride, solvolysis of, 110
Classical carbanions, 109
Classification of S_E mechanisms, 218
Conducted-tour mechanism, 58, 87
Conductivity, 11
Conformational analysis, 62
Conformational changes of anions, 172
Conjugative effect, 2, 46
Conrotatory cyclization, 176
Conservation of orbital symmetry, 175
Contact ion-pairs, 13, 211
Coplanarity, 3, 172
Correlation diagram, 178, 180
Covalent character of bonds, 212
Cross-conjugation, 116
1-Cyano-2,3-diphenylcycloprop-1-yl anion, 186
Cyclic voltametry, 137
Cycloadditions, 154, 171
Cyclobutadiene, 7, 135
Cyclobutanone, ring contraction in, 145
Cyclobutenyl cation, 114
Cyclohepta-1,3-diene, 182
Cyclohepta-1,4-diene, 185
Cycloheptadienyl anion, 182, 191
Cycloheptatrienyl anion, 136
heptaphenyl, 137

Cyclohexadienyl anion, 127, 182
Cyclononatetraenide anion, 7
cis,cis,cis-Cyclonona-1,4,7-triene, 116
1,3-Cyclooctadiene, 172
1,4-Cyclooctadiene, 172
Cyclooctadienyl anion, 172
Cyclooctatetraene, 7
Cyclooctatetraene dianion, 7, 126
Cyclopentadiene, 6
Cyclopentadienyl cation, 130
Cyclopentenyl carbonium ion, 187
Cyclopropane, acidity of, 6, 185
Cyclopropanimines, 147
Cyclopropanone, cycloaddition
    reactions of, 151
    ring opening of, 145
    trans-2,3-di-t-butyl, 151
Cyclopropenyl anion, 132, 134
Cyclopropenyl cation, 7
    bishomo, 114
    monohomo, 114
    trishomo, 114
Cyclopropyl anions, 45, 134, 185

Degenerate rearrangement, 129, 188, 202
Delocalization energy, 93, 129
Delocalized carbonium ions, 111
Demercuration, 217
Demetallation, 217
Destabilization, 129
Deuteration of carbanions, 195
Deuterioenol, 74
Deuterium exchange, 41, 59
Deuterium labelling, 189
Deuterium oxide, 74, 81
Deuteroxide ion, 17, 43, 80, 82
Diastereotopic protons, 64
Diatropic species, 132
Diborane, 215
α,α-Dibromoketones, 153
Dichotomy, 218
Dielectric constant, 11
Diels-Alder adducts, 89
Dienylic carbonium ion, 187
Dihydrobullvalene, 199

2,5-Dihydrofuran, ionization of, 183
Dimerisation of boron alkyls, 215
Dimethylberyllium, 214
Dimethylcadmium, 215
1,2-Dimethylcyclohexadiene, 174
1,6-Dimethylcyclohexadiene, 176
2,2-Dimethylcyclopropanone, 147
Dimethylformamide, H_ data, 16
Dimethylmagnesium, 214
Dimethylmercury, 216
Dimethylsulfinyl carbanion, 17
Dimethyl sulfoxide, 14, 16, 56, 225
Dimethylzinc, 215
1,3(cis,trans)-Diphenyl-2-azaallyl anion,
    186
6,6-Diphenylbicyclo[3.1.0]hex-2-ene,
    181
Dipolar aprotic media, 15, 17, 226
Dipolar ionic intermediate, 145, 150, 158
Dipole moment, effect of fluorine
    substitution, 53
Disrotatory cyclization, 150, 176, 177
Dissociation-recombination of radical
    ion-pairs, 194, 197
Dissymmetric carbanions, 59
cis-Divinylcyclopropane isomerisation,
    199
Double bond—no bond resonance, 49
Dynamic nuclear magnetic resonance, 40

Eclipsing interactions, 134
E1CB mechanism of elimination, 210
Electrocyclic rearrangements, 172
Electrocyclization, of carbonium ions,
    187
    selection rules, 175
Electron-deficient molecules, 215
Electron diffraction, 115
Electronegativity, 212
Electrophilic aliphatic substitution, 216
Electrophilic aromatic substitution, 223
Electrophilic mechanism of alkene
    halogenation, 77
Electroreduction, 153
Electrostatic bonding, 212

Elimination, 1,3-, 145, 147
Elimination-addition, 194
Eliminations, olefin-forming, 210
Empirical resonance energy, 131
Enantiomers, 119
Enolate ions, expulsion of halide ion, 149
    formation of cyclopropanone, 149
    in ionization of organomercurials, 225
    rearrangement, 143
Enolization, in bullvalone, 200
    steric inhibition thereof, 88, 157
Enols, 71, 201
Enzyme catalysis, 79
Episulfone, 147
Epoxide, ring opening of, 155
α-Epoxy ketones, 155
Equilibrating classical ions, 111
Equilibrium acidity, 10
Esters, tautomerism in, 91
Ethylacetoacetate, 71

Favorskii rearrangement, 144
Field effect, 2, 128
Fluctuating bonds, in bullvalene, 202
Fluorenyl anion, 49
Fluorenylcesium, 13
Fluorenyllithium, 13
Fluoroalkanes, 48
9-Fluorofluorene, 50
Fluxional molecules, 202
Fragmentations, 171
Frontier orbital approach, 179
Furan, adducts, 154

Gauche effect, 62
Gegenions, 209
General acid catalysis, 77
General base catalysis, 77, 96
Germane, benzyltrimethyl, 234
    phenylethynyl, 232
    phenyltrimethyl, 235
Glycine, 79
Ground state, rearrangements, 177

Haloform reaction, 76

Halogenation of ketones, 75
    of organometallic compounds, 217, 224
Halonium ion, 217
α-Halosulfones, 147
Heat of combustion, 131
Helical conformation, 184
Hemiketal, 147
1,3,5-Heptatriene, 184
1,3,6-Heptatriene, 184
Heptatrienyl anion, 184
Heterolytic scission in organometallic compounds, 216
Hexameric methyllithium, 214
H_ function, 15
Highest occupied molecular orbital, 177
Homoallylic carbonium ions, 112
Homoallylic resonance, 112
Homoantiaromaticity, 138
Homoaromatic carbanions, 122
Homoaromatic carbonium ions, 114
Homobenzene, 116
Homoconjugated carbanions, 122
Homoconjugation, 112
Homocubane, 163
Homodienolate ion, 120
Homoenolate ions, 117
Homoenolization, 118, 161
Homo-Favorskii rearrangement, 166
Homoketonization, 164
Homolytic substitution, 224
Homotropilidene, 199
Homotropylium ion, 114
Hückel rule, 7, 109
Hückel transition state, 190
Hybridization, C–H, 6
Hybridization in cyclopropane, 6
Hydration of aldehydes, 91
Hydrobenzamide, 182
Hydrogen bonding, 14, 57, 74, 87, 95
Hydrogen-deuterium exchange,
    acetone, 80
    amides, 91
    arenes, 19
    benzobicyclooctadiene, 126

benzyl methyl sulfoxide, 64
bicyclooctadiene, 124
bullvalene, 201
bullvalone, 200
camphenilone, 118
cycloheptatriene, 136
cyclohexane, 19
cyclopropane derivatives, 90, 133
cyclopropene derivatives, 133
cis-1,2-dichloroethylene, 43
dimethyl sulfone, 59
dimethyl sulfoxide, 59
2,2-diphenylcyclopropyl cyanide, 44
2-halocyclohexanones, 148
homocubane systems, 163
ketones, 18
methane, 19
nitriles, 99
nitro alkanes, 95
2,2,4,6,6-pentamethylcyclohexyl-
    ideneacetonitrile, 43
phenoxide ion, 94
phenylacetic acid, 92
2-phenylbutane, 55
2-phenylbutyronitrile, 57
phosphinates, 59
polynitrobenzenes, 19
sulfonamides, 59
tetramethylphosphonium ion, 60
xylenes, 19
[1,j]Hydrogen migrations, 188
Hydrogen-tritium exchange,
    fluorenes, 19
    fluorobenzenes, 19
    ketones, 18
    toluene, 19
    xylenes, 19
Hydronium ion catalysis of enolization,
    80
α-Hydroxycarboxylic acids, 155
Hyperconjugative effect, 49, 159
Hypoiodite, 76

Iminolactone, 160
Indanone, 152

Inductive effect, 2, 47, 90
Infrared, characterization of enol, 74, 154
Intermolecular mechanisms, 194
Internal return, 27, 149
Intra-annular valence bond isomerisation,
    174
Intramolecular catalysis, 70, 79
Intramolecular cycloaddition, 174
Intrinsic acidity, 21
Inverse isotope effect, 28, 233
Inversion in S_E2 reactions, 219
Ion cyclotron resonance, and gas phase
    acidity, 21
Ionic character, 212
Ion-pairing, 13, 211
Isobutyraldehyde-2-d, 91
Isoinversion, 58
Isomarine, 183
Isomerisation, in heterocyclic systems,
    174
    of 1,3,5-cyclooctatriene, 174
Isoracemization, 57
Isotope effect, 23
    enolization, 85
    imine equilibrium, 100
    in hydrogen migration, 188
    malononitrile, 97
    nitroalkanes, 96
    organosilicon compounds, 233
    2,4,6-trinitrotoluene, 96
Isotope exchange, base catalyzed, 16
    bullvalone, 200
    α-carbethoxybenzylmercuric bromide,
        225
    dimethylmercury, 216
    di-sec-butylmercury, 222
    p-nitrobenzylmercuric bromide, 227
Isotope tracer, in anion migration, 194
Isotopic labelling, 189, 191
I-strain, 44, 90, 134

$J(^{13}C-H)$, 50

Keto-enol tautomerism, 70
Keto-esters, tautomerism, 71, 73

Ketones, α-epoxy, 155
    equilibrium enol content, 73
    halogenation of, 75
    α-halogeno, 144
    β-halogeno, 166
7-Ketonorbornane, 89
Kinetic acidity, 16
Kinetic control, 83
Kinetic isotope effect, 23
Kolbe reaction, 93

α-Lactam, 158
γ-Lactone, 156
Lead, organometallic derivatives of, 217
Levelling effect, on acidity, 10
Ligand displacement, 218
N-Lithio-2,3-cis-diphenylaziridine, 186
Lithium cyclohexylamide-cyclohexyl-
    amine, 12, 29, 55, 128

Magnetic anisotropy, as criterion of
    aromaticity, 132
Malononitrile, proton abstraction, 97
Mass spectrometry for D determination,
    81
Medium effect, on acidity, 23
Mercury isotopes, 216
Metal alkyls, 40, 213
Metal-halogen interchange, 216
Metal-metal interchange, 217
1-Methyl-3,3-diphenylcyclopropane-1,2-
    trans-dicarboxylate, 185
Methyllithium, structure, 214
p-Methylpyridinium ion, 228
Microwave spectrum in tunneling, 38
Mirror-plane, 176
Möbius system, 190
Molecular orbitals in 1,3,5-hexatriene,
    178
Multi-centered reactions, 171
Multi-centre molecular orbitals, 214

Negative hyperconjugation, 49
Nitrenes, 192
Nitrenium ions, 192

Nitriles, tautomerism in, 46, 97
p-Nitrobenzyl anion, 227
p-Nitrobenzylmercuric bromide, 227
Nitrogen trifluoride, inversion barrier,
    48
Nitromethane, pK_a, 5
p-Nitrophenoxide ion, 82
NMR, ^{13}C, 165
    and homoconjugation, 114
    in keto-enol equilibria, 72
    of bullvalene, 202
    of delocalized carbanions, 125, 172,
        184
    of metal alkyls, 213
    of methyllithium, 213
Nonclassical carbanions, 109
Nonclassical carbonium ions, 110
    in solvolysis of organomercurials, 217
Non-linear transition states, 27
Norbornene, 89, 123
Norbornenyl carbanion, 123
Norbornyl carbonium ion, 113
Norbornyl mercuric perchlorate, 217
Nortricyclanone, 89
Nortricyclene, 123
Nucleophilic aliphatic substitution, 218
Nucleophilic catalysis, 226
Nucleophilic coordination, in S_E reactions,
    221
Nucleophilic substitution at silicon, 229

Octamethylsemibullvalene, 200
Olefins, isomerisation of, 98, 172
s Orbital effect, 42, 90
Orbital overlap, in electrocyclic closure,
    177
d Orbitals, 48, 60
Orbital symmetry, 171
Organomercurials, electrophilic
    substitution, 224
    solvolysis, 217
Organometalloid compounds, 211
Organosilicon compounds, 228
Organotin compounds, 228
    cleavage by bromine, 219

Oxygen-17, 72
Oxygen acids, 21

Paratropic species, 132, 137
Participation reactions, stereochemistry, 111
Pentacoordinate intermediates, in substitution at germanium, 229
in substitution at silicon, 229
in substitution at tin, 229
Pentadienylic-pentenylic transformations, 181
t-Pentyl alcohol-O-d, 181
Perfluoromethyl anions, stereochemistry, 52
Pericyclic reactions, rules for, 175
Phase dislocation, in Möbius system, 190
Phenylallyl anion, 233
Phenylallylsilane, 233
Phenylallylstannane, 233
2-Phenylbutane-2-d, 55
2-Phenylbutyronitrile, racemization, 97
Phenyl anion, 1
Phenylethynylsilane, 232
Phenylethynyl carbanion, 232
Phenylethynylgermane, 232
Phenyllithium, 182
Phenyl migration, 164, 193
Phenylnitromethane, ionization, 94
Phosphine, trialkyl, 120
Phosphorus, adjacent carbanions, 59
Photochemical processes, stereochemistry of, 176
Photoelectron spectroscopy, 115, 116
Pi-bonding energy, 129
Pi-inductive mechanism, 53
Pi-participation in halide displacement, 150
pK_a, acetic acid, 2
acetone, 5
acetonitrile, 5
acetylacetone, 5
acetylene, 6
ammonia, 11
benzene, 20

9-t-butylfluorene, 23
9-carbomethoxyfluorene, 9
9-cyanofluorene, 9
cyclobutenylcyclopentadiene, 136
cycloheptatriene, 7
cyclohexane, 4, 19
cyclopentadiene, 6
cyclopropane, 6, 19
dependence on σ parameter, 50
dimethyl sulfoxide, 11
dinitromethane, 5
diphenylmethane, 3
enols, 72
ethane, 6
ethylene, 20
fluoradene, 8
fluorene, 8
fluoronitromethanes, 51
indene, 8
malononitrile, 5
methane, 2, 19
nitromethane, 5
pentacyanocyclopentadiene, 9
phenol, 2
9-phenylfluorene, 8
propene, 7
toluene, 3, 19
trifluoromethane, 48
triphenylcyclopropene, 135
triphenylmethane, 3
triptycene, 4
tri(trifluoromethyl)methane, 48
Planar carbanions, 35
Polarizability, 48
Polymeric structures, of lithium alkyls, 213
Potassium amide, 172
Potassium t-pentoxide, 181
Potential energy-reaction coordinate profile, 24, 84, 85, 100, 112, 231
Potential energy surface, 24
Potentiometric measurement of pK_a, 11
Protodemetallation, 223
Protodetritiation, 49
Protolysis, of organometallics, 216

of 4-pyridiomethylmercuric bromide, 227
Protonation of delocalized anions, 172
1,3-Proton shift, 69
Proton tunneling, 26, 85, 96
Prototropy, 69
Pseudoaxial, 150
Pyramidal configuration, at carbon, 35 52
  at nitrogen, 37, 161
Pyramidal inversion, 38
Pyridine, H₋ data, 16
Pyridine complexes of halogens, 224
4-Pyridiomethide anion, 228
4-Pyridiomethylmercuric bromide, protolysis of, 227
Pyrrolidine, tetraphenyl, 186
Pyruvic acid, iodination, 79

Quantum level, 60
Quantum mechanical tunneling, 26
Quantum number, vibrational, 24
Quaternary ammonium salts, 60
Quaternary phosphonium salts, 60

Racemization, amides, 91
  camphenilone, 118
  trans-2,3-di-t-butylcyclopropanone, 151
  enolization, 86
  esters, 91
  ketones, 44, 86
  nitriles, 97
  nitro alkanes, 95
  S_E1 reactions, 218, 224
Radical ions, 127, 197
Radioactivity, 82, 101
Radio-iodine, 82
Radio-mercury-203, 222, 225
Rate enhancement, in proton abstraction, 124
  in solvolysis, 110
Reaction coordinate, 26
Rearrangement, benzylic acid, 145, 155
  carbanions, 193

carbonium ions, 192
Cope, 198
enolate ions, 143
α-epoxy ketones, 155
Favorskii, 144
halogeno amides, 157
halogeno ketones, 144
homo-enolate ions, 161
homo-Favorskii, 166
semi-benzylic acid, 145
Sommelet-Hauser, 197
Stevens, 197
Zimmerman-Grovenstein, 193
Re-distribution reactions, 215, 220
Reduced mass, 24
Regiospecific ring opening, 147
Reimer-Tieman reaction, 93
Reresolution, 58
Resonance destabilization, 130
Resonance energy, 124, 129
  thermochemical estimation, 131
Retention, in alkyl migration, 192
  in isotopic exchange, 41, 55, 61
  in S_E2 reactions, 219
Ring current, 125, 127, 132
Ring-strain, 44, 134, 157
Rule(s), Bredt's, 88
  Woodward-Hoffman, 171

Scrambling of isotopic label, 128, 146, 201
Secondary isotope effects, 26, 86
Selection rules for sigmatropic migrations, 189
Semibullvalene, 199
Shielding effect, 219
Sigma complex intermediates, 96, 223
Sigma-rho correlations, 234
Sigmatropic rearrangements, 188
Sign inversion, in p-orbital array, 190
Silane, benzyltrimethyl, 234
  phenyltrimethyl, 235
Siliconium ions, 218
Silyl anions, 218
Singlet state, 130, 137

S_E1 mechanism, 218
$S_E1(N)$ mechanism, 221
S_E2 mechanism, 218
S_EC mechanism, 220
S_Ei mechanism, 220
S_F2 mechanism, 220
S_N1 mechanism, 219
S_N2 mechanism, 219
Sodium hydride, 185
Sodium-potassium alloy, 125
Solute-solute interactions, 23
Solute-solvent interactions, 23
Solvation effects, 78
Solvent isotope effect, 28, 233
Solvent-separated ion-pairs, 13, 211
Spiroconjugation, 116
Spiro[2.4]hepta-4,6-diene, 115
Spiro-α-lactam, 160
Squalene synthesis, 197
Standard state, 22
Stannane, benzyltrimethyl, 234
    phenyltrimethyl, 235
Steady-state concentration, 209
Stereochemical consequences in keto-
    enol tautomerism, 86
Stereochemistry, in electrophilic
    displacement, 219
    in Favorskii rearrangement, 152
    in pericyclic reactions, 171
    of substitution at silicon, 233
Stereoisomers, configurational integrity
    thereof, 40
Stereomutation, 40
Stereoselectivity, in pericyclic reactions,
    171
    of hydrogen exchange, 119, 126
Stereospecificity in solvolyses, 111
Steric inhibition of resonance, 4, 90
Steroids, 152, 158
trans-Stilbene, 186
Strain in cyclopropyl systems, 90, 157,
    199
Substitution, at aromatic carbon, 223
    electrophilic, 218
    nucleophilic, 218

α-Sulfinyl carbanions, 62
α-Sulfonyl carbanions, 62
Sulfur heterocycles, 183
Suprafacial migration, 189
Symmetrical carbanions, 54
Symmetrical transition state, 27, 188
Symmetry elements, 176
Synartesis, 111

Tautomeric equilibria, 69
Tautomerism, 69
    aldehydes, 91
    amides, 91
    α,β- and β,γ-unsaturated derivatives,
        98
    anhydrides, 92
    esters, 91
    ketones, 70
    methyleneazomethine, 100
    nitriles, 97
    nitro-aci-nitro, 94
    propenes, 98
Temperature-jump, 72
Terminus, migration, 188
Termolecular reaction, 78
Tetraalkylstannanes, 229
Tetrahedral intermediate, 145
Tetrahydrofuran, 172
Tetramer, methyllithium, 213
Tetramethylene sulfone, 16
Thallium, organometallic derivatives of,
    217
Theory, orbital symmetry-controlled
    reactions, 171
Thermal isomerisation, 172, 175
Thermodynamic acidity constant, 21
Thermodynamic control, 83
Thermodynamic stability of isomers, 73,
    176, 183
Topomerisation, constitutional, 202
Transition metal, 211
Transition state, in S_E1 reactions, 218
    in S_E2 reactions, 218, 219
    in S_Ei reactions, 220
    in 1,2-rearrangements, 192

in 1,3-rearrangements, 192
in [2,3] rearrangements, 195
three-centered, 26
Transmission of electronic effects, 234
4-Trichloromethylcyclohexadienone, 120
Triiodide ion, 76
Tri-isobutylaluminium structure, 215
Trimethylaluminium structure, 214
Trimethylboron structure, 215
2,4,6-Trinitrobenzyl anion, 96
2,4,6-Trinitrotoluene, 96
1,2,3-Triphenylcyclopropyl anion, 186
Triphenylmethyl carbanion, 3
Triphenylmethyl carbonium ion, 4
Triphenylmethyllithium tetramethylene-
diamine, 4
Triphenylmethyl radical, 4
Triplet state, 130
Triptycene, ionization of, 4
Tritiated water, 81
Tritium exchange, 28, 81
Tritium isotope effect, 25
Tropenide anion, 137
Tropylium ion, 137, 187
Tunneling, electron-pair, 38
proton, 26

Unimolecular substitution,
electrophilic, 218
nucleophilic, 218
Unshared electron pairs, 35, 62
Unsymmetrical transition states, 27, 85

Valence bond isomerisation, 174, 199
Valence isomers of bullvalene, 202
Valence shell electron pair repulsion, 36
Vibrational energy, 24
Vibrational frequency, 24, 38
Vibration, imaginary, 26
Vinyl anion, 43

Woodward-Hoffman theory, 171

X-ray crystal structures, 4, 161, 213

Ylides, 9
Ylides, nitrogen, 197
phosphorus, 9, 196
sulfur, 9, 197

Zero-order rate dependence, 75, 218
Zero-point energy, 24
Zwitter-ion in Favorskii rearrangement,
151